# Physical and Chemical Data for Selected Elements

| Atomic Number | Element | Density at 20°C (g/cm³) | Crystal Structure at 20°C | Atomic Number | Element | Density at 20°C (g/cm³) | Crystal Structure at 20°C |
|---|---|---|---|---|---|---|---|
| 1 | Hydrogen | | | 43 | Technetium | | hex. |
| 2 | Helium | | | 44 | Ruthenium | 12.36 | hex. |
| 3 | Lithium | 0.533 | bcc | 45 | Rhodium | 12.42 | fcc |
| 4 | Beryllium | 1.85 | hex. | 46 | Palladium | 12.00 | fcc |
| 5 | Boron | 2.47 | | 47 | Silver | 10.50 | fcc |
| 6 | Carbon | 2.27 | hex. | 48 | Cadmium | 8.65 | hex. |
| | | 3.516 | dia. cub. | 49 | Indium | 7.29 | tet. |
| 7 | Nitrogen | | | 50 | Tin (white) | 7.29 | tet. |
| 8 | Oxygen | | | | (grey) | 5.76 | dia. cub. |
| 9 | Fluorine | | | 51 | Antimony | 6.69 | rhomb. |
| 10 | Neon | | | 52 | Tellurium | 6.25 | hex. |
| 11 | Sodium | 0.966 | bcc | 53 | Iodine | 4.95 | ortho. |
| 12 | Magnesium | 1.74 | hex. | 54 | Xenon | | |
| 13 | Aluminum | 2.70 | fcc | 55 | Cesium | 1.91 | bcc |
| 14 | Silicon | 2.33 | dia. cub. | | | (−10°) | |
| 15 | Phosphorus | 1.82 | ortho. | 56 | Barium | 3.59 | bcc |
| | | (white) | | 57 | Lanthanum | 6.17 | hex. |
| 16 | Sulfur | 2.09 | ortho. | 58 | Cerium | 6.77 | fcc |
| 17 | Chlorine | | | 59 | Praseodymium | 6.78 | hex. |
| 18 | Argon | | | 60 | Neodymium | 7.00 | hex. |
| 19 | Potassium | 0.862 | bcc | 61 | Promethium | | hex. |
| 20 | Calcium | 1.53 | fcc | 62 | Samarium | 7.54 | rhomb. |
| 21 | Scandium | 2.99 | fcc | 63 | Europium | 5.25 | bcc |
| 22 | Titanium | 4.51 | hex. | 64 | Gadolinium | 7.87 | hex. |
| 23 | Vanadium | 6.09 | bcc | 65 | Terbium | 8.27 | hex. |
| 24 | Chromium | 7.19 | bcc | 66 | Dysprosium | 8.53 | hex. |
| 25 | Manganese | 7.47 | cubic | 67 | Holmium | 8.80 | hex. |
| 26 | Iron | 7.87 | bcc | 68 | Erbium | 9.04 | hex. |
| 27 | Cobalt | 8.8 | hex. | 69 | Thulium | 9.33 | hex. |
| 28 | Nickel | 8.91 | fcc | 70 | Ytterbium | 6.97 | fcc |
| 29 | Copper | 8.93 | fcc | 71 | Lutetium | 9.84 | hex. |
| 30 | Zinc | 7.13 | hex. | 72 | Hafnium | 13.28 | hex. |
| 31 | Gallium | 5.91 | ortho. | 73 | Tantalum | 16.67 | bcc |
| 32 | Germanium | 5.32 | dia. cub. | 74 | Tungsten | 19.25 | bcc |
| 33 | Arsenic | 5.78 | rhomb. | 75 | Rhenium | 21.02 | hex. |
| 34 | Selenium | 4.81 | hex. | 76 | Osmium | 22.58 | hex. |
| 35 | Bromine | | | 77 | Iridium | 22.55 | fcc |
| 36 | Krypton | | | 78 | Platinum | 21.44 | fcc |
| 37 | Rubidium | 1.53 | bcc | 79 | Gold | 19.28 | fcc |
| 38 | Strontium | 2.58 | fcc | 80 | Mercury | | |
| 39 | Yttrium | 4.48 | hex. | 81 | Thallium | 11.87 | hex. |
| 40 | Zirconium | 6.51 | hex. | 82 | Lead | 11.34 | fcc |
| 41 | Niobium | 8.58 | bcc | 83 | Bismuth | 9.80 | rhomb. |
| 42 | Molybdenum | 10.22 | bcc | 92 | Uranium | 19.05 | ortho. |

SOURCE: J. F. Shackelford. *Introduction to Materials Science for Engineers*, 2nd ed., Macmillan, New York, 1988.

# Electronic Materials Science:

**For Integrated Circuits in Si and GaAs**

# Electronic Materials Science:
## For Integrated Circuits in Si and GaAs

# James W. Mayer

*Department of Materials Science and Engineering*
*Cornell University*
*Ithaca, New York*

# S.S. Lau

*Department of Electrical and Computer Engineering*
*University of California, San Diego*
*La Jolla, California*

**Macmillan Publishing Company**
NEW YORK

**Collier Macmillan Publishers**
LONDON

**Editor:** David Johnstone
**Production Supervisor:** John Travis
**Production Manager:** Nick Sklitsis
**Text Designer:** Natasha Sylvester
**Cover Designer:** Natasha Sylvester
**Illustrations** provided by the authors

This book was set in Times Roman by Waldman Graphics,
and printed and bound by Quinn Woodbine.
The cover was printed by Phoenix Color.

Macmillan Publishing Company
866 Third Avenue, New York, New York 10022

Collier Macmillan Canada, Inc.

**Library of Congress Cataloging in Publication Data**

Mayer, James W. (date)
  Electronic materials science for integrated circuits in Si and GaAs.

  Bibliography: p.
  Includes index.
  1. Transistor circuits—Design and construction.
2. Transistor circuits—Materials.   3. Integrated
circuits—Design and construction.   4. Integrated
circuits—Materials.   I. Lau, S. S.   II. Title.
TK7871.9.M388   1988      621.381′71      89-2844
ISBN 0-02-378140-8

Printing: 1 2 3 4 5 6 7 8      Year: 0 1 2 3 4 5 6 7 8 9

**This book is dedicated to our families.**

# Preface

## Objectives

The principles and concepts of device operation and fabrication—how transistors work and how they are made—are the focus of this book. It begins with the electrical and structural properties of silicon and gallium arsenide and the operation of $p$-$n$ junctions and transistors. This sets the stage for the diffusion, ion implementation, oxidation, and metallization process steps that are required for closely-packed transistors and interconnects. The book then treats epitaxial layers and heterojunctions used for light-emitting diodes and high-speed devices for opto-electronic integrated circuits.

The book is written for the materials scientist or engineer who is interested in the operation of $p$-$n$ junctions and transistors and the processing methods that have evolved for integrated circuits. It is also written for the electronics specialist concerned with advances in materials processing; but is particularly addressed to the students in engineering, applied physics, chemistry and materials science who will be responsible for the next generation of integrated systems.

Both of us are heavily engaged in research and teaching programs involving electronic materials and fabrication processes. We were colleagues at the California Institute of Technology and continue joint research programs at Cornell University and the University of California, San Diego. We are convinced that an understanding of both electrical behavior and materials science are required for integrated circuits. Hence this book.

## Book Topics

Chapter One is an overview of the fabrication of integrated circuits from growth of single crystals to final packaging of the circuit. It gives a perspective on the sizes of transistors and the methods used to form patterns on the metal and oxide layers. This chapter provides a review of the units used throughout the book.

The operation of integrated circuits is based on the transport of charge. We start in Chapter Two with current flow so that the electrical properties of metals and semiconductors can be evaluated. Silcon (Si) and aluminum (Al) are adjacent elements in the periodic table whose resistivities differ by ten orders of magnitude. In Chapter Three we explain the origin of this difference, and develop the semiconductor concepts required to understand the operation of transistors. We treat the electron as a particle that moves under the action of a field and postpone the wave mechanical treatment to Chapter Twelve. With the exception of the density of states in the conduction band, the electron as a particle approach is used to treat current flow, chapter four, in *p-n* junctions and metal-semiconductor barriers and Chapter Five, in bipolar transistors and field effect transistors.

Chapters Two through Five cover the motion of charge carriers in junctions and transistors. At this point we consider how to fabricate the circuit elements. We start Chapter Six with the structure of solids with emphasis on silicon and gallium arsenide, and discuss vacancies, interstitials and other defects. The diffusion process and introduction of dopants into semiconductors is treated in Chapter Seven. Ion implantation is used industry-wide for the controlled introduction of dopants and is presented in Chapter Eight.

Integrated circuit production processes are based on the contol by ion implantation and diffusion of dopants in the outer micron of the surface of the semiconductor. The connection to the individual devices and the interconnection between them requires the thin film technology discussed in Chapters Nine through Eleven. Chapter Nine covers thermal oxidation of silicon and the deposition and etching of oxide and nitride films. Chapter Ten is on metallization for silicon and gallium arsenide and the interpretation of binary and ternary phase diagrams. Thin film reaction kinetics are discussed in Chapter Eleven based on the formation of silicides and aluminides.

Gallium arsenide emits light and has a higher electron mobility than silicon. To understand the different behavior of silicon and gallium arsenide, we describe the electron as a wave and energy bands in Chapter Twelve. This chapter is a prelude to the description of band-gap engineering by heteroepitaxial growth of semiconductor layers with controlled composition and energy band gap. Chapter Thirteen describes heterojunction devices and epitaxial layers which are the key elements in lasers and high speed transistors. Chapter Fourteen describes heteroepitaxial growth and lattice matching in III-V semiconductors.

Chapter Fifteen tells how the semiconductor chips are mounted, bonded and packaged. It also covers soft errors and reliability.

## Course Organization

We have used this book in our undergraduate level course, primarily for sophomores and juniors, at Cornell University and portions of the material in undergraduate courses at Caltech and the University of California, San Diego. The topics are presented in a planned sequence to progress from electron transport and transistor operation to fabrication of devices. We treat the electron as a particle for description of a junction and then use wave behavior to discuss indirect and direct band-gaps.

The individual chapters and the entire book as well are written so that the basic material is presented first and the advanced concepts later. This allows flexibility in course presentation as some portions of the chapters or the chapters themselves can be omitted.

The problem sets are designed for large classes and for the most part have numerical answers. There is redundancy in the problems so some can be discussed in class and others assigned for home work. Answers are provided in the text. A solutions manual can be obtained from Macmillan Publishing Company.

## Acknowledgments

We would like to thank our colleagues who read the manuscript and made suggestions for revisions: Evan Colgan, IBM E. Fishkill; Karen Kavanagh, University of California, San Diego; Tom Kuech, IBM Watson Research Center; Paul Yu, University of California, San Diego. We would also acknowledge with pleasure our colleagues who provided photographs and data: Jun Amano, H.P. Labs; John Argyle, AT&T Allentown; M. Hirara, NEC Corporation; Joe McPherson, Texas Instruments, Kunio Nakamura, NEC Corporation; Simon Nieh, Caltech; Greg Olson, Hughes Research Laboratories; Richard Pashley, Intel; Scott Pesarcik, Cornell; John Poate, AT&T Bell Labs; Wen Tseng, Naval Research Labs; Chuck Woychik, IBM Endicott; and Jerry Woodall, IBM Watson Research Center. Professors John Weaver, Univ. of Minnesota, Eugene Haller, Univ. of California, Berkeley, Arthur S. Nowick, Columbia University, and G.P. Rodrigue, Georgia Tech provided useful comments as reviewers.

We are indebted to Magdalena Barandiaran, Michell Parks and Coraleen Rooney for typing the manuscript and to Ali Avcisoy and Jane Jorgensen for their drawings and art work. David Johnstone, Engineering Editor at Macmillan Publishers was a great help in guiding us through the various stages of production. John Travis of Macmillan was a master at handling the proofs.

Finally, we wish to thank our undergraduate students who have used the book and the undergraduate and graduate teaching assistants as well as our colleagues for their comments. Their suggestions provided invaluable help in the many revisions.

J.W.M.
S.S.L.

# Contents

CHAPTER 3
# Bands, Bonds and Semiconductors

51

CHAPTER 4
# Barriers and Junctions

82

CHAPTER 5
# Transistors: Bipolar and Field Effect

111

CHAPTER 6

# Crystallography and Crystalline Defects

142

CHAPTER 7

# Diffusion in Solids

183

CHAPTER 8

# Ion Implantation 222

CHAPTER 9

# Thermal Oxidation of Silicon and Chemical Vapor Deposition of Insulating Films 251

CHAPTER 10

# Metallization and Phase Diagrams 276

# List of Tables

# Symbols Frequently Used in Text[†]

| Symbol | Name and Units |
|---|---|
| $a$ | Lattice parameter, lattice constant  *145* |
| $C$ | Capacitance (F)  *41* |
| $D_n, D_p$ | Diffusion coefficient electron, holes ($cm^2/s$)  *43* |
| $D$ | Diffusion coefficient for impurities ($cm^2/s$)  *184* |
| $\mathscr{E}$ | Electric field (V/cm)  *29* |
| $E_F$ | Energy at the Fermi level (eV)  *67* |
| $E_G$ | Energy gap in semiconductors (eV)  *63* |
| $F$ | Flux of carriers, particles (no./$cm^2 \cdot s$)  *43, 184* |
| $I, J$ | Current (A), current density (A/$cm^2$)  *28* |
| $kT$ | Thermal energy (0.026 eV at 300 K)  *30* |
| $K_{th}$ | Thermal conductivity (W/cm $\cdot$ K)  *45* |
| $L_n, L_p$ | Diffusion length (cm) (electrons, holes)  *74* |
| $m, m^*$ | Free electron mass, effective mass (kg)  *66* |
| $n_n, n_p$ | Electron concentration, in $n$-type, in $p$-type ($cm^{-3}$)  *70* |
| $n_i$ | Intrinsic concentration of electrons ($cm^{-3}$)  *69* |
| $N_A, N_D$ | Concentration of acceptors, donors ($cm^{-3}$)  *58* |
| $N_C, N_V$ | Density of states in conduction and valence bands ($cm^{-3}$)  *66* |
| $p_p, p_n$ | Hole concentration in $p$-type, in $n$-type ($cm^{-3}$)  *70* |
| $R$ | Resistance ($\Omega$)  *32* |
| $v, v_d$ | Velocity, drift velocity (cm/s)  *28* |
| $V_O$ | Junction potential (V)  *82* |
| $Y$ | Young's modulus (N/$m^2$)  *20* |
| $\epsilon, \epsilon_r$ | Permittivity (F/cm), relative dielectric constant  *41* |
| $\varepsilon_s$ | Strain, $\varepsilon_s = \sigma_s/Y$  *20* |
| $\lambda$ | Characteristic diffusion length (cm)  *186* |
| $\mu_n, \mu_p$ | Mobility of electrons, holes ($cm^2/V \cdot s$)  *31, 36* |
| $\rho$ | Resistivity ($\Omega \cdot cm$)  *32* |
| $\sigma$ | Conductivity [($\Omega \cdot cm)^{-1}$]  *32* |
| $\sigma_s$ | Stress (N/$m^2$)  *20* |
| $\tau_n, \tau_p$ | Lifetime of electrons, holes (s)  *72* |

[†]Other symbols are defined as they are used in the text. See also the table "Physical Constants, Conversions and Useful Combinations" inside back cover.

# CHAPTER 1

# Integrated Circuits

## 1.1 Introduction

The worldwide prevalence of computers and information processing is a direct result of the widespread availability of the integrated circuits contained on small chips of silicon. The crucial point in the development of modern electronics was the joining—integration—of individual resistors, capacitors, and transistors on the surface of a semiconductor. The individual components exist, but rather than appearing as discrete components, they are formed in or deposited on the semiconductor and are connected by deposited metal layers. Figure 1.1a and b show optical micrographs of a *static random-access memory* (SRAM); part (a) is the overview of the 1.6 × 1.6 mm Si chip and part (b) is a portion outlined in the square in (a). Figure 1.1c and d are magnified views where a scanning electron microscope was used to obtain details of the memory array. It is in Fig. 1.1d that we see the electronic materials—metals, oxides, and semiconductors—that are of concern in this book.

The memory cell in Fig. 1.1d is made from silicon (Si), a semiconductor that has dominated integrated-circuit design and fabrication. Two-component semiconductors such as gallium arsenide (GaAs) are also being used in integrated circuits. Figure 1.2 is a schematic drawing of a metal–semiconductor field-effect transistor made on gallium arsenide. Transistors are essentially three-terminal (contact) structures that contain a source (or emitter) of charge carriers—electrons, in this case— and a drain (or collector) of charge carriers as well as a third contact, a gate (or base for a bipolar transistor) to modulate the flow of carriers. In the field-effect transistor shown in Fig. 1.2, the width of the high-field region under the gate controls the flow of carriers. Transistors are used as amplifiers or switches and their speed is determined by how fast the electrons move and how far the electrons must travel in transversing the gate or base region. Gallium arsenide has the advantage that its electrons have a higher mobility than electrons in silicon by a factor of about 6. Silicon, on the other hand, is less expensive and is easier to process than gallium arsenide. In both semiconductors, the goal is to make the devices as small as possible to decrease carrier travel distances (to increase speed of operation) and to pack as many devices as possible onto a given area—semiconductor real estate is expensive.

The goals of electronic materials science in integrated circuits are to understand the principles of device operation and the processes used to form the devices in the semiconductors so that well-designed circuits can be made with controlled, reproducible procedures. As the dimensions of the devices are scaled down to smaller and smaller sizes, the electronics engineers and process engineers—the distinction is blurred in materials science—must interact in design so that the structure can be produced with a high yield, or fraction, of chips fabricated without a fatal flaw.

## 1.2  Sizes of Transistors

A bit is either of the digits 0 and 1 used in the binary notation, so that "bits" are the basic unit of information in a computer system. In integrated-circuit memory

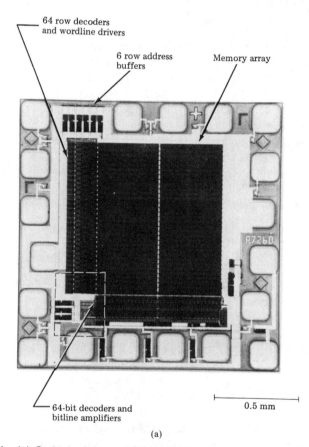

(a)

**FIGURE 1.1**  (a) Optical micrograph of an Si chip (1.6 × 1.6 mm) for an n-channel 4096-bit SRAM. Square area indicated in lower left-hand corner is magnified in (b). Area at the right portion of (b) is again magnified by scanning electron micrograph in (c) and a part of this is again magnified in (d), showing a single memory cell. From Marcus and Sheng, 1983.

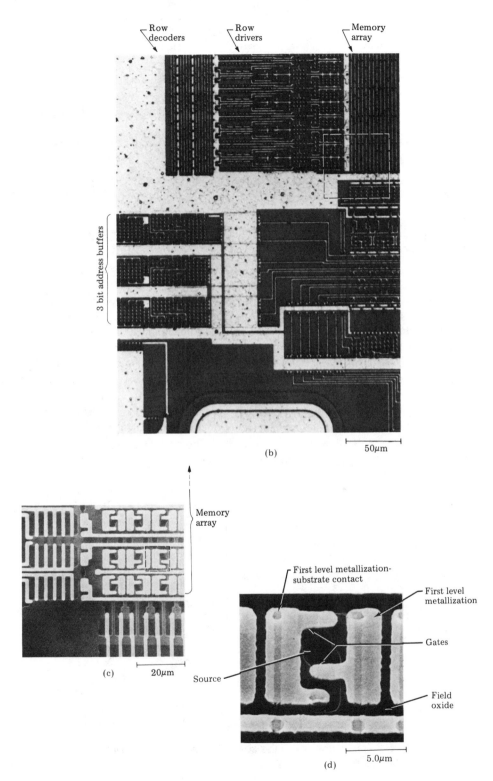

Row
decoders

Row
drivers

Memory
array

3 bit address buffers

(b)

50μm

Memory
array

(c)

20μm

First level metallization-
substrate contact

First level
metallization

Gates

Source

Field
oxide

(d)

5.0μm

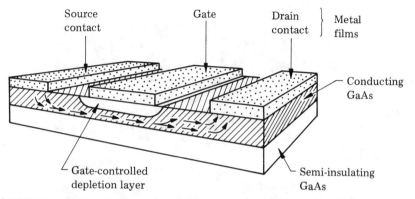

Source contact

Gate

Drain contact } Metal films

Conducting GaAs

Gate-controlled depletion layer

Semi-insulating GaAs

**FIGURE 1.2** Schematic drawing of a metal–semiconductor field-effect transistor (MESFET) showing electron flow from source to drain.

elements such as shown in Fig. 1.1, the bit is the circuit element used to store and transfer the electrical charge that represents one unit of information. The memory bit is generally a field-effect transistor and capacitor combination where the capacitor stores the charge and the transistor switches (or transfers) the charge.

The static random-access memory of Fig. 1.1 contains 4096 bits—a 4096-bit SRAM—in an area of $0.75 \times 0.9$ mm ($0.675$ mm$^2$ or $6.75 \times 10^{-3}$ cm$^2$) on a $1.6 \times 1.6$ mm chip. The average area per bit is $6.75 \times 10^{-3}$ cm$^2$ divided by 4096 (the number of bits) or $1.65 \times 10^{-6}$ cm$^2$. The lateral dimensions of semiconductor devices are given in micrometers or microns ($\mu$m); 1 micrometer equals $10^{-6}$ meter (m) or $10^{-4}$ centimeter (cm). The gate length, the extent of the gate between source and drain, is 1 $\mu$m in the memory cell shown in Fig. 1.1.

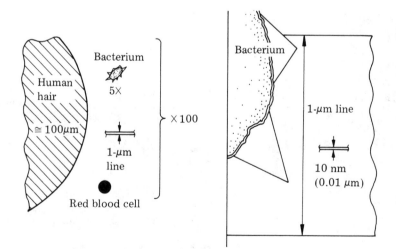

Human hair $\cong 100\mu$m

Bacterium 5×

1-$\mu$m line

Red blood cell

×100

Bacterium

1-$\mu$m line

10 nm (0.01 $\mu$m)

**FIGURE 1.3** Schematic drawing of the sizes of objects. (From National Nanofabrication Facility, Cornell University.)

One micrometer is small compared to most biological features, as shown in Fig. 1.3. The diameter of human hair is huge compared to 1 μm. Bacterium and red blood cells are several times larger than the width of a 1-μm line. The resolution of optical microscopes is about 0.25 μm, so that electron microscopes are used to reveal the details of integrated-circuit devices. One can proceed to finer dimensions, and lines as small as 20 angstroms (Å) have been made with electron beams; 1 angstrom = $10^{-8}$ cm. The spacing between atoms in a solid is about 2 to 4 Å. In this book we will use the meter rather than the angstrom as the length unit.

1 nanometer (nm) = 10 angstroms = $10^{-9}$ meter (m)

1 micrometer (μm) = $10^{-4}$ centimeter = $10^{-6}$ meter

1 centimeter (cm) = $10^{-2}$ meter.

If one looks at a field-effect transistor from the side, as shown in Fig. 1.4, the depth of the source and drain regions into the semiconductor is about 1 μm. As a general rule, all the electronic processes of charge transport in integrated circuits take place within the top 1 μm of the surface. The active volume of a typical device is area times thickness, where the area is about 100 μm² and the thickness is 1 μm, so that

$$\text{volume} \simeq 1 \times 100 \times 10^{-12} \text{ cm}^3 = 10^{-10} \text{ cm}^3.$$

The number $N_A$ of atoms per mole, Avogadro's number, is

$$N_A = 6.0220 \times 10^{23}. \tag{1.1}$$

The atomic density $N$ of atoms/cm³ is related to $N_A$ by

$$N = N_A \frac{\rho_m}{A} \tag{1.2}$$

where the mass density $\rho_m$ is in gram/cm³ and the $A$ is the atomic mass number (i.e., the number of protons and neutrons). The atomic mass of elements listed in periodic tables is the average mass, the average of the stable isotopes weighted by

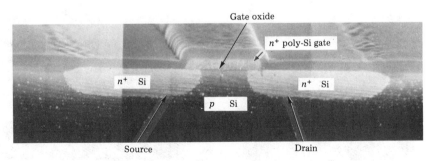

**FIGURE 1.4**    Scanning electron micrograph of the side view of a metal–oxide–semiconductor field-effect transistor (MOSFET) with a polycrystalline silicon (poly Si) gate and phosphorus-implanted ($n^+$) source and drain (junction depth of 1.14 μm) on p-type Si [From W.F. Tseng and B.R. Wilkins, *J. Electrochem. Soc.* *134,* 1258 (1987).]

their abundance. Silicon, for example, has an atomic mass of 28.086 and a mass density of 2.33 g, so that the atomic density of Si, $N_{Si}$, is

$$N_{Si} = \frac{6.022 \times 10^{23} \times 2.33}{28.086} = 4.996 \times 10^{22} \text{ atoms/cm}^3$$

$$= 5 \times 10^{22} \text{ atoms/cm}^3.$$

Therefore, the number of atoms in a device

$$N_{device} = N_{Si} \times \text{volume} \simeq 5 \times 10^{22} \times 10^{-10} = 5 \times 10^{10} \text{ atoms}$$

or

$$\frac{N_{device}}{N_A} \simeq \frac{5 \times 10^{10}}{6 \times 10^{23}} \simeq 10^{-13} \text{ mole (mol) of Si.}$$

The sizes and active volumes of electronic devices in integrated circuits are small. The recurrent goals in the electronic circuit industry are the further reduction in size to increase the packing density and the improvement in the control of process steps to increase the yield of devices. Both the size reduction and process control require knowledge of the properties of electronic materials—the topic of this book. For example, to achieve 1 million bits (1 Mb) in an array, not only is the length between source and drain reduced to less than 1 μm, but the capacitor can be folded into a trench as shown in Fig. 1.5 for a 4-Mb DRAM. Clearly, the fabrication of such devices requires exceptional control.

Such control is achieved by use of vacuum technologies and controlled ambients at various stages in the processing of integrated circuits. In vacuum technology, the basic relation is the ideal gas law:

$$PV = \bar{v}RT = N_m kT \tag{1.3}$$

where  $P, V$ = pressure [newtons (N)/m²] and volume (m³)
$T$ = absolute temperature, Kelvin (K) (K = °C + 273.16)

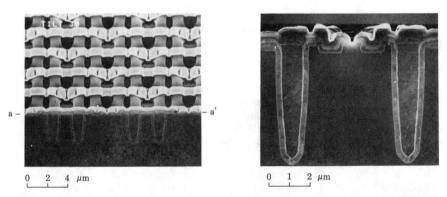

0  2  4 μm

0  1  2 μm

**FIGURE 1.5** Scanning electron micrograph of the memory array of a 4-Mb DRAM with 10-μm² cell size made with capacitors formed in trenches etched in the silicon. (From K. Nakamura, Nippon Electric Co., Japan.)

$N_m$ = total number of molecules

$\bar{\nu}$ = number of moles = $N_m/N_A$

$R$ = gas constant = 8.31 joules (J)/K · mol

$k$ = Boltzmann's constant = $R/N_A$ = 1.38 × $10^{-23}$ J/K.

The Standard International (SI) unit of pressure is N/m² (see Table 1.1), which is also 1 pascal (Pa). The torr is the unit of pressure based on the height of a column of liquid:

$$1 \text{ torr} = 1 \text{ mm Hg} = 133.3 \text{ N/m}^2$$

and

$$1 \text{ atmosphere (atm)} = 760 \text{ torr} = 1.013 \times 10^5 \text{ N/m}^2.$$

The number of molecules per unit volume at 1 atm and 25°C is

$$\frac{N_m}{V} = \frac{P}{kT} = \frac{1.013 \times 10^5}{1.38 \times 10^{-23} \times 298.16} = 2.46 \times \frac{10^{25}}{\text{m}^3} = 2.46 \times \frac{10^{19}}{\text{cm}^3}$$

In vacuum process technology, vacuums of $10^{-4}$ Pa (7.5 × $10^{-7}$ torr) are common, with the number of molecules per unit volume:

$$\frac{N_m}{V} = \frac{P}{kT} = \frac{10^{-4}}{1.38 \times 10^{-23} \times 298.16} = 2.43 \times \frac{10^{16}}{\text{m}^3} = 2.43 \times \frac{10^{10}}{\text{cm}^3}.$$

The average kinetic energy $\bar{E}_k$ of the molecules is

$$\bar{E}_k = \frac{3}{2} \frac{R}{N_A} T = \frac{3}{2} kT = \frac{1}{2} m_m v_a^2 \tag{1.4}$$

where $m_m$ is the mass of the molecular and $v_a$ is the average velocity, and the rate $r_c$ of molecular collisions per unit area per second is

$$r_c = \frac{N_m}{V} \frac{v_a}{4.} \tag{1.5}$$

where the factor of 4 is a geometrical correction. At a pressure of $10^{-4}$ Pa of nitrogen gas ($M$ = molecular weight = 28 g/mol = 28 × $10^{-3}$ kg/mol) at 25°C,

**TABLE 1.1**  Common Units and Conversions Used in Pressures and Gases

| |
| --- |
| 1 newton · meter (N · m) = 1 joule (J) = $10^7$ dyne (dyn) · cm = $10^7$ ergs |
| $1 \dfrac{\text{N}}{\text{m}^2}$ = 1 pascal (Pa) = 7.5 × $10^{-3}$ torr |
| $10^6$ N/m² = 10.2 kg/cm² = 146 lb/in² = 146 psi |
| 1 torr = 133.3 N/m² = 133.3 Pa |
| 1 atm = 760 torr = 1.013 × $10^5$ N/m² |
| 1 bar = $10^5$ N/m² = 750 torr |
| 1 dyn/cm² = 0.1 N/m² |

$$v_a = \sqrt{\frac{2\overline{E_k}}{m_m}} = \sqrt{\frac{3RT}{M}} = 421 \text{ m/s}$$

so that

$$r_c = 2.43 \times \frac{10^{16} \times 421}{4} = \frac{2.55 \times 10^{18}}{\text{m}^2 \cdot \text{s}} = 2.55 \times 10^{14} \frac{1}{\text{cm}^2 \cdot \text{s}}.$$

This rate of collisions can be put in perspective if we consider the average number of atoms on the surface of silicon, 1 monolayer, and assume that silicon has a simple cubic structure (see Section 6.2 for an exact treatment) so that

$$1 \text{ monolayer} \approx (N_{\text{Si}})^{2/3} = \left(\frac{5 \times 10^{28}}{\text{m}^3}\right)^{2/3} = \frac{13.6 \times 10^{18}}{\text{m}^2} = \frac{13.6 \times 10^{14}}{\text{cm}^2}.$$

Consequently, at a nitrogen pressure of $10^{-4}$ Pa, the time required for 1 monolayer of nitrogen atoms ($2 \times N_m$) to form on a silicon surface assuming that all molecules stick is

$$\frac{1 \text{ monolayer}}{2r_c} = \frac{13.6 \times 10^{18}}{2 \times 2.55 \times 10^{18}} = 2.67 \text{ s.}$$

This time is sufficiently short compared to normal processing times so that pressures below $10^{-4}$ Pa are required in critical vacuum processing steps in which the surface of the semiconductor must be free from contamination from the gas in the ambient.

In SI (Standard International) units, the unit of force is the newton (N), roughly the weight of an apple. The newton is the force required to accelerate 1 kilogram of matter by 1 meter per second in a time interval of 1 second ($F = ma$). This concept allows direct conversion to dynes, the force required to accelerate 1 gram of matter by 1 centimeter per second:

$$1 \text{ N} = \frac{1 \text{ kg} \cdot \text{m}}{\text{s}^2} = \frac{10^3 \text{ g} \times 10^2 \text{ cm}}{\text{s}^2} = \frac{10^5 \text{ g} \cdot \text{cm}}{\text{s}^2} = 10^5 \text{ dyn.}$$

The pascal (Pa) equals 1 N/m², so that

$$1 \text{ Pa} = \frac{1 \text{ N}}{\text{m}^2} = \frac{10^5 \text{ dyn}}{(10^2 \text{ cm})^2} = 10 \text{ dyn/cm}^2.$$

Conversions among the various units are given in Table 1.1.

## 1.3 The Time Domain

In the design of integrated circuits, the goal is to pack devices more closely together and speed up their performance without sacrifices in cost or reliability. In the time domain,

$$1 \text{ hertz (Hz)} = 1 \frac{\text{cycle}}{\text{second}} = \frac{2\pi \text{ radians}}{\text{second}}$$

and high-frequency operation of devices can be in the range 10 to 100 gigahertz:

$$1 \text{ gigahertz (GHz)} = 10^9 \text{ hertz.}$$

The velocity $c$ of light in vacuum is

$$c = 2.998 \times 10^8 \text{ meters/second} \simeq 3 \times 10^8 \text{ m/s} \tag{1.6}$$

and the wavelength $\lambda$ of light is

$$\lambda = \frac{c}{\nu} \tag{1.7}$$

where the frequency $\nu$ is units of inverse seconds, $(\text{seconds})^{-1}$ or $\text{s}^{-1}$. The angular frequency $\omega$ in radians per second is

$$\omega = 2\pi\nu. \tag{1.8}$$

Visible light with wavelengths between 400 nm (0.4 $\mu$m) and 700 nm (0.7 $\mu$m) have frequencies of 7.5 and $4.3 \times 10^{14}$ Hz, respectively. For high-speed communications, wavelengths can lie in the millimeter (mm) range.

The transit time for a field-effect transistor is the time required for an electron to move from the source to the drain. For a gate length $L$, the transit time $T_R$ is

$$T_R = \frac{\text{length}}{\text{velocity}} = \frac{L}{v} . \tag{1.9}$$

For a device to operate at 1 gigahertz, transit times of 1 nanosecond are required, where

$$1 \text{ nanosecond (ns)} = 10^{-9} \text{ second}$$

and

$$1 \text{ picosecond (ps)} = 10^{-12} \text{ second.}$$

With a gate length of 1 $\mu$m and $T_R = 10^{-9}$ s, the velocity

$$v = \frac{L}{T_R} = \frac{10^{-4} \text{ cm}}{10^{-9} \text{ s}} = 10^5 \text{ cm/s,}$$

a value that is easy to achieve in silicon. However, at 100-fold higher frequencies, a velocity of $10^7$ cm/s is required—a value near the maximum velocity at which electrons can move in semiconductors. In these high-speed applications, gallium-arsenide devices have an advantage because of the higher electron mobility in gallium arsenide than in silicon. Of course, in silicon integrated-circuit technology the push is to reduce the gate length $L$. In the present technology, the minimum gate length

$$L_{\text{min}} \simeq 0.1 \ \mu\text{m.}$$

## 1.4 Interconnection and Metallization

In integrated circuits one usually connects the various electronic devices with metal lines (Fig. 1.6) in order to carry current or transport the charge. These metal lines are separated from the substrate, except in the contact area, by insulating layers of dielectric material, usually silicon dioxide, $SiO_2$. The metal films are deposited on the substrate, patterned using lithographic techniques and etched to remove the metal except in the desired lines.

Interconnect lines are deceptively simple—especially the first-level metallization (i.e., contact to the Si) shown in Figs. 1.1d and 1.6, where aluminum is often used. The lines, however, must retain their structural integrity without deformation, loss of adhesion, or penetration into the silicon during the temperature cycles encountered in subsequent processing or the current densities involved in actual device operation. The interconnects should be designed to have the minimum electrical resistance so that the voltage $V$ is minimized for a given current $I$. The power, $P$, in joules per second or watts (W) dissipated in an interconnect is

$$P = IV \tag{1.10}$$

where the current $I$ is in amperes and the voltage $V$ is in volts.

These constraints become major issues when the next metal levels are added to provide the connections to other circuit elements (Fig. 1.7). Rather than aluminum in contact with silicon, one can have a metal–silicon compound, a silicide; the gate might be polycrystalline silicon (poly Si); and the connections between the metal lines or silicides are made by metal plugs in the holes or vias in the oxide insulating

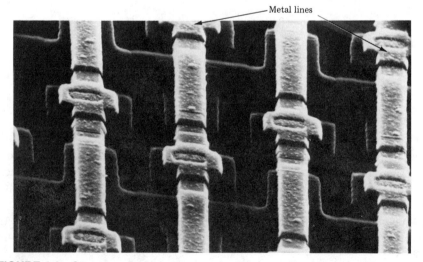

**FIGURE 1.6** Scanning electron micrograph of the metallization and oxides on an array of field-effect transistors.

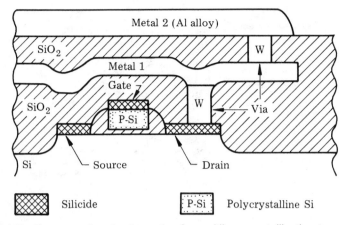

**FIGURE 1.7** Cross-sectional schematic of a multilayer metallization to a polycrystalline-silicon-gate field-effect transistor with silicide (metal–silicon compound) contacts to the source and drain regions. Tungsten is deposited in the openings (vias) in the oxide layer to provide vertical interconnects.

layers. The proper selection of metals and silicides is a crucial issue in integrated circuits; the properties of the materials are described in subsequent chapters.

The metal lines and insulating layers as well as the device itself have capacitance $C$ associated with them. The capacitance and the resistance $R$ constitute a time constant $\tau_{RC}$ that determines the flow of charge in high-frequency operation,

$$\tau_{RC} = RC \text{ (seconds)} \tag{1.11}$$

where the resistance $R$ is in ohms and the capacitance $C$ is in farads. The choice of materials in the metallization is often dictated by the requirement for a low $RC$ time constant. As the circuit moves through the fabrication process a memory array (Fig. 1.8a) becomes covered with metals, oxides, and protective coatings, as shown in part (b) of the figure.

## 1.5 Lithography and Patterning

The original definition of lithography is the print-making process of treating a stone surface so that certain areas accept ink and others reject ink. In intergrated-circuit technology, lithography is the process of forming patterns on the surface of a semiconductor wafer.

The sequence of steps required to form a pattern on a silicon wafer is shown in Fig. 1.9. Silicon with a layer of silicon dioxide ($SiO_2$) is coated with a layer of photoresist, a light-sensitive organic film similar to the emulsion on a photographic film. The resist is then exposed to ultraviolet light that passes through the clear portions of a "mask," a glass plate with precise pattern of opaque material that block the light. In the case of a positive resist, the areas of opaque material on the

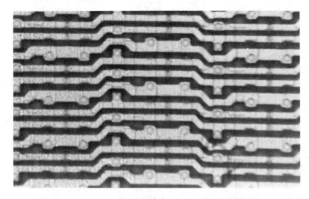

(a)

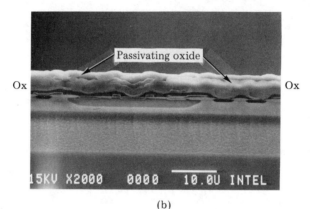

(b)

**FIGURE 1.8** Scanning electron micrographs of an Intel static random-access memory (SRAM) 16K (4096 × 4 bit organization) with complementary metal–oxide–semiconductor (CMOS) field-effect transistors: (a) top view of a memory array; (b) cross-sectional view after the upper, passivating oxide layers have been deposited. (Supplied by R.D. Pashley, Intel Corporation.)

surface of the mask are located where $SiO_2$ is to remain on the surface. After exposure to light in selected areas, the photoresist long-chain molecules have been broken (scission process) and can be removed in an organic solvent. The remaining photoresist after hardening is resistant to hydrofluoric acid (HF), which etches the $SiO_2$ but does not etch the surface of silicon. By immersing the photoresist-covered surface in HF, the $SiO_2$ layer is removed in selected areas not covered by the resist. The final step is to remove (strip off) the photoresist. The geometric pattern on the mask has now been transferred to the silicon dioxide layer, a true example of the classic lithography process.

There are many variations on this lithography process. We have described a *positive* resist in which exposure to light breaks the polymer bonds so that the exposed resist is removed in a solvent. There are *negative* resists, where exposure to light causes cross-linking of the polymer bonds, which makes the exposed resist

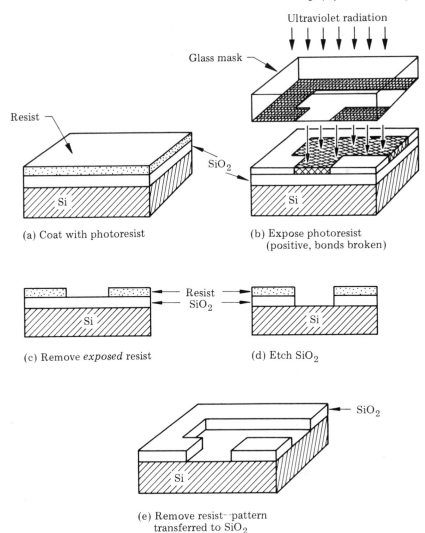

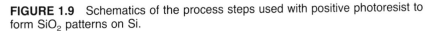

**FIGURE 1.9** Schematics of the process steps used with positive photoresist to form $SiO_2$ patterns on Si.

harder to remove than that in the unexposed areas. A negative resist remains on the surface area in regions where it was exposed. There are electron-beam (e-beam) resists which are exposed by a finely focused electron beam. These resists are often used in generating the patterns on the mask. A computer-generated program is used in a scanning electron microscope to sweep the electron beam across the resist-covered glass plate to form a precise pattern. X-ray lithography is also under development using a synchrotron as a source of x-rays. As device dimensions are scaled down to $\frac{1}{4}$-$\mu$m gate lengths, higher-energy photons, x-rays rather than ultraviolet (UV) light, with shorter wavelengths, are used to obtain the required line resolution. Finely focused ion beams in an ion microscope can also be used to define patterns in a resist.

A *lift-off* process (Fig. 1.10) is often used to define metal lines. The substrate is covered with photoresist, which is then exposed in patterns so that openings are made where the metal is to remain. A metal film is then deposited over the surface and the resist is removed. Portions of the metal film on top of the resist will be removed, leaving the metal in patterns on the substrate.

We have illustrated pattern definition by use of chemical etching in which a liquid or a gas is used to remove any material not protected by a hardened resist. Dry etching with ions can also be used to etch unprotected material. For example,

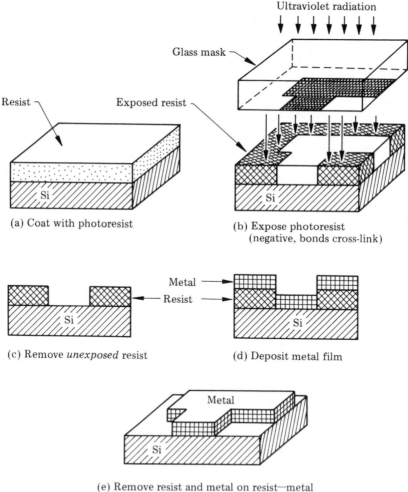

(a) Coat with photoresist

(b) Expose photoresist
(negative, bonds cross-link)

(c) Remove *unexposed* resist

(d) Deposit metal film

(e) Remove resist and metal on resist—metal
pattern remains on Si

**FIGURE 1.10** Schematics of the steps used in the lift-off process to form metal patterns on Și.

Resist pattern on $SiO_2$ on Si

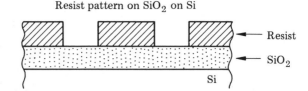

(a) Wet oxide etch (HF solution)          (b) Dry oxide etch (fluorine ions)

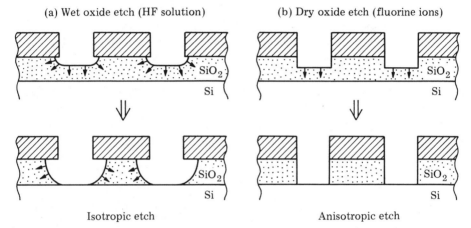

Isotropic etch                          Anisotropic etch

**FIGURE 1.11**  Formation of grooves in $SiO_2$ by use of resist patterns and (a) isotropic, wet-etch (HF) solutions or (b) anisotropic, dry-etch (fluorine ion) plasmas.

*plasma etching* uses fluorine or chlorine ions, which attack $SiO_2$. Reactive-ion beam etching (RIBE) uses reactive ions, which are accelerated through a potential before bombarding the surface. These dry-etching processes are widely used in integrated-circuit processing. Highly directional, highly anisotropic, etching profiles can be achieved as shown in Fig. 1.11. Straight-wall cuts can be made in oxides or deep grooves formed in the semiconductor as shown for the trench capacitor in Fig. 1.5. Finally, the resist can be removed in a dry process, *resist ashing*, by oxidizing (burning) the resist in an oxygen plasma system.

Lithography, from masks to etches, is a major part of the fabrication of integrated circuits. A complete integrated circuit generally requires between 10 and 20 lithography processing steps. It is a specialized topic, with conferences and books devoted to it, so we will not cover the topic further here except where properties of the electronic materials are directly involved.

## 1.6 Ingots and Wafers

Integrated circuits originate with the growth of single crystals of silicon and gallium arsenide. The crystals are grown from the melt with the melting temperatures

$$T_m(Si) = 1414°C$$
$$T_m(GaAs) = 1238°C$$

(1.12)

well above 1000 degrees Celsius (°C). In the Czochralski technique (Fig. 1.12) the melt is contained in a crucible, quartz (crystalline $SiO_2$) for silicon or graphite (carbon) for gallium arsenide, and is kept in a molten state by radio-frequency (RF) inductive heating of the crucible. Arsenic is a volatile element. Gallium arsenide will decompose with the loss of gaseous arsenic unless an arsenic overpressure of about 1 atm is maintained during growth. Liquid-encapsulated Czochralski (LEC) uses a capping layer of an inert liquid (usually boron trioxide, $B_2O_3$) to cover the exposed melt. For growth, a seed crystal is inserted into the melt and then slowly withdrawn. Crystal growth occurs by freezing at the interface between the solid seed and melt. Crystal growth proceeds by the successive additions of layers of atomic planes at the liquid–solid interface. The growth rate is about 10 μm/s.

After growth and surface finishing, the ingot is cut on a diamond-tipped saw into slices or wafers 100 to 250 μm thick (depending on the diameter of the wafer). The wafers are evaluated for dislocation content and resistivity. The wafers are etch polished on one surface and then placed in individual slots in plastic trays for delivery to the fabrication lines.

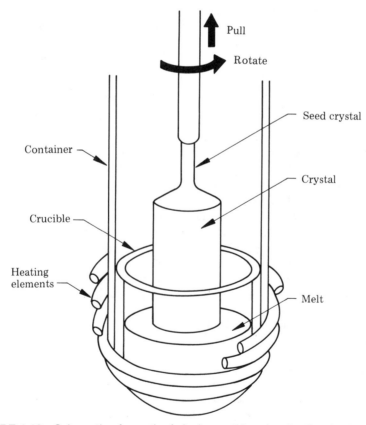

**FIGURE 1.12** Schematic of growth of single crystals using the Czochralski method.

Crystal growth is a highly specialized process, and generally single-crystal ingots are not grown by the group or even the company that carries out the circuit fabrication. In this book we focus on the properties of the electronic materials rather than the crystal growth processes.

# 1.7 Oxidation and Doping of Silicon

The growth of large perfect single crystals of silicon—over 1 m long and 20 cm in diameter—is the ultimate in the art of manufactured crystals. High-purity silicon with only one electrically active impurity for every $10^9$ silicon atoms can be grown. The population of China is about 1 billion, so the impurity concentration of such silicon is equivalent to one foreigner in all of China.

Such high-purity silicon is impressive as a technological achievement but it is useless in integrated circuits. We must destroy the purity by introducing controlled amounts of electrically active impurities—called dopants—into the silicon. Dopants determine current flow in semiconductors, and the introduction of electron-rich dopants (donors) and electron-deficient dopants (acceptors) determines the formation of $p$-$n$ junctions, which are the heart of transistors.

After the wafers have been cut from the ingot and etch polished, one of the first steps in integrated-circuit processing is growth of an oxide layer on the silicon. Controlled thicknesses of silicon dioxide ($SiO_2$) are grown by placing the wafers in a quartz-tube furnace at temperatures of 900 to 1100°C in an oxidizing ambient, generally dry oxygen or water vapor (Fig. 1.13). The oxidizing species (oxygen) diffuses through the growing oxide layer to reach the silicon surface, where the reaction is

$$Si + O_2 \longrightarrow SiO_2 . \tag{1.13}$$

The oxide layer contains $2.2 \times 10^{22}$ molecules/cm³ of $SiO_2$ (density of 2.2 g/cm³) and consumes an amount of silicon that is 0.45 times the thickness of the oxide layer.

Oxide layers can be grown in selected areas of the wafer by depositing and patterning nitride layers as shown in Fig. 1.13b. The nitride layers act as a barrier, masking portions of the silicon from the penetration of the oxidizing species. During thermal processing the oxide layer is formed in the exposed regions, with some lateral penetration beneath the nitride mask.

The oxide layers themselves act as barriers for the introduction of dopants into the silicon. Perhaps the simplest example of the masking properties of $SiO_2$ is shown in Fig. 1.14, where dopants such as arsenic are introduced by ion implantation. The energetic As ions penetrate the exposed portion of silicon but are blocked from the silicon in areas covered by $SiO_2$. The energetic dopant ions typically penetrate less than 0.1 μm into the silicon, depending on the energy of the ion beam. To distribute the dopants deeper into the silicon, the high-temperature process step of diffusion can be used.

(a) Thermal oxidation of Si–SiO₂ growth

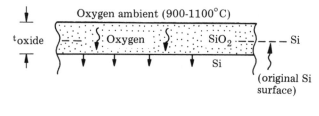

(b) Selective oxidation

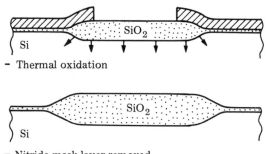

(c) Cross-section TEM

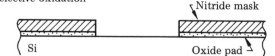

**FIGURE 1.13**   (a) Thermal oxidation of silicon by growth of SiO₂; (b) selective oxidation by use of nitride masks; (c) cross-sectional view of a 1.0-μm thick oxide grown with a nitride mask. [From Marcus and Sheng, (1983).]

Oxide and nitride layers are deposited on silicon during later stages of circuit fabrication. These films are used as insulating layers between metal lines as shown in Fig. 1.7, or as an insulating protective cover, as shown in Fig. 1.8. The deposited oxide layers can also be used as masks for selected-area ion implantation or diffusion. The layers are generally deposited by chemical vapor deposition (CVD) techniques in which silane (SiH₄) and oxygen react to form SiO₂:

$$SiH_4 + O_2 \longrightarrow SiO_2 + 2H_2 . \tag{1.14}$$

Ion beam-100 keV As$^+$

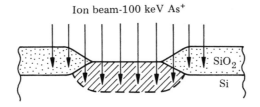

(a) Ion implantation of dopants (As)

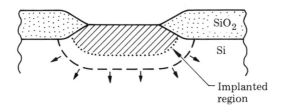

(b) Drive-in diffusion-950-1050°C

**FIGURE 1.14**    (a) Ion implantation of dopants (As) and (b) subsequent drive-in diffusion (950–1050°C) of the dopants.

The reactions can take place at 450°C or as low as 250°C, depending on the mixtures and pressures, and whether plasma enhancement is used. Silicon nitride films can be grown by chemical vapor deposition, which involves the reaction of silane gas and ammonia:

$$3SiH_4 + 4NH_3 \longrightarrow Si_3N_4 + 12H_2 . \tag{1.15}$$

Oxidation of Si and deposition of insulating layers are described in Chapter 9.

## 1.8  Chips and Bonds

The integrated circuits—or rather, the integrated systems of microprocessors, memories, and other devices which contain tens of thousands or even millions of transistors—are laid out in rectangular arrays on the wafer. Each rectangle defines the boundary of the chip that contains the complete integrated circuit. The wafers are now typically 150 mm (6 in.) in diameter and the chips may be 5 to 10 mm on an edge, depending on the number of bits on each chip, about 64,000 (64 k) or 1 million (1 M) in this example. The surface of the wafer is then cut along the boundary lines between the chips. Each chip is then attached to a metallized substrate (chip carrier), which is often part of the final assembled component or package. Heating is involved in the attachment (bonding of chip to substrate) since alloying or soldering processes are used and the metallized substrate must be chosen to have the same linear thermal expansion coefficient as the silicon to prevent thermal fracture. Attachment and bonding are described in Chapter 15.

The linear thermal expansion coefficient $\alpha$ is defined as the rate of fractional change of length $l$ with temperature $T$:

$$\alpha = \frac{1}{l}\frac{dl}{dT} \tag{1.16}$$

and values of $\alpha$ near room temperature are $2.6 \times 10^{-6}$ per degree Celsius for silicon and $6.86 \times 10^{-6}/°C$ for gallium arsenide. Values of $\alpha$ for Al and Au are 23.6 and $14.2 \times 10^{-6}/°C$, respectively. To provide better-matched thermal coefficients, alloys such as Kovar (an alloy of Fe, Ni, and Co with $\alpha \simeq 5 \times 10^{-6}/°C$) are used as the metallized substrate.

The circuits on the chip must be connected with the outside world. The package frame has metal lines, leads, which provide this access to the outside world. At the periphery of the chip (Fig. 1.15) are rectangular pads to which are connected all the memory elements—resistors, capacitors, and transistors—in the chip. These pads are aluminum layers that are deposited during the metallization and patterning process. The chip circuit-to-package connections are made by gold (Au) wires that are bonded to the pads and to the metal lines on the package frame (Fig. 1.16). These Au bond wires typically have diameters of 25 to 75 $\mu$m (0.001 to 0.003 in., called 1- or 3-mil wires) and lengths of 3 to 4 mm. They are more flexible than a rigid type of connection. The fabrication of the package imparts mechanical stresses in the wire bond and pull tests (a 5-g weight attached to the wire) can be used to ensure good interface bonding.

Stress $\sigma_s$ is the load or force $F$ per area $A$,

$$\sigma_s = \frac{F}{A} \tag{1.17}$$

which for 5 g ($5 \times 10^{-3}$ kg) on a 25-$\mu$m wire (area equals $\pi d^2/4 = 4.9 \times 10^{-6}$ cm$^2$) is a stress of $1.02 \times 10^3$ kg/cm$^2$. In SI (Standard International) units the unit of force is the newton:

$$1 \text{ N} = 0.102 \text{ kg}_{force} \tag{1.18}$$

and the SI unit of stress is newtons per square meter:

$$10^6 \text{ N/m}^2 = 10.2 \text{ kg/cm}^2. \tag{1.19}$$

As 1 kg/cm$^2$ = $9.8 \times 10^4$ N/m$^2$, the stress in the Au wire is $1.0 \times 10^8$ N/m$^2$.

Strain $\varepsilon_s$ is the change in length $dl$ over the length:

$$\varepsilon_s = \frac{dl}{l} \tag{1.20}$$

and the ratio of stress to strain is Young's modulus $Y$, a property of the material,

$$Y = \frac{\sigma_s}{\varepsilon_s} \tag{1.21}$$

in units of N/m$^2$ or dyn/cm$^2$. The statement that stress is directly proportional to

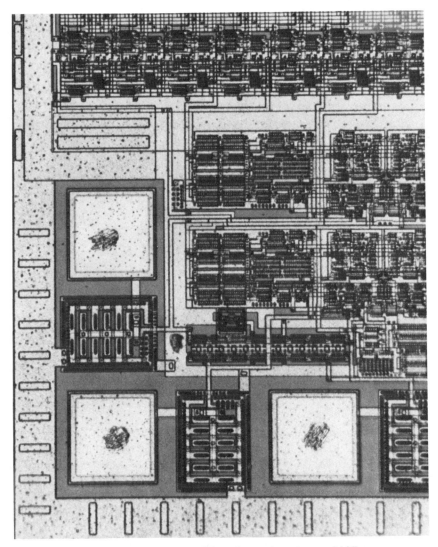

**FIGURE 1.15**  Top view of an Intel 16K static random-access 5167 memory (SRAM) where the squares at the side and lower edges are 100-μm Al thin-film pads contacted by wire bonds. (From R.D. Pashley, Intel Corporation.)

strain is Hooke's law—a law that holds fairly well for small strains of less than 1%. Young's modulus for most materials is around $10^{11}$ N/m$^2$ ($10^{12}$ dyn/cm$^2$). For Si it is $1.9 \times 10^{11}$ N/m$^2$; for gold, about $8 \times 10^{10}$ N/m$^2$; and for Al, $7 \times 10^{10}$ N/m$^2$. The tensile strength is the stress (N/m$^2$) required to break the wire. Gold has a tensile strength of about $1.3 \times 10^8$ N/m$^2$. During the pull test the weak point is usually not the wire but the bond to the aluminum pad.

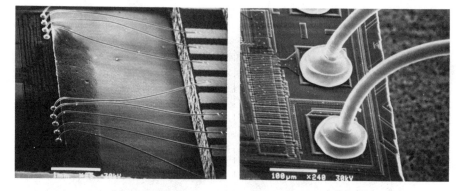

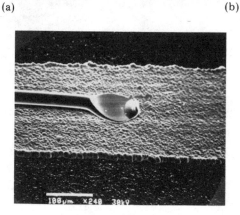

**FIGURE 1.16** Scanning electron micrographs: (a) Au-wire bonds between chip and chip carrier; (b) ball bond to the chip pad; (c) wedge bond to the metal lines. [From C. Woychik and R. Senger, in *Principles of Electronic Packaging* edited by D.P. Seraphim, R.C. Lasky and C-Y. Li, McGraw-Hill, NY (1989).]

Of course, during the initial elastic deformation as the wire stretches, its diameter decreases. The Poisson ratio $\mu$ is

$$\mu = \frac{\text{lateral strain}}{\text{longitudinal strain}} . \tag{1.22}$$

When the volume of the sample remains constant during plastic deformation, $\mu = \frac{1}{2}$. In silicon, Poisson's ratio is 0.27 (Beadle et al., 1985). The shear modulus $G$ is often given where $G = C_{44}$ in terms of tabulated elastic constants, $C$ (Beadle et al., 1985). For an isotropic medium $Y = 2G(1 + \mu)$ with $\mu = C_{12}/(C_{12} + C_{11})$.

There are other methods for connection to the outside world. One technique is to place solder balls (Fig. 1.17) on the aluminum pads that are coated with thin films of chrome and copper–tin to provide a reliable joint. The solders are tin–lead alloys (soft solders) that have the strength and ductility to meet the assembly requirements. The use of solder balls [about 0.005 in. (125 $\mu$m) in diameter] allows

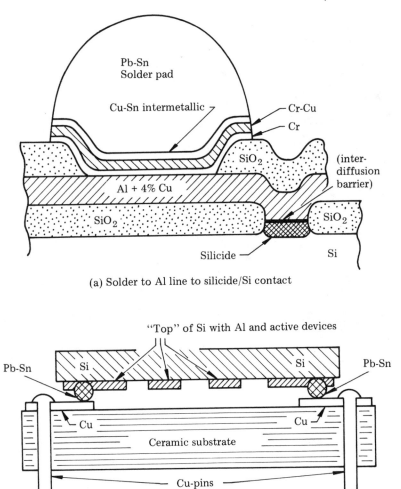

(a) Solder to Al line to silicide/Si contact

(b) Flip-chip with Pb-Sn solder to Cu

**FIGURE 1.17** (a) Schematic of Pb–Sn solder ball on Al metallization to a Si device showing the silicide contact and the diffusion barrier to block the reaction between Al and silicide; (b) cross section of the flip-chip method of contacting the Si chip to the carrier.

the connection to the copper leads on the chip carrier to be made without wire bonds. The chips are placed face down—the flip chip—on the metal leads. Surface tension of the molten solder joint during heat treatment aligns the chip and provides a mechanically sound joint upon solidification of the solder. The use of the flip chip allows a greater number of output leads in the package. The copper leads can be printed on a plastic (polyimide, usually) carrier which can contain over 100 output leads.

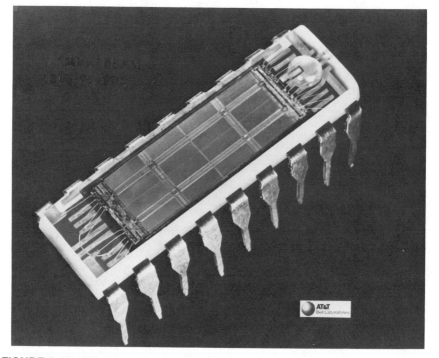

**FIGURE 1.18** Plastic-encapsulated 1-Mb dynamic random-access memory. (From J. Argyle, AT & T Bell Labs, Allentown, PA.)

After wire-bonding or soldering (flip-chip) connections are made, the unit is sealed in the final package. Figure 1.18 shows a memory array in final form after packaging in plastic materials. The package can then be connected to a circuit board. Assembly and packaging are described in Chapter 15.

## GENERAL REFERENCES

BARBE, D. F., ed., *Very Large Scale Integration (VLSI),* 2nd ed., Springer-Verlag, Berlin, 1982.

BEADLE, W. E., J. C. C. TSAI, and R. D. PLUMMER, eds., *Quick Reference Manual for Silicon Integrated Circuit Technology,* Wiley, New York, 1985.

BRODIE, I., and J. J. MURAY, *The Physics of Microfabrication,* Plenum Press, New York, 1982.

COLCLASER, R. A., *Microelectronics: Processing and Device Design,* Wiley, New York, 1980.

GHANDHI, S. K., *VLSI Fabrication Principles,* Wiley, New York, 1983.

JAEGER, R. C., *Introduction to Microelectronic Fabrication,* Addison-Wesley, Reading, Mass., 1988.

MALY, W., *Atlas of IC Technologies,* Benjamin/Cummings, Menlo Park, Calif., 1987.

MARCUS, R. B., and T. T. SHENG, *Transmission Electron Microscopy of Silicon VLSI Circuits and Structures,* Wiley, New York, 1983.

MEAD, C., and L. CONWAY, *Introduction to VLSI Systems,* Addison-Wesley, Reading, Mass., 1980.

PRINCE, P., and G. DUE-GUNDERSEN, *Semiconductor Memories,* Wiley, New York, 1983.

SERAPHIM, D. P., R. C. LASKY, and C.-Y. LI, *Principles of Electronic Packaging,* McGraw-Hill, New York, 1989.

SHACKELFORD, J. F., *Introduction to Materials Science For Engineers,* Macmillan, New York, 1988.

STREETMAN, B. G., *Solid State Electronic Devices,* 2nd ed., Prentice-Hall, Englewood Cliffs, N.J., 1980.

SZE, S. M., ed., *VLSI Technology,* McGraw-Hill, New York, 1983.

WOLF, S., and R. N. TAUBER, *Silicon Processing for the VLSI Era,* Lattice Press, Sunset Beach, Calif., 1986.

## General Journal References

*IEEE Spectrum* (Institute of Electrical and Electronics Engineers, New York)
*Solid State Technology* (Penwell Publishing, New York)

## PROBLEMS

**1.1. (a)** The insulator $Si_3N_4$ has a mass density of 3.1 $g/cm^3$. What is the molecular density (molecules/$cm^3$)? If you grew a $10^{-5}$-cm-thick layer of $Si_3N_4$ on a Si wafer, how many atoms of Si/$cm^2$ would be consumed, and how thick (cm) would be the layer of Si consumed?

**(b)** Aluminum has an atomic density of $6.02 \times 10^{22}$ atoms/$cm^3$. What is the mass density (g/$cm^3$)? If the Al atoms are all stacked uniformly in a cubic array, how many atoms/$cm^2$ are in the top atomic layer?

**1.2. (a)** An ingot of Si 100 mm in diameter is grown at a rate of 10 μm/s. How many atoms per second recrystallize?

**(b)** A 1 × 1 cm seed crystal 10 cm long supports the ingot. If the ingot is 1 m long, what is the stress in the seed crystal? If Young's modulus is $10^{12}$ $dyn/cm^2$, what is the strain?

**1.3. (a)** Show that a material that maintains constant volume during plastic deformation has Poisson's ratio = $\frac{1}{2}$ for elastic flow.

**(b)** Poisson's ratio for Si is 0.27. What does that imply?

**1.4.** The linear expansion coefficient α of Si at 100°C is $3 \times 10^{-6}$/°C.

**(a)** How much would a 10-mm chip increase in length if heated by 300°C, assuming that α is constant?

**(b)** In the flip-chip design, the Si chips are mounted on a sintered alumina ceramic substrate ($\alpha = 6 \times 10^{-6}$/°C). If one end of the chip is fixed, how much does the other end move relative to the substrate for a 300°C temperature change? Compare this distance to the diameter $\alpha$ of the solder ball ($d = 0.005$ in. $= 5$ mils).

**1.5.** Describe where the photomask would be opaque if you used a positive resist in defining a metal line $L$ in the lift-off process.

**1.6.** At the maximum electron drift velocity ($10^7$ cm/s), what is the transit time in a $\frac{1}{4}$-$\mu$m-gate-length field-effect transistor? How far would light travel in that time?

**1.7.** How many transistors could you place on the head of a pin?

# 2

# Current Flow and Capacitance

## 2.1 Introduction

Control of the flow of electrical charge underlies the discussion of the fabrication and operation of integrated circuits. One of the elements in an integrated circuit is the metal thin-film line that connects various junction regions on the semiconductor chip. It is shown in simplest form in Fig. 2.1 and idealized as a rectangular bar of length $L$ and area $A$.

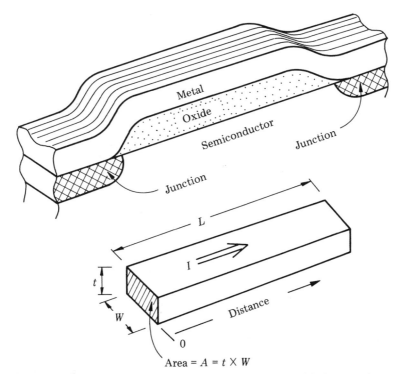

**FIGURE 2.1** Metal line connecting two junctions over an oxide layer and an idealized representation of a metal-line interconnect.

An electric current is the flow of electric charge, in this case the flow of electrons. If the number of electrons per unit volume is $n$, the current flow is

$$I(\text{coulombs/second}) \ (C/s) = q n v_d A \tag{2.1}$$

where $v_d$ is the velocity component directed along the bar, the drift velocity, and $q$ is the charge on the electrons,

$$q = -e = -1.6 \times 10^{-19} \ C \tag{2.2}$$

where we use the symbol $e$ to denote the absolute magnitude of the charge on the electron. The fact that $q$ is negative means that the flow of electrons is in the direction opposite to that of the conventional current (which is the direction of flow of positive charge), so that the velocity $v_d$ in eq. (2.1) is also negative. The unit of current is the ampere (A), the number of coulombs per second.

To describe the current flow in a cross section of area $A$ we define the current density

$$J = \frac{I}{A} \quad A/cm^2. \tag{2.3a}$$

Although Standard International (SI) units give length in meters, we will use the conventional units of semiconductor electronics, with length in centimeters ($10^{-2}$ m) or in nanometers ($10^{-9}$ m). Using eq. (2.1), the current density is

$$J = n e v_d \tag{2.3b}$$

where $n$ is the number of electrons/cm$^3$, $e$ the absolute charge on the electron, and $v_d$ the velocity in centimeters per second (cm/s). Here the subscript $d$ is used to indicate that it is the drift velocity of the electrons.

In this chapter we are concerned with the concentration of charge carriers and their drift velocity and the relation of these two parameters to the materials used in integrated circuits: metals, insulators, and semiconductors.

## 2.2 Electric Fields, Voltages, and Energies

Current depends on the velocity of electrons, so the appropriate concern is to determine the force on the electrons. The electrical force $F$ between two charged particles with $q_1$ and $q_2$ units of electronic charge separated by a distance $r$ is given by Coulomb's law,

$$F = k_C \frac{q_1 q_2}{r^2} \tag{2.4}$$

where the Coulomb law constant $k_C = 8.988 \times 10^9$ m per farad in the SI system; 1 farad (F) = 1 A $\cdot$ s/volt (V) = 1 C/V.

For charges of the opposite sign the force is attractive, and for charges of the same sign the force is repulsive. Therefore, a force can be applied to the conduction electrons merely by placing other charges around in such a way that either by attraction or repulsion a coulomb force is applied to the electrons to make them move and thus have velocity. An approach known as the concept of a "field" is of great value in actually making analyses of forces of this type. An electrostatic field is the condition or influence of one or all of the various charges on the charge $q_1$ of interest. It will replace the sum of all of the $q_2/r^2$ terms with a single vector quantity. The electric field is usually given by the symbol $E$, where the boldface type indicates a vector quantity. The force $\mathbf{F}$ on particle 1 can then be written as

$$\mathbf{F} = q_1\mathbf{E} \tag{2.5}$$

The electric field can be calculated for various characteristic charge distributions once and then used in many different kinds of problems to calculate the force in different situations.

In the present case we have a uniform bar of length $L$ and apply a potential $V$. The unit of potential is the volt, and the difference in potential between two points is 1 volt if 1 joule (J) of energy is required to transport 1 coulomb of electric charge from one point to another. Energy or work is force times distance and the unit of electric field strength $\mathscr{E}$ is volts per meter or, in semiconductor electronics usage, volts per centimeter.

The electric field is given by

$$E = -\operatorname{grad} V = -\nabla V \tag{2.6}$$

which for the one-dimensional example given in Fig. 2.2 is

$$\mathscr{E} = -\frac{dV}{dx} \tag{2.7}$$

or for a uniform bar,

$$\mathscr{E} = -\frac{V}{L} \quad \text{V/cm.} \tag{2.8}$$

The direction of the electric field as well as that for the current is from positive to negative, whereas the force on the electrons is toward positive potential.

The SI system of units gives the joule as the derived unit of energy; however we use the electron volt (eV), which is the traditional unit of energy in electronic

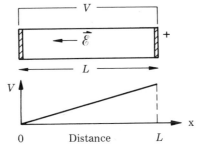

**FIGURE 2.2** Electric field and potential distribution in a metal line of length $L$ with an applied voltage $V$.

materials. Since the charge on the electron is $1.602 \times 10^{-19}$ C and a joule is a coulomb-volt,

$$1 \text{ eV} = 1.602 \times 10^{-19} \text{ J.} \tag{2.9}$$

The thermal energy for a sample at a temperature $T$ in the unit kelvin (K) is given by $kT$, where Boltzmann's constant $k = 8.617 \times 10^{-5}$ eV/K. At a temperature of 300 K (which is often used, although it is a bit warm, for room temperature)

$$kT = 0.02585 \text{ eV} \tag{2.10}$$

which is often approximated for convenience, as $\frac{1}{40}$ eV.

For photons, the wavelength $\lambda$ is the ratio of $c/v$, where $c$ is the speed of light and $v$ is the frequency, so that the energy $E$ is

$$E = hv = \frac{hc}{\lambda} = \frac{1.24 \text{ eV}}{\lambda \text{ (μm)}} \tag{2.11}$$

where Planck's constant $h = 4.136 \times 10^{-15}$ eV · s, and $c = 2.998 \times 10^8$ m/s.

## 2.3 Mobility and Conductivity

To determine the current flow, let us treat the electrons as if they are free to move in the solid and have a mass $m$ equivalent to that of a free electron in a vacuum, where $m = 9.1 \times 10^{-31}$ kg. Of course, the solid is electrically neutral, so that a corresponding amount of positive charge exists; we consider these positive charges to be fixed within the solid and surrounded by a cloud of free electrons. An approximation to the freedom of the electrons is to consider them as an electron gas and to think in terms of the kinetic theory of gases. Assume that we have a perfect monatomic gas. All of its thermal energy will be in the form of kinetic energy, so that with three degrees of freedom, the mean thermal velocity $v_{\text{th}}$ of the electrons is given by

$$\tfrac{1}{2} m v_{\text{th}}^2 = \tfrac{3}{2} kT. \tag{2.12}$$

With the value of $kT$ at 300 K in joules and the free electron mass in kilograms, the value of $v_{\text{th}} = 1.17 \times 10^5$ m/s. This velocity cannot contribute to a current because in the perfect gas model the electrons are visualized bouncing around at random in every direction (Fig. 2.3), so that they are moving in one direction just as much as they are moving in another. They bounce off of each other, but more

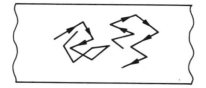

**FIGURE 2.3**  Schematic representation of the paths of two free electrons moving randomly and scattering within a solid.

often they bounce off scattering centers in the solid. In a typical conductor at room temperature the mean free path between collisions can be 100 nm, giving a mean free time $\tau$ between collisions of about $10^{-12}$s.

In our idealized sample we have applied a potential difference of $V$ volts between the ends of the sample, giving an electric field $\mathscr{E} = V/L$. The uniform electric field in the sample will give a constant force on the electrons, $F = ma = e\mathscr{E}$, causing an acceleration

$$a = \frac{e}{m}\mathscr{E}. \tag{2.13}$$

At first glance, this concept is inconsistent with the idea of constant current flow, as the electrons will move faster and faster under the action of the electric field. However, the electron is accelerated by the field *only* during the time of flight between one scattering center and the next. The acceleration is stopped when the electron scatters and must be accelerated again from zero drift velocity. The picture is that of electrons bouncing around at random but with a small component of velocity $v_d$ caused by the electric field. This velocity is smaller than the thermal velocity and is referred to as the drift velocity. It can be estimated as

$$v_d = \tfrac{1}{2}a\tau.$$

This is not quite correct, as the average times between collisions should not be used but rather, the actual times and then an average. The correct procedures give

$$v_d = a\tau. \tag{2.14}$$

The relationship between drift velocity and electric field may be obtained from eqs. (2.13) and (2.14) as

$$v_d = \frac{e\tau}{m}\mathscr{E}. \tag{2.15}$$

For field strengths of less than $10^5$ V/m, the drift velocity is orders of magnitude lower than the thermal velocity $v_{th}$, $v_d \ll v_{th}$.

The parameter $\tau$ is a material parameter for a given sample, so that we can define the mobility

$$\mu = \frac{e\tau}{m} \qquad . \tag{2.16}$$

and thus

$$v_d = \mu\mathscr{E} \tag{2.17}$$

where $\mu$ is in cm$^2$/V · s. With these definitions we can write the current density, $J = nev_d$ [eq. (2.3b)] as

$$J = ne\mu\mathscr{E} \tag{2.18}$$

or

$$J = \sigma\mathscr{E} \tag{2.19}$$

where the conductivity

$$\sigma = ne\mu \tag{2.20}$$

is a material parameter depending on $n$ and $\mu$ with units of siemens (S)/cm.

## 2.4 Resistance, Resistivity, and Ohm's Law

The conductivity $\sigma$ is an intrinsic characteristic of the material and is independent of geometry. It indicates the current density obtained for a given electric field. The inverse of the conductivity is the resistivity $\rho$:

$$\rho = \frac{1}{\sigma} \tag{2.21}$$

where the units of $\rho$ are ohm-meters ($\Omega \cdot$ m), more conventionally given as ohm-centimeters ($\Omega \cdot$ cm). The units are named after eighteenth- and nineteenth-century scientists, as indicated in Table 2.1.

In an actual experiment with the sample of Fig. 2.1, one applies a voltage $V$ across a sample of length $L$ and area $A$ and measures a current $I$. The resistance $R$ is found from Ohm's law:

$$V = IR \tag{2.22}$$

where the units of $R$ are ohms. Ohm's law represents a macroscopic measurement that is equivalent to $J = \sigma\mathscr{E}$ [eq. (2.19)], where $\mathscr{E} = V/L$ and hence

$$R = \rho\frac{L}{A}. \tag{2.23}$$

The contacts to the sample introduce a contact resistance $R_C$. In this case, as shown in Fig. 2.4a, if one applies a voltage (we use the symbol ⊣⊢ for a battery to indicate a voltage supply), measures the current $I$ through the sample, and measures the voltage $V$ across the sample (using a high-impedance voltmeter, which draws only negligible current), the total resistance $R_T$ is the sum of the contact resistance and sample resistance, $R_T = R + 2R_C$. To get around the contact resistance problem, which can be severe in the case of semiconductors, a four-point contact

**TABLE 2.1** Named SI Units

| ampere | A | current | A.M. Ampère (1775–1836) |
|---|---|---|---|
| coulomb | C | charge | Charles de Coulomb (1736–1806) |
| farad | F | capacitance | Michael Faraday (1791–1867) |
| ohm | $\Omega$ | resistance | Georg Ohm (1787–1854) |
| siemens | S | conductance | William Siemens (1823–1883) |
| volt | V | potential | Alessandro Volta (1745–1827) |
| weber | Wb | magnetic flux | Wilhelm Weber (1804–1891) |

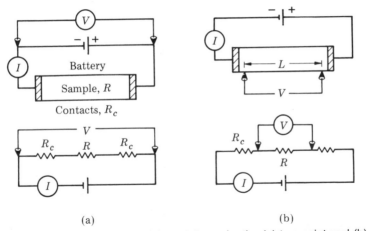

(a)　　　　　　　　　　　　　(b)

**FIGURE 2.4** Measurement of sample resistance by the (a) two-point and (b) four-point probe methods.

or probe method is used (Fig. 2.4b). Voltage probes are placed on the sample away from the current-carrying contacts and the spacing $L$ between the probes is used to determine the electric field, $\mathscr{E} = V/L$, rather than the actual sample length.

Ohm's law, or ohmic behavior, is generally followed at low electric field strengths, which implies that the conductivity $\sigma$ is a constant. For the conductivity $\sigma$ to be constant means that the drift velocity $v_d$ increases linearly with the electric field $\mathscr{E}$, $v_d = \mu\mathscr{E}$ [eq. (2.17)]. In our model of a free electron gas, it is reasonable to assume that the distance $l$ between collisions is independent of electric field. The average distance $l$ is related to the average time $\tau$ between collisions by

$$l = \tau(v_{th} + v_d).$$

For low values of the electric field, the drift velocity $v_d$ is much less than the thermal velocity $v_{th}$, so that the time between collisions,

$$\tau \simeq l/v_{th},$$

is independent of the drift velocity. The mobility $\mu$ equals $e\tau/m$ [eq. (2.16)] and thus is independent of $\mathscr{E}$, so that the conductivity $\sigma = ne\mu$ is also field independent, leading to ohmic behavior.

At high field strengths, however, in semiconductors such as silicon (Si) or gallium arsenide (GaAs), the drift velocity is not proportional to the electric field, as shown in Fig. 2.5. For most of these semiconductors the maximum drift velocity is around $10^7$ cm/s, a value close to our estimate of the thermal velocity $v_{th}$. This electric-field-independent drift velocity, called saturation velocity, naturally leads to deviations from ohmic behavior. In GaAs, the drift velocity decreases with increases in electric field above 3000 V/cm (Section 12.6). In this high-field regime, there is a negative differential mobility, a phenomenon used in high-frequency oscillators.

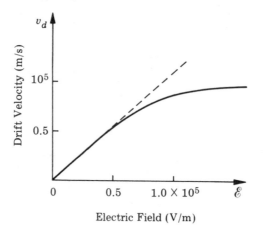

Electric Field (V/m)

**FIGURE 2.5**   Drift velocity versus electric field for electrons in a semiconductor.

## 2.5 Sheet Resistance

The resistance of a thin conductive layer, such as a diffused layer or a thin metal film, is proportional to the resistivity $\rho$ of the layer and inversely proportional to the thickness $t$. It is convenient to define a quantity $R_s$ which is equal to $\rho/t$. The quantity $R_s$ is called the sheet resistance and may be thought of as a material property for conductors which are essentially two-dimensional. A conducting thin layer consisting of a simple rectangle of length $L$ (in the direction of the current) and width $W$ has a resistance $R$ (Fig. 2.6),

$$R = \frac{\rho}{t}\frac{L}{W} = R_s\frac{L}{W}. \tag{2.24}$$

The ratio $L/W$ is referred to as the number of squares ($\square$), since it is the number of squares of size $W$ that can fit into the rectangle without overlapping. The term "squares" is dimensionless. The sheet resistance has the units of ohms, but it is convenient to refer to it as ohms per square ($\Omega/\square$). The resistance of a rectangular thin layer is therefore the sheet resistance times the number of squares ($L/W$). The concept can be generalized to include any arbitrarily shaped thin-film conductor by calling $Rt/\rho$ the effective number of squares.

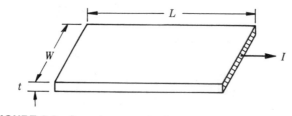

**FIGURE 2.6**   Sample geometry for a metal line with current $I$.

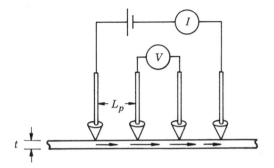

**FIGURE 2.7**  Four-point probe with probe spacing $L_p$ to measure sheet resistivity in a thin sample.

The sheet resistance can be measured by the four-point probe method illustrated in Fig. 2.7. In this method, four point contacts are made to the thin layer; current is carried through the outer two contacts and the voltage drop is measured across the remaining contacts. The sheet resistance is given by

$$R_s = K_p \frac{V}{I} \tag{2.25}$$

where $K_p$ is a constant that depends on the configuration, position, and orientation of the probes. The probes are commonly equally spaced in a straight line, with the two outer probes being the current probes. The probe spacing $L_p$ is usually small compared to the dimensions of the film and large compared to layer thickness, $t$. If the film were infinite in all planar directions, the constant $K_p$ is

$$K_p = \frac{\pi}{\ln 2} = 4.5324.$$

Correction factors are needed when the dimension of the layer becomes comparable to the probe spacing (Beadle et al., 1985). In commercially available four-point probe machines, the sheet resistance value is directly displayed.

Square samples always have resistance values numerically equal to their $R_s$ value irrespective of their sizes. A metallic layer 1 mm × 1 mm in area has a measured $R_s$ value of 10 $\Omega/\square$. The resistance of the sample 1 mm × 1 mm in area is

$$R = \frac{\rho L}{tW} = \frac{\rho}{t} \frac{1 \text{ mm}}{1 \text{ mm}} = \frac{\rho}{t} = R_s = 10 \ \Omega$$

where $\rho$ is the resistivity of the layer, $t$ the thickness of the layer, $L$ the length, and $W$ the width of the sample.

## 2.6  Electrons, Holes, and the Hall Effect

In the measurement of the resistance or conductivity $\sigma$ of the sample, one obtains the product of the carrier concentration $n$ and mobility $\mu$ from $\sigma = ne\mu$ [eq.

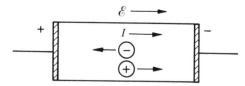

**FIGURE 2.8** Motion of electrons and holes in a sample with electric field $\mathcal{E}$ and current $I$.

(2.20)]. To understand the material properties in detail, an independent method of obtaining either $n$ or $\mu$ is required. Further, we have tacitly assumed that only negatively charged free electrons were involved in charge transport. As shown in Fig. 2.8, measurement of the direction of current flow $I$ does not allow a distinction between the motion of electrons or of positive charge carriers. A flow of current from the positive to negative contacts can be produced by negative charge carriers moving toward the positive contact or by positive charge carriers moving toward the negative contact. A method is required to distinguish between positive and negative charge carriers.

In semiconductors, as discussed in Chapter 3, both negative and positive charge carriers contribute to current flow. The positive charge carriers are called holes because they correspond to an absence of free electrons. The motion of holes in an electric field follows the relationship expected for positively charged particles, with a positive mass nearly equal to that of the electron. The symbol $p$ is used to denote the hole concentration and $\mu_p$ the mobility of holes (see Table 2.2).

When both holes and electrons are considered, the conductivity is

$$\sigma = e(n\mu_n + p\mu_p). \tag{2.26}$$

In general, however, the conductivity is dominated by either electrons or holes. In semiconductor technology, it is the controlled introduction of dopant species that determines whether the charge carriers are predominantly electrons or holes. The sign of the charge carriers and their mobility can be found from their motion in a magnetic field.

The symbol for the magnetic field is **B**, where the boldface type indicates a vector quantity. The units of magnetic field (the magnetic flux density) are webers/meter$^2$, Wb/m$^2$ (which translates to Volt $\cdot$ s/m$^2$ in the conversion to SI units). Magnetic fields (flux density) are sometimes given in tesla or gauss; 1 weber/m$^2$ = 1 tesla = $10^4$ gauss. Magnetic fields of 0.1 to 1 Wb/m$^2$ can easily be obtained in laboratory magnet setups.

**TABLE 2.2** Electron and Hole Notation

|  | **Electrons** | **Holes** |
| --- | --- | --- |
| Charge | $-e$ | $+e$ |
| Mass | $m_e$ | $m_h$ |
| Concentration (number/cm$^3$) | $n$ | $p$ |
| Mobility (cm$^2$/V $\cdot$ s) | $\mu_n$ | $\mu_p$ |

For a magnetic field **B**, the equation that gives the force, the Lorentz force, on a charged particle can be written as

$$\mathbf{F} = q(\mathbf{v} \times \mathbf{B}) \tag{2.27}$$

where **v** is the vector velocity of the particle and $q$ is its charge ($-1.6 \times 10^{-19}$ C for electrons). Equation (2.27) is a vector equation where both the magnitude and the direction of each of the components are specified. The cross product ($\mathbf{v} \times \mathbf{B}$) results in a vector perpendicular to both **v** and **B** in a right-hand direction. In a normal right-handed coordinate system, $\mathbf{x} \times \mathbf{y} = \mathbf{z}$; that is, **x** rotated into **y** obtains the direction of **z**. Here **v** rotates into **B** to obtain the direction of **F**. The magnitude is obtained as

$$|\mathbf{F}| = q|\mathbf{v}| \cdot |\mathbf{B}| \sin \theta, \tag{2.28}$$

where $\theta$ is the angle between **v** and **B** and the bars $|\ |$ indicate absolute magnitude. Actually, for all the cases of interest here the velocity will be perpendicular to the magnetic field, so that $\sin \theta$ will always be equal to unity and the force is given as

$$|\mathbf{F}| = q|\mathbf{v}| \cdot |\mathbf{B}|$$

or

$$\tag{2.29}$$

$$F = qvB.$$

The direction of the force for ($\mathbf{v} \perp \mathbf{B}$) is in the direction perpendicular to both. It is not possible to change the energy of a charged particle with a magnetic field. The particle is deflected but not accelerated since the force is always normal to the direction of motion.

A bar of conducting material is aligned along the axis of a right-handed Cartesian system with a magnetic field along the $y$ axis as in Fig. 2.9. Electrons in the bar are moving in all directions at their thermal velocity. The electrons will be deflected by the magnetic field, but just as many will be deflected in one direction as the other, so that no net effect will be observed; the motion will still be random. Consider, however, what happens when a potential source (a battery) is connected so that an electric field is created along the $x$ axis as indicated in Fig. 2.9. The electric field is defined as positive when it goes from positive potential to negative potential, so electrons will move, on an average, in the $-x$ direction. It is their

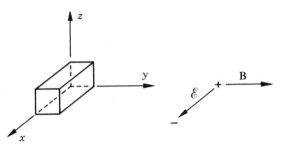

**FIGURE 2.9**  Geometry for the Hall measurement.

drift velocity coupled with the magnetic field that will cause a force on the electrons on the $z$ direction. Equation (2.27) gives $(\mathbf{v} \times \mathbf{B})$ in the negative $z$ direction, but $q$ is negative for electrons, so that the force is in the positive $z$ direction.

Conduction electrons in a magnetic field can be pictured as drifting in the negative $x$ direction caused by the electric field with a component of their drift velocity along the $z$ axis caused by the magnetic field. Quantitatively, then, $v_x$ is

$$v_x = -\mu \, \mathscr{E}_x \tag{2.30}$$

and from the magnetic field interaction [eq. (2.29)], the force $F_B$ due to the magnetic field $B$ on the moving electrons is

$$F_B = qv_xB_y = -q\mu\mathscr{E}_xB_y \tag{2.31}$$
$$= e\mu \, \mathscr{E}_xB_y$$

where $-q = e$, the magnitude of the charge on the electron. A force in the $z$ direction will cause the electrons to drift in this direction.

The deflection of the moving charge carriers is the basis for the Hall effect [named after Edwin Hall (1855–1938)]. For clarity the sample geometry is redrawn in Fig. 2.10 with the magnetic field pointing out of the page, the positive $y$ direction. The electrons are deflected upward, positive $z$ direction, but they cannot move farther than the top of the sample and will accumulate there. The sample was electrically neutral before and motion of the electrons upward leaves positive fixed charge at the bottom. Separation of charge creates a field $\mathscr{E}_H$ (the Hall field), hence an electrostatic force also in the $z$ direction. This force on the electron, however, will be along the *negative z* direction. In equilibrium, the magnetic force along the $z$ axis would be expected to be just balanced by the electrostatic force caused by the initial separation of charge. No deflection of electrons would occur after equilibrium is established. This balance of electrostatic and electromagnetic forces in the $z$ direction gives an interesting relation. Equating the magnitude of the two forces, from eqs. (2.5) and (2.31), respectively:

$$|q| \, \mu\mathscr{E}_xB_y = |q|\mathscr{E}_H. \tag{2.32}$$

Remembering that the field is the potential divided by distance [eq. (2.7)], eq. (2.32) can be rearranged to

$$\mu B_y = \frac{\mathscr{E}_H}{\mathscr{E}_x} = \frac{V_HL_x}{L_zV_x} \tag{2.33}$$

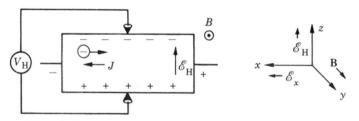

**FIGURE 2.10** Configuration for Hall effect measurements.

where $V_x$ is the applied potential across the distance $L_x$ and $V_H$ is the Hall potential generated across distance $L_z$. The Hall voltage is a real potential that can be measured with a good high-impedance voltmeter to produce a direct measure of the mobility which is independent of the number of carriers. Measurement of the Hall effect is the most direct measurement of mobility, hence one of the more important measurements in the characterization of a semiconductor.

The Hall field can also be expressed in terms of current density, $J$:

$$\mathscr{E}_H = R_H JB \quad \text{and} \quad R_H = \frac{1}{ne} \qquad (2.34)$$

where $R_H$ is the Hall coefficient. The magnitude of the Hall field is easily measured from the potential $V_H$ developed across the sample. For a drift velocity of 2 m/s and a magnetic field of 1 Wb/m$^2$ (or 1 V $\cdot$ s/m$^2$), the Hall field is 2 V/m or a potential of 0.02 V measured across a 0.01-m (1-cm) sample.

If the current in Fig. 2.10 were carried by holes, the hole drift velocity would be along the positive $x$ axis, the direction of current flow. They would be deflected upward (similar to the electrons, which have a drift velocity in the opposite direction). Consequently, the Hall field would point downward, so that the polarity of the Hall potential is reversed for holes as compared to that of electrons. Thus the polarity of the Hall voltage indicates whether the current is carried by holes or electrons.

# 2.7 Electrical Properties of Metals and Semiconductors

A typical value of mobility $\mu$ of a metal such as silver (Ag) is 50 cm$^2$/V $\cdot$ s, which gives a drift velocity of 50 cm/s for an electric field of 1 V/cm. The resistivity $\rho$ of high-conductivity metals is about 2 to 6 $\times$ 10$^{-6}$ $\Omega$ $\cdot$ cm (Table 2.3).

A typical mobility $\mu$ of a semiconductor such as silicon is 1000 cm$^2$/V $\cdot$ s, a value much higher than that in metals (Table 2.4). The resistivity is orders of magnitude higher in semiconductors than in metals. A typical value of the resistivity of silicon is 10 $\Omega$ $\cdot$ cm. Further, although the resistivity of metals is little influenced by the addition of 0.1 atomic percent (at %) of impurities, the resistivity in semiconductors is completely dominated by the addition of dopant impurities.

**TABLE 2.3**  Resistivity $\rho$ of Pure Metals at 300 K

|  |  | $\rho$ (10$^{-6}$ $\Omega$ $\cdot$ cm) | Atoms/cm$^3$ ($\times$ 10$^{22}$) |
|---|---|---|---|
| Aluminum | Al | 2.5 | 6.02 |
| Copper | Cu | 1.6 | 8.45 |
| Gold | Au | 2.0 | 5.9 |
| Silver | Ag | 1.5 | 5.85 |
| Tungsten | W | 5.6 | 6.3 |

**TABLE 2.4** Mobility of Electrons $\mu_n$ and Holes $\mu_p$ (at 300 K)

| | $\mu_n$ (cm²/V · s) | $\mu_p$ (cm²/V · s) | Atoms/cm³ (× 10²²) |
|---|---|---|---|
| Silicon (Si) | 1500 | 450 | 5.0 |
| Gallium arsenide (GaAs) | 8500 | 400 | 4.42 |
| Germanium (Ge) | 3900 | 1900 | 4.42 |

*Source:* Sze (1981).

## 2.8 Dielectrics and Capacitance

Insulators, oxide and nitride layers, are at the other extreme of electrical behavior from that of metals and conductors. Insulators are characterized by the absence of charge transport, and it is this property that makes them so important in isolating the various layers of interconnects in integrated circuits. Insulators do have positive and negative charges (think of a positively charged nucleus surrounded by an electron cloud) and when an electric field is applied there is a shift in the charge distributions. This field-induced polarization leads to the dielectric behavior of insulators and hence to their capacitance $C$. It is the combination of the resistance $R$ of an interconnect and the capacitance $C$ of the insulating layer (the oxide layer in Fig. 2.1) that leads to a finite $RC$ time constant. It is the $RC$ time constant of the interconnects rather than the intrinsic speed of the semiconductor devices that may set the high-frequency limit of the finished integrated circuit.

Consider two parallel metal plates separated by a distance $L$ with a vacuum between (Fig. 2.11a). When a voltage $V$ is applied to the plates, positive and negative charges are induced on the surface of the plates and an electric field, $\mathscr{E} = V/L$, is created in the vacuum. One can visualize the field and the equal number of positive and negative charges by connecting each positive and negative charge. These Faraday strings, as these lines are called, show that the electric field is constant; the number density of strings is constant in any cross section between the plates.

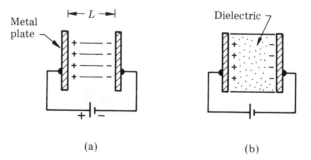

(a)          (b)

**FIGURE 2.11** Metal plates, a distance $L$ apart: (a) in vacuum; (b) with dielectric.

The charge per unit area $Q$ (on either plate) is proportional to the field $\mathscr{E}$ through the relationship

$$Q = \epsilon_0 \mathscr{E} \tag{2.35}$$

where $\epsilon_0$ is the permittivity of vacuum, $\epsilon_0 = 8.85 \times 10^{-13}$ F/m. The farad is a coulomb/volt and the charge on the electron $e$ is $1.6 \times 10^{-19}$ C, so that the permittivity can also be expressed as $\epsilon_0 = 55.3 e/\text{V} \cdot \mu\text{m}$. For a voltage $V$ between the plates, the charge per unit area is

$$Q = \frac{\epsilon_0 V}{L}. \tag{2.36}$$

The proportionality constant between the applied voltage and the charge is called the capacitance $C$, which for plates of area $A$ is

$$C = \frac{QA}{V} = \frac{\epsilon_0 A}{L} \tag{2.37}$$

where the unit of the capacitance is the farad.

When a dielectric is inserted between the plates in place of the vacuum, the capacitance is increased by a factor $\epsilon_r$:

$$C = \frac{\epsilon_r \epsilon_0 A}{L} = \frac{\epsilon A}{L} \tag{2.38}$$

where $\epsilon_r$ is the relative dielectric constant (with typical values between 4 and 13, Table 2.5) and $\epsilon$ is the permittivity, $\epsilon = \epsilon_r \epsilon_0$. The reason for the increase in capacitance is the polarizability of the dielectric, which causes charges to appear at the surface of the dielectric and hence requires fresh charges from the battery to keep the voltage constant. For Si, $\epsilon_r = 11.9$ and $\epsilon = 1.05 \times 10^{-12}$ F/cm = 1.05 pF/cm.

The capacitance can be measured directly and is one method used to determine the thickness of an insulator. If we consider two interconnect lines of area $A = 3 \times 10^{-10}$ m$^2$ (3 $\mu$m wide by 100 $\mu$m long) separated by $L = 10^{-7}$ m (100 nm or 1000 Å) of insulator with a relative dielectric constant $\epsilon_r = 4$, the capacitance is

$$C = \frac{\epsilon_r \epsilon_0 A}{L} = \frac{4 \times 8.85 \times 10^{-12} \times 3 \times 10^{-10}}{10^{-7}} = 10^{-13} \text{ F}$$

or $C = 10^{-1}$ pF. Although this may appear to be (and is) a small value, if the

**TABLE 2.5**  Values of the Relative Dielectric Constant $\epsilon_r$ at 300 K

| Material | Si | GaAs | SiO$_2$ | Si$_3$N$_4$ |
|----------|------|------|---------|-------------|
| $\epsilon_r$ | 11.9 | 13.1 | 3.9 | 7.5 |

*Source:* Sze (1981).

resistance of the interconnect line is 10 $\Omega$, the $RC$ time constant is $10^{-12}$ s or 1 ps. With a large number of such interconnects in the integrated circuit, the response time of the system can be degraded.

## 2.9 Carrier Diffusion in Semiconductors

The description of carrier diffusion processes in semiconductors is similar to the spread of a drop of ink in a glass of water or molecular diffusion of a gas in a container. Imagine a bar of $n$-type Si irradiated with a flash of light at time $t_0$ (Fig. 2.12). The excess carriers, shown in the lower portion of the figure, will begin to spread out due to their random paths as they collide with lattice atoms. The carrier distribution for times greater than $t_0$ ($t_0 < t_1 < t_2$) are indicated. There is a net flow of carriers away from the initial point of illumination.

In our description of the diffusion process, we assume that there is no applied electric field and neglect any field due to the space charge associated with the diffusing carriers themselves. The latter is a good assumption as long as the concentration of dopant atoms (or majority carriers) is much greater than the concentration of carriers created by the light flash. In this case, the majority carriers need to make only a small change in their local concentration to accommodate any charge unbalance caused by the diffusing carriers.

Let us consider that the particles have a mean free path $l$ between collisions and that their concentration follows the distribution shown in Fig. 2.13, where we assume a linear one-dimensional distribution. (The discussion does not depend on whether there is a uniform gradient where $dn/dx$ is constant, but it is easier to visualize.) The carrier concentration distribution at time $t_1$ is composed of a series

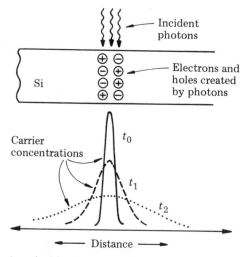

**FIGURE 2.12** Photons incident on silicon create hole–electron pairs. The lower portion shows the carrier distribution broadening with time.

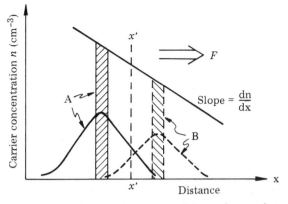

**FIGURE 2.13**   The carrier concentration versus distance has a slope $dn/dx$ and a diffusion flux $F$, illustrated by the broadening of two distributions, A and B.

of slabs of carriers such as that shown at A and B. At a later time $t_2$, the carrier distributions in the individual blocks will have broadened due to diffusion. If we consider some point $x'$ between A and B, there will be more particles from A that have passed $x'$ than particles from B. There will be net flux $F$ of carriers directed to the right from A to B. It is clear that the greater the slope, $dn/dx$, the larger the flux; that is, the flux $F$ is proportional to the gradient $dn/dx$. We introduce a proportionality constant $D$, the diffusion coefficient, and write

$$F = -D\frac{dn}{dx} \qquad (2.39)$$

where the units of $D$ are in $cm^2/s$ and the minus sign indicates that the particles diffuse from regions of high concentration to regions of low concentration; that is, $n$ is decreasing, so that $dn/dx$ is negative. Equation (2.39) is described in Chapter 7 as Fick's first law.

A flux of charged particles represents a current where the current density $J$ is given by the charge $q$ on the particles times the flux $F$, so that:

For holes:  $$J_p = -eD_p\frac{dp}{dx} \qquad (2.40)$$

and

For electrons:  $$J_n = eD_n\frac{dn}{dx} \qquad (2.41)$$

where $D_n$ and $D_p$ are the electron and hole diffusion constants, respectively. The change in sign between eqs. (2.40) and (2.41) reflects the fact that electrons carry a negative charge, as shown in Fig. 2.14. When there are both voltage gradients (electric field $\mathscr{E} = -dV/dx$) and concentration gradients present,

$$J_p = e\left(p\mu_p\mathscr{E} - D_p\frac{dp}{dx}\right) \qquad (2.42)$$

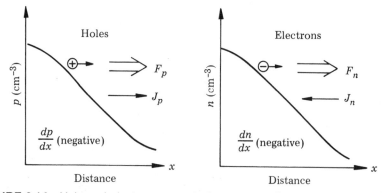

**FIGURE 2.14** Hole and electron concentrations versus distance have negative gradients and positive fluxes $F$. The direction of the current depends on the charge of the carrier.

$$J_n = e\left(n\mu_n \mathscr{E} + D_n \frac{dn}{dx}\right).\tag{2.43}$$

In most cases the field term dominates [see eq. (2.18)] and diffusion terms are only used where there are appreciable carrier gradients caused by carrier injection (forward-biased $p$-$n$ junctions) or carrier generation by light (solar cells).

Both the carrier mobility $\mu$ and diffusion constant $D$ are determined by the motion of charge carriers, which have a mean free path $l$ between collisions. The Einstein relation gives the proportionality

$$\mu = \frac{eD}{kT}.\tag{2.44}$$

At room temperature where $kT = 0.026$ eV, a semiconductor with a carrier mobility of $\mu$ of 1000 cm$^2$/V $\cdot$ s will have a carrier diffusion constant $D$ of 26 cm$^2$/s.

## 2.10 Thermal Properties

Heat is dissipated in integrated circuits due to current in resistive elements. A joule (J) is 1 coulomb $\cdot$ volt, so that in terms of the SI unit of power, the watt (W),

$$\text{joule} = \text{watt} \cdot \text{second} = \text{ampere} \cdot \text{volt} \cdot \text{second}$$

and $\tag{2.45}$

$$W = I \times IR = I^2 R$$

with $I$ in amperes and $R$ in ohms. One of the design considerations is to provide packaging material with sufficient thermal conductivity so that device operating temperatures will not be excessive.

The ratio of the heat supplied $\Delta Q$ per unit mass $M$ to the corresponding temperature rise $\Delta T$ is called the specific heat $c_p$:

$$c_p = \frac{\text{heat capacity}}{\text{unit mass}} = \frac{\Delta Q}{M \Delta T} \tag{2.46}$$

where the units of specific heat are J/g · K or cal/g · K;

$$1 \text{ calorie} = 4.186 \text{ joules.} \tag{2.47}$$

The thermal properties of several metals and semiconductors are listed in Table 2.6

The temperature rise of resistive elements would be large if it were not for the substrate, which conducts heat away. If we consider a self-supporting Al interconnect (Fig. 2.1) 100 μm ($10^{-2}$ cm) long with an area of 1 μm × 1 μm ($10^{-8}$ cm$^2$) carrying $10^{-4}$ A for just 1.0 s, the temperature rise would be about 100°C. For this example, the resistance $R$ is

$$R = \frac{\rho L}{A} = \frac{2.5 \times 10^{-6} \times 10^{-2}}{10^{-8}} = 2.5 \ \Omega$$

the thermal energy $\Delta Q$ is

$$\Delta Q = I^2 R \times 1 = 2.5 \times 10^{-8} \text{ J}$$

and the mass $M$ is 2.7 g/cm$^3$ times the volume ($10^{-10}$ cm$^3$), so that

$$\Delta T = \frac{\Delta Q}{M c_p} = \frac{2.5 \times 10^{-8}}{2.7 \times 10^{-10} \times 0.9} = 92.3°C$$

where the specific heat $c_p$ of Al is 0.9 J/g · K.

The transfer of energy arising from the temperature difference between adjacent components is called heat conduction. As in the case of carrier diffusion, the flux $F_{th}$ of energy transmitted across a unit area per unit time is proportional to the temperature gradient $dT/dx$ and we write

$$F_{th} = -K_{th} \frac{dT}{dx} \tag{2.48}$$

where the units of thermal conductivity $K_{th}$ are W/cm · K.

**TABLE 2.6**   Thermal Properties: $K_{th}$, the Thermal Conductivity, and $c_p$, the Specific Heat, with Atomic or Molecular (SiO$_2$) Concentrations $N$ and Mass Densities, $\rho_m$.

|        | $K_{th}$ (W/cm · K) | $c_p$ (J/g · K) | $N$ ($10^{22}$/cm$^3$) | $\rho_m$ (g/cm$^3$) |
|--------|---------|--------|--------|--------|
| Al     | 2.37    | 0.9    | 6.02   | 2.7    |
| Cu     | 4.01    | 0.39   | 8.45   | 8.93   |
| Si     | 1.48    | 0.7    | 5      | 2.33   |
| Ge     | 0.6     | 0.31   | 4.42   | 5.32   |
| GaAs   | 0.46    | 0.35   | 4.42   | 5.32   |
| SiO$_2$ | 0.014  | 1.0    | 2.2    | 2.2    |

For metals around room temperature where the electrons dominate in heat transport, the thermal conductivity $K_{th}$ is directly related to the electrical conductivity (the Wiedemann–Franz law),

$$\frac{K_{th}}{\sigma} = LT \tag{2.49}$$

where the Lorenz number $L$ has a value $= 2.45 \times 10^{-8}$ W $\cdot$ $\Omega/K^2$ that is independent of the particular metal. Experimental values are close to the prediction of eq. (2.49), implying that the collision processes are equivalent in the two transport phenomena (electrical and thermal).

In semiconductors it is the thermal vibration of atoms about their equilibrium positions that transmits the thermal energy. If we treat the vibrating atoms as harmonic oscillators, the average energy per atom based on a classical three-dimensional oscillator is $3kT$, where the Boltzmann constant $k = 1.38 \times 10^{-23}$ J/K. The energy per unit volume is $3NkT$ (where $N$ = atomic density, atoms/cm$^3$) and the specific heat $c_v$ per unit volume is $3Nk$ in the classical limit.

The thermal vibrations can be treated as elastic waves called phonons with energy $E_{ph} = h\nu$, in analogy with the photon of the electromagnetic wave. Thermal energy can be transported by phonons, which have a mean free path $l$ determined by scattering by other phonons and imperfections. If we treat the phonons as molecules in a gas and use the kinetic theory of gases, the thermal conductivity $K_{th}$ can be expressed as

$$K_{th} = \tfrac{1}{3} c_v v_s l \tag{2.50}$$

where the velocity of sound $v_s \simeq 5 \times 10^5$ cm/s. In Si, the specific heat/cm$^3$ $= 0.7$ (J/g $\cdot$ K) $\times$ 2.33 (g/cm$^3$) $= 1.63$ J/cm$^3$ $\cdot$ K and the mean free path for phonons is about $5 \times 10^{-6}$ cm; so that

$$K_{th} = \tfrac{1}{3} \times 1.6 \times 5 \times 10^5 \times 5 \times 10^{-6} = 1.3 \text{ W/cm} \cdot \text{K}$$

a value close to that in Table 2.6.

The velocity of sound may be treated as the propagation of an elastic wave of wavelength $\lambda$ in a sample. For longitudinal waves, where elastic displacements $u(x)$ at point $x$ occur along the sample, the strain $\varepsilon_s$ is the change in length per unit length (Chapter 1),

$$\varepsilon_s = \frac{du}{dx} . \tag{2.51}$$

The stress, force per unit area, is proportional to the strain:

$$\sigma_s = Y\varepsilon_s = Y\frac{du}{dx} \tag{2.52}$$

where Young's modulus $Y$ has values of about $10^{11}$ N/m$^2$ for most solids.

We consider the stress as a restoring force and use $F = ma = m\dfrac{d^2u}{dt^2}$. For a

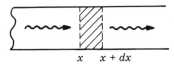

**FIGURE 2.15**  Elastic waves propagating along the x direction in a bar of area A.

segment of length $dx$ in a bar of area $A$ (Fig. 2.15), the mass of the segment is the mass density $\rho_m$ times the volume, $A\,dx$. The equation of motion of the segment is

$$(\rho_m A\,dx)\,\frac{d^2u}{dt^2} = [\sigma_s\,(x + dx) - \sigma_s\,(x)]\,A \tag{2.53}$$

which from eq. (2.52) can be written as

$$\rho_m\,\frac{d^2u}{dt^2} = Y\,\frac{d^2u}{dx^2} \tag{2.54}$$

where we have set

$$\sigma_s\,(x + dx) - \sigma_s\,(x) = d\sigma_s.$$

Equation (2.54) is the wave equation in one dimension and has a solution

$$u = u_0\,\exp\,[i(qx - \omega t)] \tag{2.55}$$

where the wave number $K = 2\pi/\lambda$ and $\omega$ is the frequency of the wave. The velocity of the wave, $v_s$, is the sound velocity and is

$$v_s = \frac{\omega}{K} \tag{2.56}$$

and from Eq. (2.54)

$$v_s = \left(\frac{Y}{\rho_m}\right)^{1/2} \tag{2.57}$$

The mass density of Si is $\rho_m = 2.33 \times 10^3$ kg/m³, and using $Y = 10^{11}$ N/m² yields

$$v_s = \left(\frac{10^{11}}{2.33 \times 10^3}\right)^{1/2} = 6.5 \times 10^3\ \text{m/s}$$

a value close to $5 \times 10^5$ cm/s, the velocity of sound in Si.

The Debye temperature, $\Theta_D$, is used to characterize thermal properties, or elastic wave propagation, in solids. The maximum frequency that can be propagated in the Debye approximation is $\omega_D$, given by

$$h\,\frac{\omega_D}{2\pi} = k\Theta_D. \tag{2.58}$$

**TABLE 2.7** Values of the Debye Temperature (Kelvin)

|             | Si  | Ge  | GaAs | Al  | Au  |
|-------------|-----|-----|------|-----|-----|
| $\Theta_D$  | 645 | 374 | 350[a] | 428 | 165 |

Source: C. Kittel, *Introduction to Solid State Physics*, 6th ed., (Wiley, New York, 1986.)

[a]From *Properties of GaAs*, EMIS Data Reviews Series 2, Inspec, London, 1986.

Inserting values for $h = 4.136 \times 10^{-15}$ eV · s and $k = 8.617 \times 10^{-5}$ eV/K gives us

$$\omega_D = 13.09 \times 10^{10}\ \Theta_D. \qquad (2.59)$$

Values of the Debye temperature are given in Table 2.7. For Si, $\Theta_D = 645$ K, so that

$$\omega_D(\text{Si}) = 8.44 \times 10^{13}/\text{s}.$$

In the Debye approximation all waves have the sound velocity, $v_s = \omega/K$ and at $\omega_D$, the wave number is $K_D$.

We can estimate the sound velocity by assuming that the maximum wave number $K_D$ is determined by the shortest wavelength, $\lambda_D$. In a solid with atom spacings of $a$, the shortest realistic wavelength is $\lambda_D = 2a$. The value of $K_D$ is

$$K_D = \frac{2\pi}{\lambda_D} = \frac{2\pi}{2a} = \frac{\pi}{a} \qquad (2.60)$$

where $a$ is given by $(1/N)^{1/3}$ for a simple cubic solid. For Si $a = 2.7 \times 10^{-10}$ m for $N = 5 \times 10^{28}$ atoms/m$^3$. In this approximation, the velocity of sound is

$$v_s = \frac{\omega_D}{K_d} = \omega_D \frac{a}{\pi} \qquad (2.61)$$

which for Si has a value

$$v_s = 8.44 \times 10^{13} \times \frac{2.7 \times 10^{-10}}{\pi} = 7.25 \times 10^3 \text{ m/s},$$

a value that is 50% higher than the true value due to our one-dimensional approach to a three-dimensional problem.

## GENERAL REFERENCES

BEADLE, W. E., J. C. C. TSAI, and R. D. PLUMMER, eds., *Quick Reference Manual for Silicon Integrated Circuit Technology*, Wiley, New York, 1985.
GHANDHI, S. K., *VLSI Fabrication Principles*, Wiley, New York, 1983.

HOWES, M. J., and D. V. MORGAN, eds., *Gallium Arsenide*, Wiley, New York, 1985.

PIERRET, R. F., *Semiconductor Fundamentals*, Addison-Wesley, Reading, Mass., 1983.

RUOFF, A. L., *Introduction to Materials Science*, Prentice-Hall, Englewood Cliffs, N.J., 1972.

SOLYMAR, L., and D. WALSH, *Lectures on the Electrical Properties of Materials*, 3rd ed., Oxford University Press, Oxford, 1984.

SZE, S. M., *Physics of Semiconductor Devices*, 2nd ed., Wiley, New York, 1981.

TIPLER, P. A., *Physics*, Worth, New York, 1976.

## PROBLEMS

**2.1.** You deposit a strip of Al (resistivity $= 2.5 \times 10^{-6} \, \Omega \cdot cm$) that is 0.1 cm long with a thickness of 1 µm and a width of 2.5 µm.

(a) What is the resistance $R$ in ohms of the strip?

(b) For a current of 10 milliamperes (mA) flowing along the strip, what voltage would you measure along the length and how many watts are dissipated?

(c) If the thickness of the Al strip is doubled, what is the resistivity? If the thickness of half the length of the strip is doubled, what is the resistance of the strip?

(d) You keep the same amount of Al in the strip but find that there are bubbles of argon (an insulating, inert gas) in the Al, so that the thickness of the strip is doubled. Would the resistance (ohms) decrease?

(e) What is the sheet resistance $R_s$ of a 1-µm layer of Al?

**2.2.** A semiconductor with an area of $10^{-4} \, cm^2$ and a length of 1 cm is held at $T = 300$ K with an electric field of 2 V/cm along the length. The resistivity is $10 \, \Omega \cdot cm$ and the electron concentration $n = 1 \times 10^{16}/cm^3$.

(a) What is the drift velocity of electrons?

(b) What is the flux of electrons (electrons/$cm^2 \cdot s$) and current density?

(c) Is the drift velocity less than, equal to, or greater than the thermal velocity?

(d) What is the value of the diffusion coefficient of electrons?

(e) What concentration gradient of electrons would be required to give the same magnitude of current density as in part (b)?

**2.3.** Cu has a density 8.93 g/$cm^3$ and a resistivity of $1.7 \times 10^{-6} \, \Omega \cdot cm$. You deposit a Cu film that is $10^{-4}$ cm thick on an insulating substrate.

(a) What is the number of Cu atoms/$cm^2$?

(b) If there are $10^{23}$ electrons/$cm^3$ that are "free" to conduct current, what is the electron mobility in $cm^2/V \cdot s$?

(c) You "pattern" the film into a strip with a total length of 0.2 cm with one-half the length $5 \times 10^{-4}$ cm wide and the other half $10 \times 10^{-4}$ cm wide. What is the resistance $R$?

**(d)** What is the current density $J$ (A/cm$^2$) and the electric field $E$ (V/cm) in the region $5 \times 10^{-4}$ cm wide if 0.1 V is applied across the entire length?

**(e)** What is the sheet resistance $R_s$ of the deposited Cu film and what is the value if the layer is $10^{-5}$ cm thick?

**2.4.** The mobility of electrons in GaAs is 8500 cm$^2$/V $\cdot$ s at 300 K.

**(a)** Using the free-electron model, what is the average time $\tau$ between collisions and the average distance the electron travels between collisions assuming thermal velocity at 300 K?

**(b)** For an electric field of $10^3$ V/cm, how much energy (eV) is gained from the electric field by an electron in $\tau$ seconds? Using your value of $\tau$ in part (a), compare the energy gained with the value of $kT$ at 300 K.

**2.5.** You make a Hall measurement with magnetic field at right angles (normal) to the sample with applied electric field $\mathscr{E}$ along the sample. Do the current $I$ and Hall voltage increase, decrease, equal zero, remain the same, or change polarity (sign) if you change one parameter at a time and

**(a)** the carrier mobility decreases?

**(b)** the number of carriers decreases?

**(c)** the magnetic field is directed along the sample?

**(d)** the applied electric field decreases?

**2.6.** You make a Hall effect sample from a semiconductor bar with a cross-sectional area of $10^{-2}$ cm$^2$. With a constant current supply, you have a current of 0.4 mA ($4 \times 10^{-4}$ A) along the sample.

**(a)** You measure a voltage $V_x$ of 1.0 V across the resistivity probes that are separated by 0.5 cm along the sample length. What is the resistivity ($\Omega \cdot$ cm)?

**(b)** For a magnetic field of 7500 gauss (G) normal to the sample, you measure a Hall voltage $V_H$ of 0.1 V across probes separated by 0.2 cm. What is the mobility in cm$^2$/V $\cdot$ s?

**(c)** What is the carrier concentration?

**(d)** What are the values of $V_x$ and $V_H$ if the carrier concentration is doubled?

**2.7.** You have a capacitor of silicon dioxide, SiO$_2$, with two plates of area $A = 10^{-6}$ cm$^2$ separated by a distance $d$ of $5 \times 10^{-5}$ cm.

**(a)** What is the capacitance in farads, and what is the charge on the plates if 0.5 V is applied?

**(b)** An $\alpha$-particle with energy 5.4 MeV ($5.4 \times 10^6$ eV) creates hole–electron pairs with an average energy of 3.6 eV required to form a pair. How many pairs are created, and what voltage is induced if the charge is on the plates?

**(c)** For a memory cell in an FET you want to design as small a capacitor as possible but big enough so that the charge created by an $\alpha$-particle will not cause errors with your 5 V operating signal. You choose a 50 femtofarad (fF) capacitor ($50 \times 10^{-15}$ F). What voltage is induced by the charge created by the $\alpha$-particle? How much area is required for the capacitor if the $SiO_2$ thickness is $10^{-4}$ cm (1 $\mu$m)?

**2.8.** You have a semiconductor with a carrier mobility of 1500 cm$^2$/V · s at 300 K. You have a sample $10^{-4}$ cm thick and inject $10^{15}$ carriers/cm$^3$ at one side ($x = 0$) and collect all the carriers at the other side ($x = L$), so that the carrier concentration is zero at $x = L$.

(a) If the electric field is zero and the concentration gradient is constant, what is the current density (A/cm$^2$)?

(b) At any time, how many carriers/cm$^2$ are in the semiconductor? If the injection is turned off at $t = 0$, estimate the time required for the carriers to diffuse out of the semiconductor.

(c) If the carrier concentration was uniform, $10^{15}$ carriers/cm$^3$, what applied voltage is required to maintain the same current density as in part (a)?

**2.9.** Your silicon chip is 5 mm $\times$ 5 mm and 250 $\mu$m thick mounted on a heat sink that maintains the bottom of the chip at 300 K. In the outer micron of the chip, the active devices have a total current of 10 mA ($10^{-2}$ A) at 10 V.

(a) If the chip were self-supported without a substrate, what would be the temperature after 1 min of operation (assuming no radiative heat loss)?

(b) If the chip is attached to the substrate, what is the heat flux through the silicon, and what is the temperature at the top surface assuming that the temperature gradient is constant?

**2.10.** You have a *n*-type semiconductor and make measurements in the high-electric-field region, where the electron drift velocity is constant (independent of electric field). You measure current $I$, voltage $V$ along the sample length $L$ (along the direction of current), and Hall voltage $V_H$ across the sample (magnetic field perpendicular to current flow and Hall field).

(a) You increase $V$.

    (1) Does the resistance $R$ increase, decrease, or remain constant?

    (2) Does $V_H$ increase, decrease, or remain constant?

(b) You increase $L$ with $V$ constant.

    (1) Does the resistance $R$ increase, decrease, or remain constant?

    (2) Does $V_H$ increase, decrease, or remain constant?

**2.11.** Consider carrier diffusion.

(a) You have a sample of *n*-type Si and heat it so that the number of electrons is 10 times greater at the hot end than at the cold end. You measure the voltage across the heated sample. Is the hot end positive, negative, or the same (zero volts) compared to the cold end?

(b) You have an *n*-type sample with one end heavily doped so that the carrier concentration is 10 times greater than the lightly doped end. You measure the voltage across the sample. Is the heavily doped end positive, negative, or the same (zero volts) as the lightly doped end?

# 3

# Bonds, Bands, and Semiconductors

## 3.1 Introduction

A portion of the periodic table is shown in Fig. 3.1. When one considers the two adjacent elements discussed in Chapter 2, aluminum (Al) and silicon (Si), it is the eleven-order-of-magnitude difference in resistivity that stands out in contrast to the many similar features (Table 3.1). The atomic density $N$ (atoms/cm$^3$) of Al is $6 \times 10^{22}$, a value close to that of Si ($5 \times 10^{22}$ cm$^3$); the mass densities (g/cm$^3$) are similar, 2.7 for Al and 2.33 for Si, and the two even feel the same. A review of the atomic physics of the electron configurations and binding energies of electrons in individual atoms shows that these are also similar in both materials. As indicated in Fig. 3.1, aluminum has an atomic number $Z$ of 13 and for Si, $Z = 14$, which means that the number of atomic electrons is 13 and 14, respectively. As shown in Fig. 3.2, the 10 most tightly bound electrons (the inner $K$- and $L$-shell electrons, to use x-ray transition notation) form the closed-shell configuration

| III | IV | V |
|---|---|---|
| 5<br>B<br>Boron | 6<br>C<br>Carbon | 7<br>N<br>Nitrogen |
| 13<br>Al<br>Aluminum | 14<br>Si<br>Silicon | 15<br>P<br>Phosphorus |
| 31<br>Ga<br>Gallium | 32<br>Ge<br>Germanium | 33<br>As<br>Arsenic |

**FIGURE 3.1** Portion of the periodic table showing column III, IV, and V elements and their atomic number $Z$ (the number of electrons).

**TABLE 3.1** Comparison of the Properties of Al and Si at 300 K[a]

|  | $N$ <br> ($\times\ 10^{22}/\text{cm}^3$) | $\rho_m$ <br> g/cm³ | $\rho$ <br> ($\Omega \cdot$ cm) | $c_p$ <br> (J/g $\cdot$ K) |
|---|---|---|---|---|
| Al | 6.02 | 2.7 | $2.5 \times 10^{-6}$ | 0.9 |
| Si | 5.0 | 2.33 | $2.3 \times 10^{5}$ | 0.7 |

[a]$N$, atomic concentration; $\rho_m$, mass density; $\rho$, resistivity; $c_p$, specific heat.

of neon ($Z = 10$). The electron binding energies (the energy required to remove an electron from the atom and bring it to rest well away from the atom) of these inner-shell electrons are within 10 to 15% of each other for Al and Si. These differences are due primarily to the differences in atomic number, which indicate 13 protons in the nucleus of Al and 14 in that of silicon. There are three outer-shell ($M$-shell) electrons in Al and 4 in Si—all with about the same binding energies. With the exception of electrical resistivity, the macroscopic and atomistic properties of Al and Si appear to be similar.

The question then arises as to nonelectrical methods to distinguish between disks of Al and Si, as the materials are odorless, tasteless, and have about the same weight and color. If one heats the materials, the difference is quickly evident, as Al melts at 660°C (not even red-hot), whereas Si melts at 1414°C (white-hot). An even simpler method is to bend the two disks; Si is brittle and fractures and Al is ductile and bends easily. One difference, then, between Al and Si is in the way the atoms are held together or bonded.

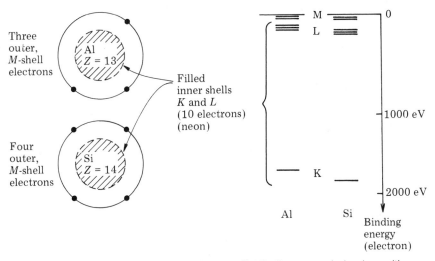

**FIGURE 3.2** Representation of the electron distribution around aluminum (three outer electrons) and silicon (four outer electrons) and the electron binding energies.

## 3.2 Bonds and the Covalent Bond

Most solids have an atomic density, $N$, between 3 and 6 $\times$ $10^{22}$ atoms/cm$^3$ with the exception of the less dense alkali metals (lithium, potassium, etc.). Consequently, it is not how close the atoms are to each other but how they are bonded together that is important.

Bonding may be understood in terms of the forces between atoms. When two atoms are brought together, there are both attractive (unlike charges) and repulsive (like charges) forces which determine the potential energy of the interaction (Fig. 3.3). The longer-range forces are attractive and are a result of electrostatic interactions between the atoms. At smaller distances repulsive forces between the positive charges of the protons in the nuclei of the two atoms begin to dominate and the net energy of interaction becomes positive (because work must be done to bring the atoms together). The minimum in energy, the binding energy, corresponds to the interatomic separation distance (actually, the distance between the small, point-like, positive nuclei as compared to the larger cloud of electrons around the nuclei). The net force, the derivative of the energy, is zero at this point, with the attractive and repulsive forces just balancing.

The bonding in a metal such as Al can be thought of in terms of all of the atoms in the solid being taken collectively, with the outer electrons (three in the case of Al) from all the atoms belonging to the solid as a whole. The metal can be pictured (Fig. 3.4) as a framework or lattice of positive metal ions embedded in an electron gas that permeates the entire solid. The lattice is held together by the resulting electrostatic interaction between the ions and electrons.

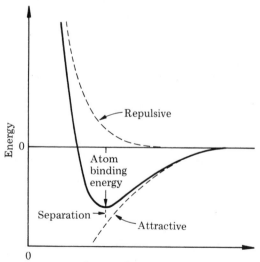

**FIGURE 3.3**   Schematic diagram of energy as a function of internuclear separation.

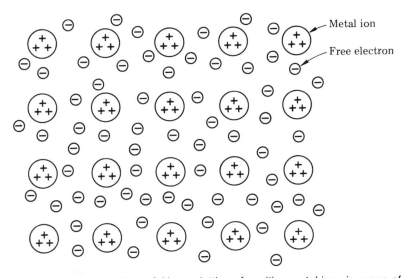

**FIGURE 3.4**  Representation of Al as a lattice of positive metal ions in a sea of free electrons.

Each metal atom contributes electrons to the free-electron gas. The electrons are distributed so that each atom on the average has enough electrons to neutralize its charge. At some instant in time, the electron gas in the metal might look as shown in Fig. 3.4. The electrons that are free to move are distributed uniformly through the metal even though at random. The negative charge carried by the electrons is neutralized by the positive charge of the immobile metal ions. This pictorial representation of the electron gas model of a metal is misleading in an important aspect. Here, as well as other places, it must be remembered that the quantum-mechanical nature of electrons keep them from being localized to this extent. Their influence is spread out over many atomic distances. For most metals, however, this electron gas model of Fig. 3.4 is quite useful, so it will be used here.

The freedom of the electrons to move through the metal and the large number of electrons gives rise to the high electrical conductivity. This freedom is not true in silicon, where the bonding is of a different type—the covalent bond.

Covalent bonding between two silicon atoms is visualized as a sharing of electrons supplied by both atoms. The bond comes about because shared electrons orbit around both atoms. This overlap of the bonding orbitals lowers the energy of the system. For the purpose of bonding, the atoms of silicon may be visualized as having 10 inner, closed-shell electrons (which do not participate in bonding), and four outer electrons that do. When a silicon atom is brought together with four other silicon atoms (Fig. 3.5), it shares its four outer electrons (heavy, curved lines in Fig. 3.5) with each of the four atoms. These atoms, in turn, share one of their four outer electrons (light, curved lines) with the central Si atom. Covalent bonds are quite directional (the four electrons are arranged symmetrically) as illustrated

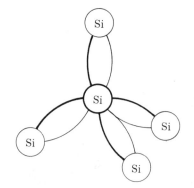

**FIGURE 3.5** Covalent bonds of silicon.

by the tetrahedral configuration. Again, we emphasize that these lines are just a pictorial representation of a more diffuse electron distribution.

Figure 3.6 is a two-dimensional representation of the silicon lattice. The covalent bonds are shown by the curved lines connecting the Si atoms. Although this is a line-and-ball representation, it shows that all the outer or valence electrons are tied up in covalent bonds. The inner 10 electrons are tightly bound (binding energies greater than 100 eV) and the outer four form the covalent bonds. Consequently, there are essentially no free electrons available for charge transport.

If we shine light of a fixed energy on the sample, we can break the bonds if the photon energy is greater than the bond energy (about 1.1 eV for Si), which corresponds to the infrared portion of the spectrum. When a bond is broken, the liberated electron is now free to move within the crystal. The empty site, or hole, left by the escaping electron can be occupied by a nearby electron. Consequently, the hole can also migrate through the crystal by exchanges with the bound electrons. Both electrons and holes can transport charge leading to a current, a photocurrent.

If the sample is irradiated with low-energy photons which are not energetic enough to break the covalent bonds, the photons will be transmitted through the

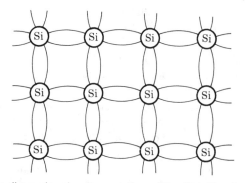

**FIGURE 3.6** Two-dimensional representation of the Si lattice for high-purity Si, where all the outer electrons are in covalent bonds.

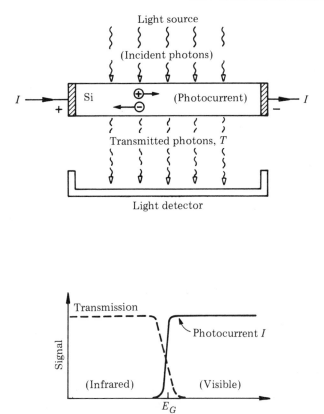

**FIGURE 3.7** Photons incident on a silicon sample, with the lower portion display-ing the signal from the transmitted photons (not absorbed) and the photocurrent versus photon energy.

sample unattenuated (Fig. 3.7). There is no mechanism for the absorption of light as well as no photocurrent. As the energy of the photons is increased, a threshold is reached where the interaction of light with the Si lattice can break the covalent bonds. In Si this occurs at an energy $E_G$ of about 1.1 eV, at a wavelength $\lambda$ of 1.1 μm [from eq. (2.11), $E = 1.24$ eV$/\lambda(\mu m)$]. This energy lies in the infrared but close to the visible range.

As the photon energy is increased above this energy $E_G$, the number of trans-mitted photons decreases and the photocurrent increases due to the absorption of light in the bond-breaking process. Each method, transmission or photoconduct-ance, can be used to measure the energy required to free an electron from the covalent bonds.

For a silicon sample at room temperature, there is enough thermal energy con-tained in the lattice to break the bonds. This thermal energy is in the form of lattice vibrations or phonons which are responsible for thermal conductivity (Chapter 2).

The average phonon energy is a few $kT$ ($kT$ = 0.026 eV at room temperature). Bond breaking by phonons is a very rare event since $kT$ is much less than the bond energies and a multiple phonon interaction is required. At thermal equilibrium there is only one broken bond for every $10^{12}$ silicon atoms. Thus pure silicon can conduct current due to thermal generation of electrons and holes through the bond-breaking process. The conductivity of pure Si lies between that of a good metallic conductor and that of an insulator—hence the name "semiconductor."

## 3.3  Dopants: Donors and Acceptors

The salient feature of integrated-circuit technology is the control of the conductivity of silicon that is achieved by the controlled introduction of electrically active elements into the lattice. As shown in the periodic table (Fig. 3.1), the elements are from column III [such as boron (B)] and column V [such as phosphorus (P) or arsenic (As)]. These elements, called dopant species, occupy substitutional sites in the Si lattice; that is, they take the place of a Si atom (the dopant is "substituted" for a Si atom). The influence of a column V element, As, is shown schematically in Fig. 3.8. Arsenic has five outer electrons, but only four are used to form covalent bonds with the neighboring Si atoms. These four are tightly bound in the lattice (1 eV binding energy as that of the covalent electrons from Si), but the extra electron is very weakly bound. The binding energy is so weak (<0.1 eV) that at room temperature there is enough thermal energy to break the bond (ionize the neutral dopant atom) and the electron is free to migrate through the lattice. Values of the ionization energy are given in Table 3.2 (see also problem 3.4). In this case the As atom donates an electron to the lattice and is called a donor,

$$N_D^0 \rightarrow N_D^+ + e^- \tag{3.1}$$

where $N_D^0$ is the concentration (atoms/cm$^3$) of neutral (with all five outer electrons) donors and $N_D^+$ the concentration of donors that have donated an electron and hence are positively charged.

When a column III element such as boron is substituted in the Si lattice, there is one missing electron, a missing covalent bond. A nearby covalently bound

**TABLE 3.2**  Values of the Dopant Ionization Energy (eV) for Donors and Acceptors in Si, Ge, and GaAs

|  | Si | Ge | GaAs |
|---|---|---|---|
| Donors | P: 0.045 | P: 0.012 | Si: 0.006 |
|  | As: 0.054 | As: 0.013 | Se: 0.006 |
| Acceptors | B: 0.045 | B: 0.01 | Si: 0.035 |
|  | Al: 0.067 | Al: 0.01 | Zn: 0.031 |

*Source:* Milnes (1973).

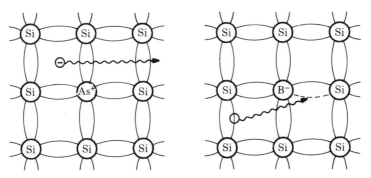

**FIGURE 3.8** Two-dimensional representation of the silicon lattice containing (left) an arsenic atom which donates a free electron, and a boron atom (right) which accepts a nearby electron to complete the covalent bond.

electron can migrate to the boron atom, completing the four covalent bonds and leaving a positively charged hole, $h^+$, behind:

$$N_A^0 \rightarrow N_A^- + h^+ \tag{3.2}$$

where $N_A^0$ is the concentration (atoms/cm$^3$) of neutral acceptors and $N_A^-$ is the concentration of acceptors that have accepted an electron and hence are negatively charged. Since the silicon lattice as a whole is neutral, the expression for charge neutrality is

$$p + N_D^+ = n + N_A^- \tag{3.3}$$

where we have used $p$ and $n$ to denote the concentration of holes and electrons, respectively. Generally, either large concentrations of donors are added, $N_D \gg N_A$, or large concentrations of acceptors, $N_A \gg N_D$, so that

$$n = N_D^+ \quad \text{or} \quad p = N_A^- . \tag{3.4}$$

We have presented the relatively simple case of dopants in Si. A compound semiconductor such as a III–V compound, gallium arsenide (GaAs), also forms a lattice with covalent bonding. The similarity to Si can be inferred from Fig. 3.9, which shows a portion of the periodic table and a representation of the atomic configuration of the electrons. All five elements have 28 electrons in filled inner shells ($K$, $L$, and $M$ shells) and electrons in the outer, less tightly bound shell ($N$ shell). The column III element Ga has three outer electrons, and the column V element As has five outer electrons. When these two elements are arranged in a lattice they share outer electrons to form covalent bonds with Ga atoms providing three electrons and As atoms, five electrons. There is, in effect, a Ga sublattice and an As sublattice (Fig. 3.10). If a column VI element such as selenium (Se) or tellurium (Te) is substituted on an As site, it acts as a donor; if a column II element such as zinc (Zn) is on a Ga site, it acts as an acceptor. Silicon with four electrons can act as a donor on a Ga site or as an acceptor on an As site. Because of the dual role of Si it is called an amphoteric dopant.

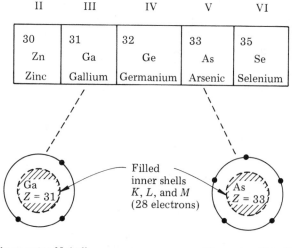

**FIGURE 3.9**    Portion of the periodic table and representation of the electron distribution around gallium (three outer electrons) and arsenic (five outer electrons).

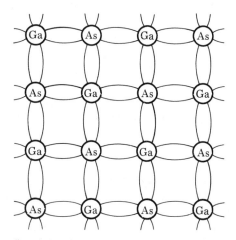

**FIGURE 3.10**    Two-dimensional representation of the gallium-arsenide (GaAs) lattice for high-purity GaAs, showing that all the outer electrons are in covalent bonds.

## 3.4 Mobility: Influence of Dopants

The introduction of dopants into a lattice also influences the mobility of the charge carriers. When we introduced the concept of mobility in Chapter 2 we noted that the mobility $\mu$ was determined by the mean free time $\tau$ between collisions; $\mu = e\tau/m$ [eq. (2.16)]. In a perfect lattice, the electron could travel long distances before encountering a scattering or collision site. However, at room temperature, the lattice is vibrating ever so slightly due to thermal energy. This vibrational energy comes in packets called phonons and has wave properties very much like photons, the energy packet for light. The phonons have a complicated wave structure, as they represent the vibrational displacement of the lattice atoms. Every once in a while, this wave structure happens to reinforce and give a particular atom a larger displacement than normal. That one atom for that instance provides a scattering center and determines $\tau_l$, the lattice scattering time. Since the speed of sound (phonon velocity) is about $5 \times 10^5$ cm/s (a value much less than the thermal velocity of electrons), the displaced lattice atoms appear frozen in place to the electron during its time between collisions.

The dopant atoms also act as scattering centers. In this case the coulomb force introduced by the presence of a charged center fixed in the lattice can deflect or scatter the carriers with a scattering time $\tau_i$. Since the smallest time dominates,

$$\frac{1}{\tau} = \frac{1}{\tau_l} + \frac{1}{\tau_i}$$

or
(3.5)

$$\frac{1}{\mu} = \frac{1}{\mu_l} + \frac{1}{\mu_i}$$

where $\mu_l$ and $\mu_i$ are the contribution to the mobility determined by the vibrating lattice and the impurities. Figure 3.11 shows the mobility of electrons and holes in silicon as a function of dopant or impurity concentration $N_i$. At low impurity

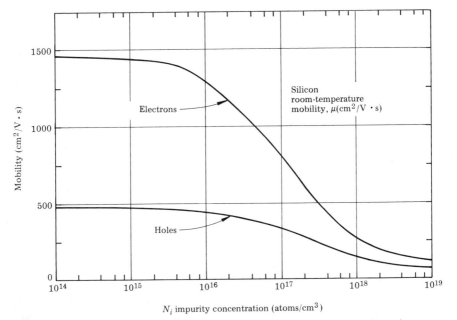

**FIGURE 3.11** Electron and hole mobilities in silicon at 300K versus impurity concentration. (Adapted from W.E. Beadle, J.C.C. Tsai, and R.D. Plummer, eds., *Quick Reference Manual for Silicon Integrated Circuit Technology*, Wiley, New York, 1985).

concentrations, the mobility is determined by lattice scattering. If we assume an electron mobility of 1500 $cm^2/V \cdot s$, the mean time between collisions is about 0.85 picosecond (ps) ($\tau = \mu \, m/e$). For a thermal velocity of $1.1 \times 10^7$ cm/s, the mean free path $l$ between collisions is about $10^{-5}$ cm. For silicon with an atomic density $N$ of $5 \times 10^{22}$ atoms/$cm^3$, the average spacing $\overline{d}$ between atoms is $1/(N)^{1/3}$, about $3 \times 10^{-8}$ cm. The electron passes $l/\overline{d}$ atoms, or about 300 lattice atoms between collisions.

As the dopant concentration is increased, the presence of the charged impurity centers will decrease the carrier mobility. We can estimate the critical impurity concentration by assuming that their influence on scattering sets in only when the average distance between the impurities is *less* than the mean free path $l$ between collisions when lattice scattering dominates. A value of $l$ of $10^{-5}$ cm corresponds to a concentration $N_i$ of $10^{15}/cm^3$ ($N_i = 1/l^3$). From the data shown in Fig. 3.11, a concentration of a few times $10^{15}/cm^3$ is a reasonable estimate of the concentration at which the presence of charged centers can influence the carrier mobility.

## 3.5 Energy Bands in Semiconductors

The energy levels of electrons in isolated Si atoms have well-defined values, as indicated in Fig. 3.2. To describe the electronic structure, we should properly describe the electron on the basis of its wave behavior and obtain a quantum-mechanical description of the solid. At present, we will continue to describe the electron as a particle occupying energy levels whose values and spacing are determined by the wave behavior of the electron. Of course, even the energy levels of the electrons in an isolated atom (Fig. 3.2) are wave-function solutions of the Schrödinger equation. The allowed, discrete energies, corresponding to particular wave functions, are designated by quantum numbers.

As the isolated atoms are brought together to form a solid, the orbitals of the outer electrons of the different atoms begin to overlap and interact with each other. This interaction causes a shift in the energy levels and a band of levels is formed extended in width over a few eV. In a solid where the number of atoms is of the order of $10^{23}/cm^3$, the number of levels within the band becomes of this order and their spacing is very close. We have shown two bands in Fig. 3.12 for high-purity silicon (with no impurities and at low temperatures) where there are no free electrons. All the outer electrons are tied up in covalent bands and we designate the lower band in Fig. 3.12 as the valence band. All the lower-band energy levels are completely occupied with electrons. One of the consequences of the quantum-mechanical treatment is that no two electrons can occupy the same electronic state at the same time (the Pauli exclusion principle). Because the electrons have two spin states, there are two electrons in each level.

With the valence band completely full and separated in energy from the next band, there are no empty states available for electrons to move to if a modest electric field is applied to the sample of silicon. Thus the electrons in the full valence band do not conduct current, a situation that we predicted from the simple geometrical picture of the Si lattice shown in Fig. 3.6.

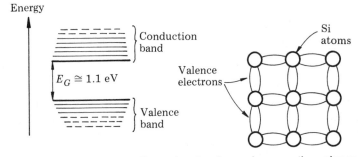

**FIGURE 3.12** Energy levels in silicon showing the region near the valence-band and conduction-band edges, and the associated Si lattice showing the outer, valence electrons in covalent bonds.

The band of states lying higher in energy is separated in energy from the valence band by an energy $E_G$. The notation $E_G$ indicates "energy gap" and $E_G$ has a value of 1.1 eV in Si and 1.4 eV in GaAs. If photons of energy greater than $E_G$, say 1.2 eV, irradiate the Si sample, the photons can interact with the valence electrons. The photon delivers all its energy to the covalent electron and breaks the bond, thereby freeing an electron (Fig. 3.13). The electron is now free to transport current under the influence of an applied electric field. On an energy band diagram, the interaction is shown as a vertical line where the electron makes a transition from the full valence band to the empty conduction band. Within the conduction band, there are a large number of energy states, closely spaced as in the valence band, but empty. The electron in the conduction band has a sufficient number of available levels, so it can gain energy from an electric field and hence act as a current conductor. This is an energy-band description of the photoconductivity process shown in Fig. 3.7. It is obvious that the onset of photoconductivity and the decrease of light transmission occur when the photon energy exceeds $E_G$.

When donors are introduced, the electron concentration is increased and we assume that $n = N_D$; the semiconductor is referred to as an "$n$-type" semiconductor. The energy required to remove an electron from the donor, the ionization

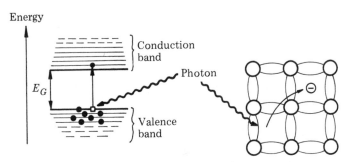

**FIGURE 3.13** Photoconduction process showing an incident photon interacting with a valence electron to break the bond and to produce a free electron in the conduction band.

energy or binding energy, is much less than $E_G$. To establish an energy scale in Fig. 3.14, we set the zero of energy at the top of the valence band, denoted $E_V$. We are concerned with differences in energies, so assignment of the reference level is arbitrary. With this convention, the bottom of the conduction band $E_C$ is located at an energy $E_G$; $E_C - E_V = E_G$. The location of the donor states is set at $E_D$; $E_C - E_D$ is the donor ionization energy. The donor level is shown in Fig. 3.14a to be empty of electrons. There are so many empty states available in the conduction band and sufficient thermal energy $kT$ is available that all the donors are ionized and $n = N_D$.

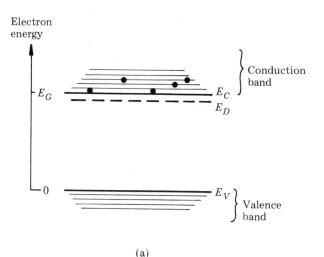

(a)

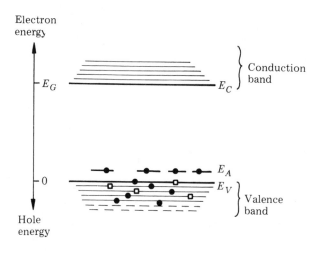

(b)

**FIGURE 3.14** Energy levels in a semiconductor: (a) electrons in the conduction band from ionized donor levels at $E_D$; (b) holes in an otherwise full valence band due to electrons on ionized acceptor levels at $E_A$.

The energy-level diagram is shown in Fig. 3.14b for a $p$-type semiconductor where acceptors are added, and we assume that the hole concentration, $p$, equals the concentration of acceptors, $N_A$. The location of the acceptor states is set at $E_A$; $E_A - E_V$ is the acceptor ionization energy. In this diagram the acceptors are shown to be occupied with electrons. There are so many valence electrons around the acceptors that there is a high probability that an electron will complete the covalent bond arrangement around the acceptor and leave a broken band or hold behind; $p = N_A$.

In Fig. 3.14 we have shown free electrons occupying states near the bottom of the conduction band and holes occupying states near the top of the valence band. Electrons will occupy the lowest levels available in the conduction and valence bands following the Pauli exclusion principle that only one electron is allowed in a given state with a set of quantum numbers. Since holes represent the absence of electrons, it follows that holes will be located at the top of the valence band. We could also arrive at the same conclusion by considering the energies of the electrons and holes. In Fig. 3.14 the electron energy is directed toward the top of the page, so that electrons occupying higher levels have higher energies than those near the bottom of the band. The hole energy is directed toward the bottom of the page so that the lowest-energy state for holes is at the top of the valence band. In analogy with bubbles in water, holes float upward.

## 3.6 Effective Density of States and the Fermi Level

The electrons have been treated as a free-electron gas in the semiconductor. We can continue this analogy to classical particles and simplify the energy diagram to a pair of lines located at $E_C$ and $E_V$. Rather than draw all the energy levels in the bands, we can condense them to a single line located at $E_C$ for the conduction band and at $E_V$ for the valence band, as shown in Fig. 3.15. This is not just an artifice but is based on a wave-mechanical description of the density of states in the bands and the probability of occupation of the levels. Although this treatment is described in detail in Chapter 12, we utilize one of the results to give the parameter, $N_S$, for the effective density of states at $E_C$ and $E_V$:

$$N_S = 2 \left( \frac{2\pi m k T}{h^2} \right)^{3/2} = 4.84 \times 10^{15} T^{3/2} \ (\text{cm}^{-3}) \tag{3.6}$$

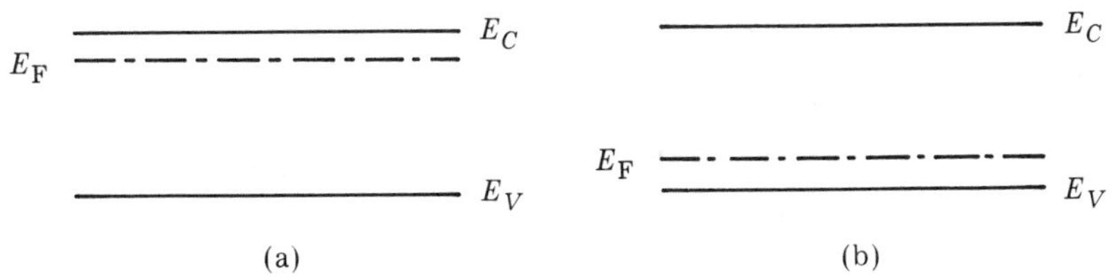

(a)   (b)

**FIGURE 3.15** Energy levels in a semiconductor. Solid lines represent the density of states at $E_C$ and $E_V$ and dashed lines the Fermi level at $E_F$ in (a) an $n$-type and (b) a $p$-type semiconductor.

**TABLE 3.3**   Density of States in the Conduction Band $N_C$ and Valence Band $N_V$, Energy Gap $E_G$, and Intrinsic Carrier Concentration $n_i$ at 300 K[a]

|  | $N_C$ (cm$^{-3}$) | $N_V$ (cm$^{-3}$) | $E_G$ (eV) | $n_i$ (cm$^{-3}$) |
|---|---|---|---|---|
| Si | $2.8 \times 10^{19}$ | $1.04 \times 10^{19}$ | 1.12 | $1.45 \times 10^{10}$[a] |
| Ge | $1.04 \times 10^{19}$ | $6.0 \times 10^{18}$ | 0.66 | $2.4 \times 10^{13}$ |
| GaAs | $4.7 \times 10^{17}$ | $7.0 \times 10^{18}$ | 1.424 | $1.79 \times 10^{6}$ |

*Source:* Sze (1981).

[a]The value for $n_i$ (Si) calculated from eq. (3.18b) differs from that given in Table 3.3. For consistency we use tabulated values.

where $m$ is the free-electron mass and $h$ is Planck's constant, $h = 6.626 \times 10^{-34}$ J $\cdot$ s. As we show in the complete treatment, an effective mass $m^*$ for electrons $(m_e^*)$ in the conduction band and holes $(m_h^*)$ in the valence band should be used. These distinctions are important when describing actual semiconductors such as Si or GaAs, where we use the notation $N_C$ and $N_V$ to give the effective density of states in the conduction and valence bands, respectively, using effective mass values (see Table 3.3). Using the free-electron mass the value of $N_S$ at 300 K ($kT = 0.026$ eV) is

$$N_S = 2.52 \times 10^{19}/\text{cm}^3.$$

Clearly, this value is orders of magnitude less than the actual number of all the levels in the conduction band, about $10^{23}/\text{cm}^3$. For dopant concentrations much less than $N_S$, there are so few electrons in the conduction band that we can consider them as classical particles occupying a single level at the bottom of the conduction band at $E = E_C$. The parameter $N_S$ reflects the number of states available at $E_C$ in the classical approximation where electrons follow Maxwell–Boltzmann statistics. The probability $F_{\text{M-B}}$ that a state of energy $E_C$ is occupied is

$$F_{\text{M-B}} = \exp\left(-\frac{E_C}{kT}\right).$$

In the semiconductor systems with dopant concentrations less than $N_S$, the value of $F_{\text{M-B}}$ is so small that we do not have to invoke the Pauli exclusion principle (one electron per set of quantum numbers) or the Fermi–Dirac probability distribution that is required in semiconductors with large dopant concentrations (Chapter 12).

It is useful when describing the occupation of states with electrons or holes to have the Fermi energy $E_F$ (also called the Fermi level) as a reference compared to $E_C$ or $E_V$. The probability of occupation is expressed

$$F_{\text{M-B}} = \exp\left(-\frac{E_C - E_F}{kT}\right) \tag{3.7}$$

so that in a system with an electron concentration $n$,

$$n = N_S F_{\text{M-B}}(E) \tag{3.8}$$

$$= N_S \exp\left(-\frac{E_C - E_F}{kT}\right).$$

For example, if $n = N_S/1000$, the Fermi level will be located about $0.18$ eV below the conduction band level at 300 K ($kT = 0.026$ eV), as shown schematically in Fig. 3.15a for the $n$-type semiconductor; and similarly for a $p$-type semiconductor (Fig. 3.15b):

$$p = N_S \exp\left(-\frac{E_F - E_V}{kT}\right) = N_S \exp\left(-\frac{E_F}{kT}\right) \tag{3.9}$$

and the Fermi level would lie about $0.18$ eV above the valence band energy $E_V$ for $p = N_S/1000$. We have set $E_V = 0$.

So far we have treated the Fermi energy in terms of a reference energy for the probability of occupation of a set of levels. In more general terms it is the electrochemical potential of the electrons [indicated by $\bar{\mu}$ in standard chemical texts such as Castellan (1971)] or partial molar free energy. For example, in a dilute system obeying Maxwell–Boltzmann statistics where there are $n$ electrons in $N$ possible energy states,

$$\bar{\mu} = \text{const.} + kT \ln \frac{n}{N}$$

which would translate in our notation to

$$E_F = E_C + kT \ln \frac{n}{N_S}. \tag{3.10}$$

An important aspect of the Fermi energy in view of the properties of an electrochemical potential is that the Fermi level must have a constant value through a system of electronic materials in equilibrium. This means that if we join $n$- and $p$-type semiconductors together as shown in Fig. 3.16, the Fermi levels are aligned, not the conduction and valence band energies $E_C$ and $E_V$. The offset in the con-

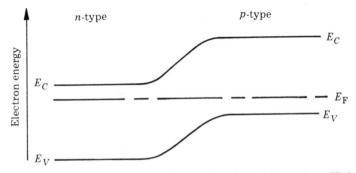

**FIGURE 3.16** Energy-band diagram of an $n$-$p$ junction at thermal equilibrium where the Fermi level is flat.

duction bands leads to the formation of internal energy barriers which are the heart of the *p-n* junction (Chapter 4).

## 3.7 Recombination, Generation, and Intrinsic Carrier Concentrations

In any semiconductor there are both electrons and holes present, irrespective of whether the semiconductor is doped with acceptors or donors or is undoped. The number of electrons and holes depends on temperature and the concentration of impurities (donors or acceptors).

As we noted before, electrons and holes can be created (or generated) by thermal energy. This occurs because the lattice atoms vibrate (fluctuate) about their ideal position when at elevated temperatures. Occasionally, a bond is broken and an electron and hole are created; equal numbers of holes and electrons are generated (Fig. 3.17). The symbol $G$ is used to denote the rate per unit volume at which electrons and holes are generated:

$$\text{thermal generation rate} = \frac{\text{electron–hole pairs}}{\text{cm}^3 \cdot \text{s}} = G. \qquad (3.11)$$

Naturally, there must be another mechanism leading to the removal of electrons or holes, or we would be overwhelmed by charge carriers. There is the inverse process, recombination, which acts to re-form broken bonds or in terms of the band picture acts to return free electrons to the holes in the valence band (Fig. 3.17).

Under conditions of thermal equilibrium the recombination rate, $R$, must equal the generation rate, $G$:

$$\text{generation rate, } G = \text{recombination rate, } R. \qquad (3.12)$$

We can specify the general expression for $R$ if we recognize that for recombination to occur there must not only be a free electron, but also an empty state, or hole, in the valence band into which the electron can move. This is the mass-action law, in analogy with chemistry, which states that the recombination rate is proportional to the number, $n$, of electrons available for recombination times the number, $p$, of empty states available for occupancy; that is,

$$R = rnp \qquad (3.13)$$

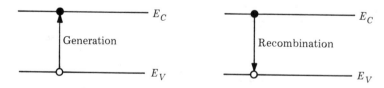

**FIGURE 3.17** Generation and recombination processes in a semiconductor.

where $r$ is a proportionality constant the value of which depends on the semiconductor material, the impurities, and temperature.

The generation rate, $G$, is a function of temperature, and since $G = R$, it follows that the product of electrons and holes will be a constant at any given temperature:

$$np = \text{const.} \tag{3.14}$$

In a crystal without any impurities, the number of electrons is equal to the number of holes (we use the subscript $i$ to denote intrinsic carrier concentrations):

$$n_i = p_i \tag{3.15}$$

and from eq. (3.14) it follows that

$$np = n_i p_i = n_i^2 . \tag{3.16}$$

The value of the intrinsic carrier concentration $n_i$ can be determined from eqs. (3.8) and (3.9):

$$
n_i^2 = np = \left[ N_S \exp\left( -\frac{E_C - E_F}{kT} \right) \right] \left[ N_S \exp\left( -\frac{E_F - E_V}{kT} \right) \right]
$$

$$
= N_S^2 \exp\left( -\frac{E_C - E_V}{kT} \right) \tag{3.17}
$$

$$
= N_S^2 e^{-E_G/kT}
$$

or

$$n_i = N_S e^{-E_G/2kT}. \tag{3.18a}$$

In a semiconductor with a band gap of 1.1 eV at 300 K where $N_S = 2.5 \times 10^{19}/\text{cm}^3$, the value of $n_i = 1.6 \times 10^{10}$ carriers/cm$^3$. In real semiconductors, one should use the electron and hole effective masses to determine the effective density of states $N_C$ and $N_V$ (Table 3.3) in the conduction and valence band, respectively, so that

$$n_i = (N_C N_V)^{1/2} e^{-E_G/2kT} = N_C \exp\left( -\frac{E_C - E_F}{kT} \right), \tag{3.18b}$$

$$E_F = \frac{E_G}{2} + \frac{kT}{2} \ln\frac{N_V}{N_C}. \tag{3.18c}$$

The intrinsic conductivity $\sigma_i$ is

$$\sigma_i = e n_i (\mu_n + \mu_p) \tag{3.19}$$

where $e$ is the charge on the electron and $\mu_n$ and $\mu_p$ are the electron and hole mobilities. The intrinsic resistivity $\rho_i$ ($\rho_i = 1/\sigma_i$) of a material with $n_i = 1.6 \times 10^{10}/\text{cm}^3$ and $\mu_n + \mu_p = 2000$ has a value of $2.0 \times 10^5 \ \Omega \cdot \text{cm}$; a value that is 11 orders of magnitude greater than the resistivity of aluminum, $\rho = 3 \times 10^{-6} \ \Omega \cdot \text{cm}$.

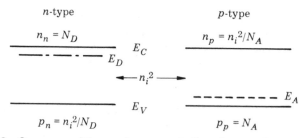

**FIGURE 3.18** Summary of the carrier concentrations in $n$- and $p$-type semiconductors assuming that all dopants are ionized and that $np = n_i^2$.

If the semiconductor is doped with donors, the electron concentration $n$ will be much greater than the hole concentration, $p$ (i.e., $n \gg p$). At this point we assume that all the donors are ionized (i.e., all the donors have given up their electrons) and that the electron concentration $n$ is equal to the donor concentration $N_D$. We follow the same assumption for materials with a concentration $N_A$ of acceptors; that is:

$$n\text{-type material: } n = N_D = n_n \tag{3.20}$$
$$p\text{-type material: } p = N_A = p_p.$$

where we use subscripts $n$ and $p$ to denote $n$- and $p$-type material, respectively.

In $n$-type material, the electrons are called majority carriers and the holes are minority carriers. In $p$-type material the holes are majority carriers and electrons are called minority carriers. Since the electron hole product is a constant, $np = n_i^2$, the minority carrier concentrations are given by:

$$n\text{-type material: } p_n = \frac{n_i^2}{N_D}$$
$$p\text{-type material: } n_p = \frac{n_i^2}{N_A}. \tag{3.21}$$

These relations are shown in Fig. 3.18.

In most cases we can neglect the concentration of minority carriers in calculating the current density, $J$. For example, in Si which contains an acceptor concentration $N_A = 10^{18}/\text{cm}^3$, the minority carrier or electron concentration is given by $n_p = n_i^2/N_A = 2.1 \times 10^{20}/10^{18} = 2.1 \times 10^2/\text{cm}^3$, whereas $p_p = N_A \simeq 10^{18}/\text{cm}^3$. Clearly, the equilibrium electron concentration can be neglected in the $p$-type semiconductor.

## 3.8 Minority-Carrier Lifetimes

In Section 3.7 we considered the case where the semiconductor was in the dark under thermal equilibrium conditions. Now let us consider the case where we create

$n_l$ and $p_l$ extra carriers/cm$^3$ by illumination with light; then

$$n = n_0 + n_l \qquad \text{and} \qquad p = p_0 + p_l \qquad (3.22)$$

where we use the subscript 0 for equilibrium conditions in the dark.

The recombination rate $R$ is given by

$$R = rnp = r(n_0 + n_l)(p_0 + p_l).$$

Since the thermal generator rate $G$ is not affected by light, $G = rn_i^2$, so that the net recombination rate $U$ is given by

$$U = R - G = r[(n_0 + n_l)(p_0 + p_l) - n_i^2] \qquad (3.23)$$

or

$$U = R - G = r(np - n_i^2). \qquad (3.24)$$

Immediately after the light is turned off, the values of $n$ and $p$ will begin to decrease. As shown in Fig. 3.19 the product $np$ will decrease or decay until it reaches the product $n_i^2$ (or $n \to n_0$ and $p \to p_0$). Since electrons recombine with holes, the rate of loss of electrons, $-dn/dt$, equals the rate of loss of holes, $-dp/dt$. That is,

$$U = -\frac{dn}{dt} = -\frac{dp}{dt} = r(np - n_i^2). \qquad (3.25)$$

For the discussion of p-n junctions and bipolar transistors, the prime concern is the decay or recombination of minority carriers. In this case, the number of carriers created by a light flash (or by injection from the emitter) is much less than the concentration of majority carriers. In n-type Si with $N_D$ donors we have

$$n \gg p$$

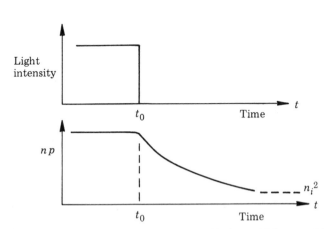

**FIGURE 3.19** Incident light creates electrons and holes, which recombine with time after the light is off. The electron-hole product, $np$, is shown versus time.

$$n = N_D + n_l \simeq N_D \tag{3.26}$$
$$p = p_n + p_l \simeq p_l$$
$$n_i^2 = N_D \cdot p_n$$

since $n_l \ll N_D$ and $p_n \ll p_l$, where $n_l$ and $p_l$ are the concentrations of excess carriers created by the flash of light ($n_l = p_l$). By use of the relations in eq. (3.26), the expression for the recombination rate of minority carriers, holes, given in eq. (3.25) can be written as

$$-\frac{dp}{dt} = rN_D (p - p_n). \tag{3.27}$$

We define the minority-carrier lifetime $\tau_p$ (the hole lifetime in $n$-type Si) by the formula

$$\tau_p = \frac{1}{rN_D} \tag{3.28}$$

so eq. (3.27) becomes

$$\frac{dp}{dt} = -\frac{p - p_n}{\tau_p} \tag{3.29}$$

which has a solution

$$p - p_n = (p - p_n)_0 \exp\left(-\frac{t}{\tau_p}\right) \tag{3.30a}$$

where $(p - p_n)_0$ represents the concentration of excess minority carriers when the light is turned off (or when injection or creation of holes stops).

A similar relation holds for electrons in $p$-type material:

$$n - n_p = (n - n_p)_0 \exp\left(-\frac{t}{\tau_n}\right) \tag{3.30b}$$

where $\tau_n$ is the electron lifetime or minority-carrier lifetime in $p$-type material, that is,

$$\tau_n = \frac{1}{rN_A}. \tag{3.31}$$

These expressions indicate that when excess carriers are created in $n$- or $p$-type material at concentrations much less than the doping concentration ($N_D$ or $N_A$), the excess carrier population dies away to zero exponentially with a time constant $\tau_p$ or $\tau_n$.

In silicon, recombination occurs through defect centers, which may be an impurity atom such as Au or Cu (often referred to as lifetime killers), or disordered places in the crystal lattice, such as dislocations or stacking faults. In transistor-grade Si, a reasonable estimate of lifetime can be made by choosing a minority-carrier lifetime $\tau$ equal to 10 $\mu$s.

## **3.9** Carrier Diffusion Lengths

The operation of a *p-n* junction or a bipolar transistor involves injection of carriers from one region into another. In the example shown in Fig. 3.20, holes are injected into an *n*-type region. The number of injected holes is usually much less than the electron concentration (the majority carriers) but much greater than the concentration of thermally generated holes (the minority carriers). If we assume that the electric field $\mathscr{E}$ is negligible in the *n*-type region, the injected holes will diffuse deeper into the *n*-type region and recombine with electrons. The concentration of injected holes will decrease with distance into the *n*-type material as shown in Fig. 3.20.

If we maintain a constant concentration of injected holes, $p$ $(x = 0)$ at $x = 0$, holes will diffuse and recombine in the *n*-type region and a steady concentration profile, $p(x)$, will be established. There will be a flow or flux, $F(x)$, of holes and if the electric field is zero, $\mathscr{E} = 0$, diffusion dominates (Chapter 2) and

$$F(x) = -D_p \frac{dp}{dx} \tag{3.32}$$

where $D_p$ is the hole diffusion coefficient in units of cm$^2$/s. We can neglect the thermally generated holes and characterize the recombination per unit time of the holes with a lifetime $\tau_p$ by

$$-\frac{dp}{dt} = \frac{p(x)}{\tau_p}$$

The diffusion-recombination process can be described by visualizing a slab of material $\Delta x$ thick with holes diffusing in from one side, some recombining, and

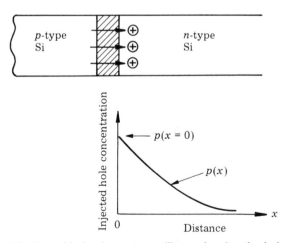

**FIGURE 3.20** Injection of holes into *n*-type silicon, showing the hole concentration versus distance.

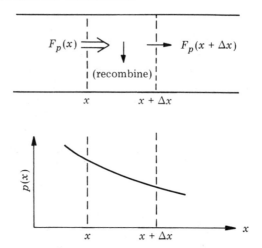

**FIGURE 3.21**   Flux $F$ of holes into and out of a slab $\Delta x$ thick where some recombination occurs.

some diffusing out (Fig. 3.21). The flux distribution can be expressed as

$$\frac{\text{flux of holes}}{\text{in at } x} = \frac{\text{recombination}}{\text{unit time}} + \frac{\text{flux of holes}}{\text{out at } x + \Delta x} \tag{3.33}$$

$$F(x) = \frac{p(x)}{\tau_p} \Delta x + F(x + \Delta x).$$

Equation (3.33) can be rewritten as

$$\frac{F(x + \Delta x) - F(x)}{\Delta x} = -\frac{p(x)}{\tau_p} \tag{3.34}$$

or

$$\frac{dF(x)}{dx} = -\frac{p(x)}{\tau_p}. \tag{3.35}$$

From eq. (3.32) we substitute for $F(x)$ with $D_p = $ constant, so that

$$D_p \frac{d^2p}{dx^2} = \frac{p}{\tau_p} \tag{3.36}$$

and rearranging gives us

$$\frac{d^2p}{dx^2} = \frac{p}{D_p\tau_p} = \frac{p}{L_p^2} \tag{3.37}$$

with

$$L_p = (D_p\tau_p)^{1/2}. \tag{3.38a}$$

The parameter $L_p$ is the hole diffusion length in $n$-type material. Similarly, we can define an electron diffusion length $L_n$ in $p$-type material,

$$L_n = (D_n \tau_n)^{1/2}. \tag{3.38b}$$

The significance of the diffusion lengths can be seen from the solution of eq. (3.37), which has the form

$$p(x) = C_1 e^{x/L_p} + C_2 e^{-x/L_p}. \tag{3.39}$$

Recombination reduces $p(x)$ to zero at large values of $x$ (remember that we have neglected the thermally generated holes), so that $C_1 = 0$. The solution to eq. (3.37) is

$$p(x) = p(x = 0)e^{-x/L_p}. \tag{3.40}$$

The injected hole concentration decreases exponentially with distance as the holes diffuse and recombine. For values of $D = 20$ cm$^2$/s and $\tau = 10^{-5}$ s, the value of $L = (D\tau)^{1/2} = 1.4 \times 10^{-2}$ cm. The diffusion length can also be used to estimate the time t for carriers to diffuse out of a region of thickness $W$, $t = W^2/D$ (see problem 2.8).

The diffusion length $L$ represents the distance at which the injected carrier concentration is reduced to a value $1/e$ of its value at the point of injection $x = 0$. We are often interested in the carrier concentration near the point of injection and can use a linear approximation, which for holes is

$$\frac{p(x)}{p(x = 0)} = e^{-x/L_p} \simeq 1 - \frac{x}{L_p}. \tag{3.41}$$

As shown in Fig. 3.22, this approximation is reasonable for $x/L_p \leq 0.4$.

In solutions of the current density where $J_p = -eD_p \, dp/dx$,

$$\frac{dp(x)}{dx} = -\frac{p(x = 0)}{L_p} e^{-x/L_p}. \tag{3.42}$$

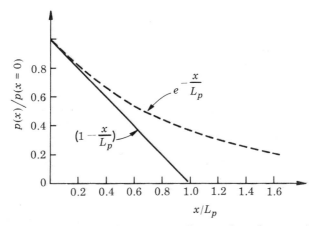

**FIGURE 3.22** Concentration of holes versus distance for a linear and exponential distribution.

If one evaluates the diffusion current at $x = 0$,

$$J_p(x = 0) = \frac{eD_p}{L_p} p(x = 0). \tag{3.43}$$

These relations are useful in derivations of operation of $p$-$n$ junctions and bipolar transistors.

## 3.10 Changes in Energy Gap with Temperature and Pressure

The energy gap $E_G$ in Si and GaAs changes with temperature and pressure. An increase in temperature causes a decrease in the band gap through the relation

$$E_G(T) = E_G(0) - \frac{AT^2}{T + B} \tag{3.44}$$

where values of $T$ are in Kelvin and Table 3.4 gives the values of the constants for Si and GaAs. From Table 3.4 and eq. (3.44), the values of the energy gap at 300 K for Si and GaAs are 1.12 eV and 1.42 eV, respectively. These values are also given in Table 3.3

The intrinsic carrier concentration $n_i(T)$ at temperature $T$ (K) can be written from eq. (3.18) as

$$n_i(T) = \text{const.} \times T^{3/2} \times \exp\left[-\frac{E_G(T)}{2kT}\right].$$

The values of $n_i(300)$ and $E_G(300)$ are well established (Table 3.3), and $n_i(T)$ can be expressed as

$$n_i(T) = n_i(300) \times \left(\frac{T}{300}\right)^{3/2} \exp\left[\frac{E_G(300)}{600k}\right] \exp\left[\frac{-E_G(T)}{2kT}\right]. \tag{3.45}$$

Values of $n_i(T)$ for Si and GaAs are given in Fig. 3.23.

The energy gaps change under applied pressure. In Si the gap decreases with pressure at a rate of about $2 \times 10^{-3}$ eV/kilobar (where 1 kbar $= 1.02 \times 10^8$ N/m²) and in GaAs the gap increases with pressure at a rate of about $15 \times 10^{-3}$ eV/kbar.

**TABLE 3.4** Values Used in Eq. (3.44) to Determine the Value of the Energy Gap $E_G(T)$ at Temperatures in Kelvin

|  | $E_G(0)$ (eV) | $A$ | $B$ |
|---|---|---|---|
| Si | 1.17 | $4.73 \times 10^{-4}$ | 636 |
| GaAs | 1.519 | $5.405 \times 10^{-4}$ | 204 |

*Source:* Sze (1981).

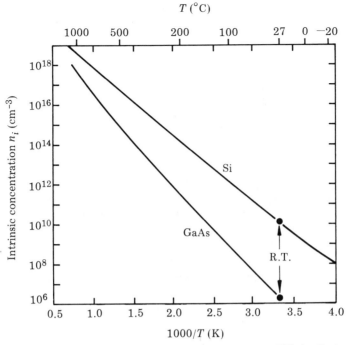

**FIGURE 3.23**  Intrinsic carrier concentration $n_i$ versus $1000/T$(K) for GaAs and Si. Room temperature is indicated by R.T. [After Sze (1981).]

Strain (see also Section 1.8) is the change in length $dl$ over length $l$,

$$\varepsilon_s = \frac{dl}{l} \tag{3.46}$$

and the stress $\sigma_s$, which is force per unit area, is related to strain by

$$\sigma_s = Y\varepsilon_s \tag{3.47}$$

where $Y$ is Young's modulus, which in Si has a value of $1.9 \times 10^3$ kbar ($1.9 \times 10^{12}$ dyn/cm²). Since the maximum elastic strain without breaking Si is about $10^{-2}$, this gives a stress of about 19 kbar, or a maximum change in the band gap $\Delta E_G$ of

$$\Delta E_G = -2 \times 10^{-3}\,\frac{eV}{kbar} \times \sigma_s$$

$$= -(2 \times 10^{-3})(19) = -3.8 \times 10^{-2}\ eV. \tag{3.48}$$

The change in resistivity $\Delta\rho$ with stress is given by

$$\frac{\Delta\rho}{\rho} = P_s\sigma_s \tag{3.49}$$

where $P_s$ is the piezoresistance constant, a tensor quantity that depends on crystal orientation, resistivity, and conductivity type ($n$- or $p$-type).

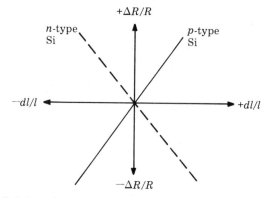

**FIGURE 3.24** Relative change in longitudinal resistance $\Delta R/R$ for $p$-type (solid line) and $n$-type (dashed line) Si versus relative changes in length. [After Heywang (1986).]

For pressure measurements (pressure transducers), we use the change in longitudinal resistance $\Delta R$ (where the resistance is measured along a length of material),

$$\frac{\Delta R}{R} = K \frac{dl}{l} \tag{3.50}$$

where $K$ is the $K$-factor or gauge-factor and depends on the orientation and resistivity. The values of $K$ for $10^{-2} \ \Omega \cdot$ cm Si are about $-60$ for (100) $n$-type Si and $+100$ for (111) $p$-type Si, with the sign convention shown in Fig. 3.24.

These changes in resistance are the basis of pressure sensors that use thin Si diaphragms that are formed by etching away a portion of the Si chip. A pressure differential across the diaphragm causes deflections that induce strains in the diaphragm, thereby modulating the resistance values. Pressure sensors are described in Chapter 15.

## GENERAL REFERENCES

BLAKEMORE, J. S., *Solid State Physics,* 2nd ed., Cambridge University Press, Cambridge, 1985.

CASTELLAN, G. W., *Physical Chemistry,* Addison-Wesley, Reading, Mass., 1971.

FELDMAN, L. C., and J. W. MAYER, *Fundamentals of Surface and Thin Film Analysis,* North-Holland, Amsterdam, 1986.

HALL, H. E., *Solid State Physics,* Wiley, New York, 1979.

HEYWANG, H., *Sensorik,* Springer-Verlag, Berlin, 1986.

KITTEL, C., *Introduction to Solid State Physics,* 5th ed., Wiley, New York, 1976.

MILNES, A. G., *Deep Impurities in Semiconductors,* Wiley, New York, 1973.

PAULING, L., *The Chemical Bond,* 3rd ed., Cornell University Press, Ithaca, N.Y., 1967.

PIERRET, R. F., *Advanced Semiconductor Fundamentals,* Addison-Wesley, Reading, Mass., 1987.

RUDDEN, M. N., and J. WILSON, *Elements of Solid State Physics,* Wiley, New York, 1980.

SZE, S. M., *Physics of Semiconductor Devices,* 2nd ed., Wiley, New York, 1981.

TIPLER, P. A., *Modern Physics,* Worth, New York, 1979.

WEIDNER, R. T., and R. L. SELLS, *Elementary Modern Physics,* 3rd ed., Allyn and Bacon, Boston, 1980.

WILLMOTT, J. C., *Atomic Physics,* Wiley, New York, 1975.

## PROBLEMS

**3.1.** Compare the properties of 1-mm-thick single-crystal slices of intrinsic silicon (Si) and high-purity aluminum (Al) at temperatures near 300 K. Indicate your answer by writing Al, Si, *same* (when the properties are the same within $\pm 20\%$) or *neither*.

    **(a)** The resistivity changes strongly with temperature.

    **(b)** Transparent to infrared radiation.

    **(c)** Has a carrier mobility greater than 100 cm$^2$/V · s.

    **(d)** Has a carrier concentration greater than $10^{21}$/cm$^3$

    **(e)** Has an atomic concentration of 5 to 6 $\times$ $10^{22}$/cm$^3$.

    **(f)** Has a density of 10 g/cm$^3$.

**3.2.** Consider the semiconductors Si (IV), GaAs (III–V), and ZnS (II–VI), where the roman numerals in parentheses indicate the columns in the periodic table occupied by the elements. Indicate which element (Si, Ga, As, Zn, S) is likely to act as a donor or acceptor in the semiconductors. For the compound semiconductors GaAs and ZnS indicate which lattice site (Ga, As, Zn, or S) the impurity is likely to occupy.

**3.3** Using the mobility values given in Table 2.4 for Si and GaAs at 300 K for carrier concentrations of $10^{15}$/cm$^3$:

    **(a)** What are the resistivities of *n*-type and *p*-type materials?

    **(b)** If the layer of Si is $10^{-4}$ cm thick, what are the sheet resistance values?

    **(c)** What is the sheet resistance of a layer of *n*-type Si, $10^{-4}$ cm thick, if half the layer thickness has a concentration of $10^{16}$ electrons/cm$^3$ with a mobility of 1400 cm$^2$/V · s and the other half, $10^{17}$ electrons/cm$^3$ with a mobility of 700 cm$^2$/V · s?

**3.4.** Silicon has an atomic density of 5 $\times$ $10^{22}$ atoms/cm$^3$. Assume that the atoms are packed in a simple cubic structure (the correct description is given in Chapter 6) with 5 $\times$ $10^{19}$ arsenic atoms (donors) uniformly distributed.

    **(a)** What is the average spacing between Si atoms, how many atoms/cm$^2$ would lie on the outermost layer (a monolayer), and what would be the average spacing between the donor atoms?

    **(b)** In the hydrogenic model of a donor electron orbiting around a positive donor, the orbit radius $r = \epsilon_r h^2/(2\pi)^2 m k_c e^2 = \epsilon h^2/\pi m e^2$. What is the value of $r$ (in Si, $\epsilon_r = 11.9$) and what concentration of donors is necessary for the orbitals to touch?

(c) In this model the donor ionization energy $E_D = (2\pi)^2 mk_c^2 e^4/2h^2\epsilon_r = me^4/8\epsilon^2 h^2$. What is $E_D$?

**3.5.** A semiconductor at 400 K with $E_G = 1.0$ eV and electron (hole) mass = $m$ (free electron mass) contains $2 \times 10^{16}/\text{cm}^3$ ionized donors.
(a) Find $n$, $n_i$, $p$, and $E_F$.
(b) If $1 \times 10^{16}/\text{cm}^3$ ionized acceptors are added, what is the value of $n$? Are the values of $n_i$, $p$, and $E_F$ changed?

**3.6.** Consider a semiconductor with a fixed number of ionized acceptors. With respect to the valence band edge, $E_V$, does the position of the Fermi level, $E_F$, move closer to, move away from, or remain at the same distance from $E_V$ if
(a) the temperature is increased?
(b) the effective mass of the holes is decreased?
(c) the energy gap is increased?
(d) donors are added?

**3.7.** Using the data in Table 3.3 at 300 K, calculate the position of the Fermi level below the conduction band for Si and GaAs for the same electron concentrations, $n = 10^{16}/\text{cm}^3$.

**3.8.** Using the data in Table 2.4 for electron mobilities in Si and GaAs at 300 K and assuming equal minority-carrier lifetimes in p-type material with $\tau_n = 10^{-7}$ s:
(a) What are the diffusions lengths $L_n$?
(b) If $10^{14}$ electrons/cm$^3$ are injected at $x = 0$, what are the diffusion current densities at $x = 0$?
(c) If there are $10^{15}$ holes/cm$^3$, what electric fields would be required to give the same current densities as in part (b)?

**3.9.** (a) You have an As-doped Si (n-type) sample and an intrinsic Si sample and you heat the intrinsic sample until the electron concentration equals that of the As-doped sample. Is the energy difference between the conduction band and Fermi level $(E_C - E_F)$ in the intrinsic sample greater than, about the same as, or less than that in the As-doped sample?
(b) You have an As-doped (n-type) sample and increase the dopant concentration so that the electron concentration increases. Does the energy $E_C - E_F$ increase, decrease, or remain about the same?
(c) You have samples of Si and GaAs with the same electron concentration, but the effective mass of electrons in GaAs is 10 times smaller than that in Si. Is the energy difference $E_C - E_F$ in GaAs greater than, less than, or about the same as that in Si?

**3.10.** (a) A semiconductor contains acceptors and the hole concentration is $1 \times 10^{16}/\text{cm}^3$. Assume that the density of states in the valence band, $N_V$, is $10^{19}/\text{cm}^3$. How far above the valence band in eV does the Fermi energy $E_F$ lie at 300 K?

**(b)** Consider a semiconductor, germanium, with a band gap $E_G = 0.66$ eV and equal density of states in the conduction and valence bands with $N_C = N_V = 10^{19}/\text{cm}^3$. The temperature is 300 K.
  **(1)** What is the value of the intrinsic carrier concentration $n_i$?
  **(2)** If the number of electrons $n$ is $10^{17}/\text{cm}^3$, what is the value of the hole concentration $p$?

**3.11.** You evaluate a semiconductor in a series of measurements at 300 K.
  **(a)** To determine the energy gap $E_G$, you shine light of various wavelengths and measure the wavelength to be 1.0 μm where the intensity of transmitted light decreases abruptly. What is $E_G$?
  **(b)** You have a Hall magnet with a magnetic field of 2000 G. You apply an electric field of 100 V/cm along the sample and measure a Hall field across the sample of 4 V/cm. What is the carrier mobility?
  **(c)** To determine the effective mass of the carriers, you use the value of mobility (assume that $\mu = 4000 \text{ cm}^2/\text{V} \cdot \text{s}$) and estimate the mean free path between collisions to be $10^{-6}$ cm at 300 K by measuring the resistivity in films of different thicknesses. Using the free-electron theory, what is the effective mass $m^*$ of the electron compared to the free-electron mass $m$?
  **(d)** Assuming that the effective mass of electrons and holes are equal with a value of $0.01m_0$, what is the effective density of states at $T = 300$ K?
  **(e)** Assuming that $N_C = N_V = 10^{17}/\text{cm}^3$ and $E_G$ is 1.0 eV, what is the intrinsic carrier concentration $n_i$ at 300 K?
  **(f)** For a value of $n_i = 10^9/\text{cm}^3$ and using 1000 cm²/V · s mobility, what is the hole concentration in $n$-type material with a resistivity of 100 Ω · cm?

# 4

CHAPTER

# Barriers and Junctions

## 4.1 Introduction

The p-n junction—located at the juncture between p-type and n-type material—is the basic building block for semiconductor devices. At the transition between n- and p-type material a potential gradient is formed to separate the regions with high and low, electron and hole concentrations. It is this potential barrier that is the key element in the description of the p-n junction.

To see how the barrier is formed, let us join a piece of n-type and p-type silicon, each with $10^{16}$ donors or acceptors, as shown in Fig. 4.1. We will assume that thermal equilibrium holds so that all the relations derived in Chapter 3 hold for the bulk n- and p-type regions separated by the potential gradient: position of the Fermi level with respect to band edges, concentrations of electrons and holes, and mass-action relation, so that in bulk regions the intrinsic carrier concentration $n_i$ is found from $n_i^2 = np$. The important concept is that in thermal equilibrium the Fermi level $E_F$ is constant across the transition from n- to p-type materials. The magnitude of the difference between conduction band edges (i.e., the barrier height $eV_0$) is given by

$$eV_0 = E_G - (E_C - E_F)_n - (E_F - E_V)_p. \tag{4.1}$$

For a semiconductor at room temperature ($kT \simeq 0.025$ eV) with $10^{16}$ carriers/cm$^3$ and densities of states $N_C$ and $N_V$ equal to $10^{19}$/cm$^3$, the values of $(E_C - E_F)_n$ and $(E_F - E_V)_p$ are 0.17 eV. Thus if the band gap $E_G$ is 1.0 eV, the potential barrier is 0.66 eV, two-thirds of the band gap.

The requirement for a potential barrier can be explained from the diagram of carrier concentrations shown in Fig. 4.1. The electron concentration in the n-region, $n_n$, equals $10^{16}$/cm$^3$ (where we assume that $n_n = N_D$ the donor concentration) while that in the p-type region $n_p = n_i^2/p_p \simeq 10^4$/cm$^3$ (if we assume that $n_i \simeq 10^{10}$/cm$^3$). As described later, the potential-barrier region has a width of about 1 μm, leading to an immense gradient in carrier concentration:

$$\frac{dn}{dx} = \frac{(10^{16} - 10^4) \text{ carriers/cm}^3}{10^{-4} \text{ cm}} = 10^{20} \frac{\text{carriers}}{\text{cm}^4}$$

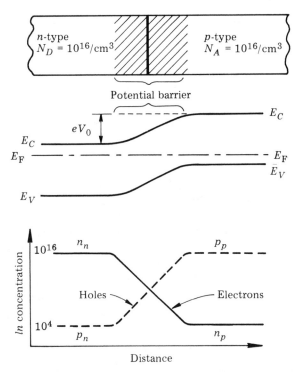

**FIGURE 4.1** Schematic of a $p$-$n$ junction with an abrupt change in doping from $N_D = 10^{16}$ donors/cm$^3$ to $N_A = 10^{16}$ acceptors/cm$^3$—the symmetrical, step junction. The energy level and carrier concentration versus distance are shown for the transition from $n$- to $p$-type. The shaded region shows the extent of the depletion region for a junction potential of $V_0$ volts.

so that the electron diffusion current density $J_n$ is hundreds of A/cm$^2$; from Chapter 2,

$$J_n = eD_n \frac{dn}{dx}$$

where $D_n = \mu_n(kT/e) \simeq \mu_n/40 \simeq 20$ to $100$ cm$^2$/s for most semiconductors.

In thermal equilibrium, this diffusion current is opposed by the field-driven current, so that the total electron current is zero:

$$0 = J_{\text{diffusion}} + J_{\text{field}}$$

$$0 = eD_n \frac{dn}{dx} + e\mu_n n\mathscr{E}.$$

The presence of a concentration gradient with no current requires an electric field. The integration of the electric field across the barrier region gives the voltage $V_0$ of the potential barrier.

The equilibrium carrier concentrations at either side of the potential barrier region are related to each other through Maxwell–Boltzmann statistics by the relationships that for electrons,

$$n_p = n_n \exp\left(-\frac{eV_0}{kT}\right) \tag{4.2}$$

and similarly for holes,

$$p_n = p_p \exp\left(-\frac{eV_0}{kT}\right). \tag{4.3}$$

The electric field $\mathscr{E}$ is established by the charge on the fixed donors and acceptors located within the potential-barrier region as shown in Fig. 4.2. In the $n$-type segment of this region, the electron concentration has decreased to a negligible fraction of the donor concentration when the potential has changed by a few $kT/e$. The same is true for holes in the $p$-type region. We will use the approximation that we can neglect the free-carrier contribution throughout the barrier region. This is a reasonable assumption except at the edges of the barrier region, where the potential varies by only a few $kT/e$. The barrier region is in effect depleted of carriers, which gives rise to the term "depletion region." Alternatively, it is called a "space-

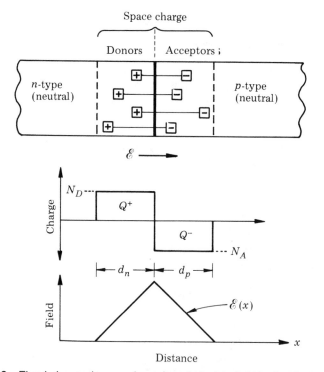

**FIGURE 4.2**   Fixed charge (space charge) and electric field in the depletion region of a symmetric, step junction under the conditions that the free carriers are swept out of the depletion region.

charge region'' because of the charge on the fixed donors and acceptors. The width $d$ of the depletion region is the sum of the depletion widths in the $n$-type, $d_n$, and $p$-type, $d_p$, regions.

The electric field is confined to the depletion region, so the amount of positive charge per unit area $Q^+$ in the $n$-type region equals the amount of negative charge $Q^-$ in the $p$-type region:

$$Q^+ = Q^- \tag{4.4}$$
$$N_d^+ d_n = N_A^- d_p.$$

In the example (Fig. 4.2) we chose $N_D = N_A$, so that in this case $d_p = d_n$.

The electric field $\mathcal{E}$ is directed from the $n$-type region to the $p$-type region (from positive donors to negative acceptors). If we visualize the electric field as lines of force (the lines joining donors and acceptors in Fig. 4.2) between the positive and negative charges, the maximum electric field will be located at the junction between the $n$- and $p$-type material.

## 4.2 Diffused Junction: Equilibrium

In general, $p$-$n$ junctions are formed by diffusing a high concentration of dopants into a less heavily doped region. The junction is located where the donor and acceptor concentrations are equal. In our discussion of the $p$-$n$ junction we assume that the doping concentration changes abruptly from donors to acceptors (the abrupt or step junction) and that one doping concentration is much greater than the other (the asymmetrical junction). Neither approximation is crucial to the concept of junction behavior, but they make the discussion simpler. We will concentrate on the case where the $n$-type region has a much higher dopant concentration than the $p$-type region. We assume that $N_D = 10^{18}/\text{cm}^3$ and $N_A = 10^{16}/\text{cm}^3$. As shown in Fig. 4.3, the depletion region will extend deeper into the $p$-type region—the region with a lower doping concentration—than into the $n$-type region—the region with the higher concentration.

The width of the space-charge region is found by equating the amount of fixed negative charge $Q^-$ (acceptors) in the $p$-material to the amount of fixed positive charge $Q^+$ (donors) in the $n$-material,

$$N_A^- d_p = N_D^+ d_n.$$

Since $N_D \gg N_A$, then $d_n \ll d_p$ and the total width $d$ of the depletion layer is given by

$$d = d_p + d_n \simeq d_p. \tag{4.5}$$

The maximum field $\mathcal{E}_{\text{max}}$ is given by the amount of fixed charge per unit area:

$$\epsilon \mathcal{E}_{\text{max}} = eN_A d_p \tag{4.6}$$

where $\epsilon$ is the permittivity ($\epsilon \simeq 10^{-12}$ F/cm in Si). This follows from the discussion in Chapter 2.

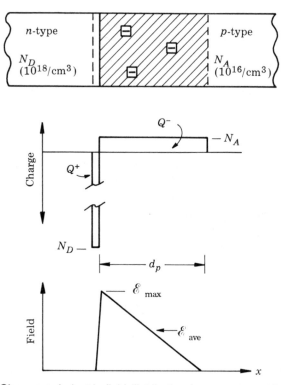

**FIGURE 4.3** Charge and electric field distribution for an asymmetric, step junction with $10^{18}$ donors/cm³ and $10^{16}$ acceptors/cm³, showing that the depletion region extends primarily into the p-type region (shaded portion).

For the simple, linear field distribution, the potential drop $V_0$ across the barrier can be related to $\mathscr{E}_{max}$ by

$$V_0 = \mathscr{E}_{ave}d_p = \tfrac{1}{2}\mathscr{E}_{max}d_p. \tag{4.7}$$

This expression can be combined with eq. (4.6) to give

$$d_p = \left(\frac{2\epsilon V_0}{eN_A}\right)^{1/2}. \tag{4.8}$$

The value of $V_0$ can be found from eq. (4.1) or (4.2). From eq. (4.2) (with $n_n = 10^{18}$/cm³ and assuming that $n_i^2 = 10^{20}$/cm⁶, hence $n_p = n_i^2/N_A = 10^4$/cm³)

$$V_0 = \frac{kT}{e} \ln \frac{n_n}{n_p} = 0.83 \text{ V}.$$

This gives a depletion layer width of $3.2 \times 10^{-5}$ cm or $0.32$ μm (with $\epsilon = 10^{-12}$ F/cm) and a maximum field strength [from eq. (4.7)] of $5.2 \times 10^4$ V/cm.

These relations between the depletion width $d_p$ and potential $V_0$ or electric field $\mathscr{E}$ were obtained from simple geometrical arguments using lines of force for the field. The potential and electric field distribution can be found explicitly from Poisson's equation:

$$\frac{d^2V}{dx^2} = -\frac{d\mathcal{E}}{dx} = -\frac{Q(x)}{\epsilon} \tag{4.9}$$

and treating only the lightly doped $p$-type layer so that $Q(x)$ is the charge in the depletion layer due to fixed ionized acceptors (the contribution of mobile carriers is again neglected), we obtain

$$\frac{d^2V}{dx^2} = \frac{eN_A^-}{\epsilon}. \tag{4.10}$$

Integration of eq. (4.10) with the boundary condition $\mathcal{E}(d_p) = 0$ gives

$$\mathcal{E}(x) = \frac{eN_A^-}{\epsilon}(d_p - x) \tag{4.11}$$

which leads to the linear variation of the electric field shown in Fig. 4.3. Integration of eq. (4.11) gives the potential distribution

$$V(x) = \frac{eN_A^-}{\epsilon}\left(d_p x - \frac{x^2}{2}\right) \tag{4.12}$$

which gives eq. (4.8) at $x = d_p$ with $V(x) = V_0$. The potential therefore increases parabolically from $x = 0$ to $x = d_p$.

## 4.3 Diffused Junction: Applied Voltage

The most striking feature of the characteristics of a $p$-$n$ junction is that current flows if $n$-type material is connected to the negative terminal of a power supply (forward bias), and essentially no current is observed with $n$-type material connected to the positive terminal (reverse bias) until a certain voltage is reached (breakdown voltage, $V_B$). This rectification behavior is shown in Fig. 4.4 with forward current obtained with negative voltage applied to $n$-type and positive voltage to $p$-type material.

The dashed line indicates the linear current–voltage behavior that would be observed if the $n$-type diffused layer were removed and replaced with an ohmic contact. One notes that even in the case of forward bias, the presence of an $n$-$p$ junction reduces the amount of current by orders of magnitude compared to the ohmic value. This reduction is due to the presence of the potential barrier. Because the barrier region reduces or blocks the current flow, under forward or reverse bias, we can consider as a first approximation that all the external voltage is applied to the barrier region alone and that the electric fields (and hence voltage differences or voltage "drops") are negligible in the $n$- and $p$-type regions outside the barrier regions. In the following sections we discuss the general characteristics of the $p$-$n$ junction under reverse and forward bias.

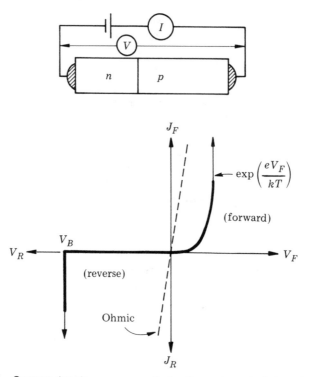

**FIGURE 4.4** Current density versus applied voltage characteristics of an *n-p* junction showing the reverse-voltage breakdown at $V_B$. The dashed line shows the linear *J–V* behavior if the *n*-type region were to be replaced by an ohmic contact.

## 4.3.1 Reverse Bias and Capacitance

Under an applied reverse-bias voltage $V_R$, the *p*-type region is negative with respect to the *n*-type region. Hence the potential barrier (shown in Fig. 4.5) is increased for electrons to move from the *n*- to the *p*-type regions. With reference to the Fermi level in the *n*-type material, the Fermi level in the *p*-type material is shifted upward in energy by an amount $eV_R$.

The potential across the depletion region is increased from $V_0$ to $V_0 + V_R$ volts and the width of the depletion region increases. Since the *n*-type region is heavily doped with respect to the *p*-type region ($N_D \gg N_A$), the depletion region extends primarily into the *p*-type region. We can again make the approximation that the width $d$ of the depletion region is given by $d = d_n + d_p \approx d_p$:

$$d \simeq d_p = \left[ \frac{2\epsilon(V_0 + V_R)}{eN_A} \right]^{1/2}. \tag{4.13}$$

The maximum in the electric field, $\mathscr{E}_{max}$, is expressed as $2\mathscr{E}_{ave}$:

$$\mathscr{E}_{max} = \frac{2(V_0 + V_R)}{d_p} = \left( \frac{2eN_A}{\epsilon} \right)^{1/2} (V_0 + V_R)^{1/2}. \tag{4.14}$$

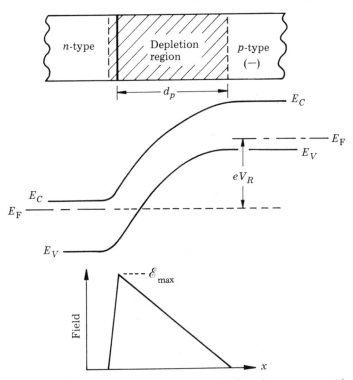

**FIGURE 4.5** Energy levels and electric field distributions in an asymmetrical, step junction ($N_D \gg N_A$) under reverse bias with $V_R$ volts applied. The shaded area shows the depletion region.

Both the width of the depletion region and the maximum in the electric field increase as $(V_0 + V_R)^{1/2}$.

The reverse-biased $p$-$n$ junction acts as a parallel-plate capacitor (Chapter 2) with a permittivity $\epsilon$ and a separation between the plates given by the width of the depletion region. If one applies a reverse bias and then superimposes a small modulation voltage, the modulation in the width of the depletion region will cause changes in the charge only at the edges of the depletion region. For a junction of area $A$, the capacitance $C$ is given by

$$C = \frac{\epsilon A}{d} \tag{4.15}$$

so that for our case of the asymmetrical step junction,

$$C = A \left[ \frac{\epsilon e N_A}{2(V_0 + V_R)} \right]^{1/2}. \tag{4.16}$$

For the case where $N_A = 10^{16}/\text{cm}^3$ and $V_0 = 0.84$ V, with $V_R = 10$ V applied, the depletion width $d_p = 1.16 \times 10^{-4}$ cm. For a 10-$\mu$m $p$-$n$ junction, $A = 10^{-6}$ cm$^2$ and the capacitance $C = 0.86 \times 10^{-14}$ F $= 0.86 \times 10^{-2}$ pF for $\epsilon = 10^{-12}$ F/cm.

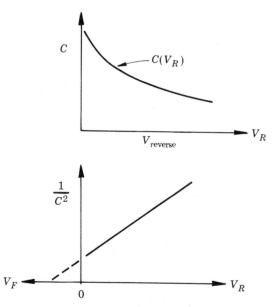

**FIGURE 4.6** Decrease in capacitance $C$ versus reverse-bias voltage in an asymmetrical, step junction. The plot of $1/C^2$ versus $V_R$ shows linear behavior with an intercept of $V_0$ on the voltage axis.

The width of the depletion region and hence the value of the capacitance is determined primarily by the dopant concentration in the region with the lowest dopant concentration—$p$-type material in this case. The capacitance decreases with applied voltage (Fig. 4.6) and measurement of the capacitance as a function of reverse bias is a method of determining the dopant concentration. The conventional way to display the $C$–$V$ data is to plot $1/C^2$ versus $V_R$ as shown in Fig. 4.6. The slope of the line gives the value of $N_A$ and the intercept at $1/C^2 = 0$ gives the value of $V_0$.

This $C$ proportional to $(V)^{-1/2}$ relation [eq. (4.16)] holds only for uniformly doped material. For linearly graded junctions where $(N_D - N_A)$ varies linearly with distance, as might be found in a deep diffused junction, the capacitance varies as the cube root of applied voltage, $C \propto V^{-1/3}$.

## 4.3.2 Reverse-Bias Breakdown Voltage $V_B$

As the reverse-bias voltage is increased, the current exhibits a sharp increase at a voltage $V_B$, called the breakdown voltage, as shown in Fig. 4.4. In asymmetrical junctions where the dopant concentration is of the order of $10^{16}/cm^3$ or less, the breakdown is caused by an "avalanche" mechanism. In this mechanism the electrons and holes that traverse the depletion region acquire sufficient energy from the electric field to create electron–hole pairs by collisions (or impact ionization) with the electrons in the covalent lattice bonds. It causes an avalanche process because the electrons and holes created in the impact ionization collisions can gain energy from the electric field in the depletion region, so that they in turn create electrons and holes. In heavily doped materials, junction breakdown can occur by another mechanism, internal field emission or Zener breakdown. At high electric field strengths in the depletion region, electrons can make a transition directly to unfilled states, again leading to sharp increases in the reverse current.

We can make a simple estimate of the breakdown voltage in the avalanche regime by considering the trajectories of electrons in the high field of the depletion region. Assume that we are in the high-field regime and that the carriers have acquired a saturated drift velocity $v_s = 1 \times 10^7$ cm/s ($= 10^5$ m/s). If we take the electron mass equal to the free-electron mass, $m = 9.1 \times 10^{-31}$ kg, we can estimate the mean free path between collisions from the relation in Chapter 2 that $\mu = e\tau/m = el/mv_s$,

$$l = \frac{mv_s\mu}{e}. \tag{4.17}$$

For our example,

$$l = \frac{9.1 \times 10^{-31} \times 10^5 \times 10^{-1}}{1.6 \times 10^{-19}} = 5.7 \times 10^{-8} \text{ m} = 5.7 \times 10^{-6} \text{ cm}$$

for a mobility $\mu = 10^3$ cm$^2$/V $\cdot$ s ($= 10^{-1}$ m$^2$/V $\cdot$ s). For avalanche to occur, the carrier must acquire enough energy from the field in traversing the distance $l$ between collisions to create an electron–hole pair. If we assume that the carrier requires an energy equal to the band gap $E_G$, the condition for avalanche breakdown is reached when the electric field reaches a value $\mathscr{E}_B$,

$$e\mathscr{E}_B l = E_G. \tag{4.18}$$

For $E_G = 1.1$ eV and $l \approx 5.7 \times 10^{-6}$ cm, the breakdown field strength $\mathscr{E}_B$ is $1.93 \times 10^5$ V/cm, a value about a factor of 2 below commonly accepted values of $\mathscr{E}_B$ ($\mathscr{E}_B = 4 \times 10^5$ V/cm). By setting $\mathscr{E}_B = \mathscr{E}_{max} = 2 V_B/d_p$ [eq. (4.13) with $V_0 \ll V_B$], we obtain

$$V_B = \frac{\mathscr{E}_B d_p}{2} = \frac{\mathscr{E}_B}{2} \left( \frac{2\epsilon V_B}{eN_A} \right)^{1/2} \tag{4.19}$$

which can be rewritten as

$$V_B = \mathscr{E}_B^2 \frac{\epsilon}{2eN_A}.$$

For $\mathscr{E}_B = 4 \times 10^5$ V/cm and $N_A = 10^{16}$/cm$^2$, a breakdown voltage of 50 V is obtained, a value close to that measured in Si $p$-$n$ junctions.

## 4.4 Forward Current

In the asymmetrical $p$-$n$ junction with a high concentration of donors diffused into a $p$-type substrate, the forward current will be determined by the number of electrons that can surmount the potential barrier and be injected into the $p$-type material.

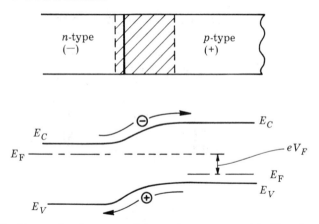

**FIGURE 4.7** Energy levels versus distance in an asymmetrical junction ($N_D \gg N_A$) under forward bias with $V_F$ volts applied. The arrows represent the injection of electrons and holes into the p-type and n-type material.

Throughout this discussion we assume that the deviation from thermal equilibrium is small, so that we can use the relations developed in Chapter 3.

As in the reverse-bias case, under forward bias the junction region is the high-resistance element and the applied voltage $V_F$ (positive voltage applied to p-type region) reduces the potential drop (Fig. 4.7) across the barrier to a value

$$V_T = V_0 - V_F. \tag{4.20}$$

With a reduction in the barrier height more electrons and holes can get over the barrier than under the case of thermal equilibrium.

The number of electrons $n$ injected into the p-type region (or stated in other terms the number of electrons that can surmount the energy barrier) is given by

$$n_p = n_n \exp\left(-\frac{eV_0 - eV_F}{kT}\right), \tag{4.21}$$

following eq. (4.2), or

$$n_p = n_{p_0} \exp\left(\frac{eV_F}{kT}\right) \tag{4.22}$$

where $n_{p_0}$ represents the number of electrons in the p-type material under thermal equilibrium conditions (the subscript zero is often omitted when describing carrier concentrations in bulk materials).

As a first approximation, we assume that the forward-bias current density $J_F$ is proportional to the number of carriers that surmount the barrier. That is, $J_F$ proportional to $n$,

$$J_F = J_0 \exp\left(\frac{eV_F}{kT}\right) \tag{4.23}$$

where $J_0$ is a parameter that depends on temperature, band gap, and dopant concentration. The value of $J_0$ is derived in the next section.

In the asymmetrical *p-n* junction in which the concentration of dopants in one region is much greater than the concentration in the other region, the current is determined primarily by carriers injected from the region with a high dopant concentration. For example, with $N_D \gg N_A$, then $n_n \gg p_p$ and the number of injected electrons is much greater than the number of injected holes $p_n$:

$$p_n = p_p \exp\left(-\frac{eV_0}{kT}\right)\exp\left(\frac{eV_F}{kT}\right) \tag{4.24}$$

as compared to

$$n_p = n_n \exp\left(-\frac{eV_0}{kT}\right)\exp\left(\frac{eV_F}{kT}\right). \tag{4.25}$$

Consequently, we can describe the major features of the forward current characteristics by considering the behavior of carriers injected from the region with the higher dopant concentration.

## 4.4.1 Forward Current, Diffusion Only

Let us consider the forward current in an abrupt *p-n* junction with the *n*-region more heavily doped than the *p*-region, as shown in Fig. 4.8. Point A represents the transition from the depletion region to the *n*-type material and point B the

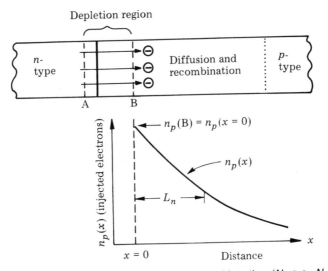

**FIGURE 4.8**  Under forward bias of an asymmetrical junction ($N_D \gg N_A$), electrons are injected into the *p*-type region, where they diffuse and recombine with a diffusion length $L_n$.

transition from the depletion region to the $p$-region. Since the $n$-type region is more heavily doped than the $p$-type region, we can consider only the electron current. Let

$$n_n = \text{concentration of electrons in } n\text{-type region } (n_n = N_D)$$
$$n_{p0} = \text{concentration of electrons in the } p\text{-region in thermal equilibrium (or at distances far from the region where electrons are injected into the } p\text{-region)}$$
$$n_p(B) = n_p(x = 0) = \text{concentration of electrons at point B, } x = 0$$
$$n_p(x) = \text{concentration of electrons at a distance } x.$$

From eq. (4.24),

$$n_p(x = 0) = n_{p0} \exp\left(\frac{eV_F}{kT}\right). \tag{4.26}$$

Now we make the assumption that for $x \geq 0$, we can neglect the influence of the electric field in the $p$-region where electrons are injected (we will justify this point later), so that the current of electrons $J_n$ is given by

$$J_n = eD_n \frac{dn}{dx} \tag{4.27}$$

and the hole current in this region is negligible.

Since the electron current $J_n(x = 0)$ at $x = 0$ dominates over the hole current $J_p(x = 0)$, the total forward current $J_F$ through the $p$-$n$ junction can be written as

$$J_F = J_n(x = 0) + J_p(x = 0) \approx J_n(x = 0). \tag{4.28}$$

We note that the total current is conserved so that $J_F =$ constant, independent of $x$. Therefore, to determine $J_F$ we only need to determine $dn/dx$ at $x = 0$.

As a first approach, let us use the linear approximation, where the concentration of injected electrons decreases linearly with distance (as we did in Chapter 3)

$$n_p(x) - n_{p0} = [n_p(x = 0) - n_{p0}]\left(1 - \frac{x}{L_n}\right) \tag{4.29}$$

where $L_n$ is the diffusion length for electrons, $L_n = (D_n \tau_n)^{1/2}$. Then

$$\frac{dn}{dx} = -\frac{n_p(x = 0) - n_{p0}}{L_n} \approx -\frac{n_p(x = 0)}{L_n} \tag{4.30}$$

where the value of $n_p(x = 0)$ is orders of magnitude greater than $n_{p0}$. Using eq. (4.27) gives us

$$J_n \approx -\frac{eD_n}{L_n}[n_p(x = 0) - n_{p0}] \approx -\frac{eD_n}{L_n}n_p(x = 0) \tag{4.31}$$

where the minus sign accounts for the fact that a flux of electrons toward positive values of $x$ is equivalent to a negative current. We use eq. (4.26) in eq. (4.31) and obtain

$$J_F = -\frac{eD_n}{L_n} n_{p_0} \exp\left(\frac{eV_F}{kT}\right)$$

$$= J_0 \exp\left(\frac{eV_F}{kT}\right) \tag{4.32}$$

where

$$J_0 = -\frac{eD_n}{L_n} n_{p_0} = -\frac{eD_n}{L_n} \frac{n_i^2}{N_A}$$

from the mass action relation $np = n_i^2$ with $p_p = N_A$. If we choose $D_n = 20 \text{ cm}^2/\text{s}$ and $\tau_n = 10^{-5}$ s, then $D_n/L_n = 1.4 \times 10^3$ cm/s. With $n_i^2 = 10^{20}/\text{cm}^6$ and $N_A = 10^{16}/\text{cm}^3$, $n_{p_0} = 10^4/\text{cm}^3$.

$$J_0 = -(1.6 \times 10^{-19} \times 1.4 \times 10^3 \times 10^4) = -2.24 \times 10^{-12} \text{ A/cm}^2$$

where the minus sign is used because of our choice of axes with electron flow toward positive $x$ equivalent to negative current. If we include the contribution of holes injected into the $n$-region, we obtain the total forward current density $J_F$,

$$J_F = -\left(\frac{eD_n}{L_n} n_{p_0} + \frac{eD_p}{L_p} p_{n_0}\right) \exp\left(\frac{eV_F}{kT}\right) \tag{4.33}$$

where $L_p$ is the diffusion length for holes. The concentration of holes in the $n$-region $(p_{n_0} = n_i^2/N_D)$ is orders of magnitude smaller than that of electrons in the $p$-region. Thus the current is determined primarily by carriers injected from the heavily doped region into the lightly doped region.

## 4.4.2 Electric Field Distribution $\mathscr{E}(x)$

We need now to justify our assumption that we can neglect the electric field in the $p$-type region near the junction in treating the diffusion of electrons. At long distances, many diffusion lengths from the depletion region, the current in the $p$-type region is carried entirely by holes moving under the influence of an electric field $\mathscr{E}(x)$ (Fig. 4.9). The electron current is negligible because at these distances all the injected electrons have recombined

$$J_F = J_p(x = \infty) = ep_p\mu_p\mathscr{E}. \tag{4.34}$$

But at $x = 0$, $J_F = J_n(x = 0) = -(eD_n/L_n)n_p(x = 0)$ from eq. (4.32). The current $J_F$ is constant, so that

$$J_F = J_n(x = 0) = J_p(x = \infty)$$

and from combining eq. (4.32) and (4.34),

$$\mathscr{E}(x = \infty) = -\frac{D_n}{L_n} \frac{n_p(x = 0)}{p_p\mu_p} \tag{4.35}$$

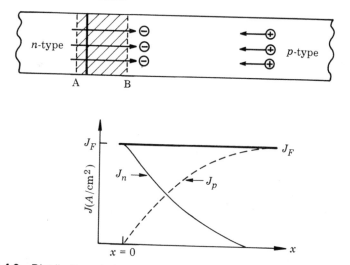

**FIGURE 4.9** Distribution of electron $J_n$ and hole $J_p$ current densities in a forward biased $n$-$p$ junction with a current density $J_F$ (solid, heavy line). At the edge of the depletion region, B, the current is given by the diffusion of electrons, while far away from the depletion region the current is given by the field-induced drift of holes in the $p$-type material.

where the minus sign just indicates the direction of the field. From the Einstein relation given in Chapter 2, $\mu = eD/kT$, we simplify eq. (4.35) and assuming that $\mu_n = \mu_p$,

$$\mathscr{E} = \frac{-kT}{e} \frac{1}{L_n} \frac{n_p(x = 0)}{p_p}. \tag{4.36}$$

Since $kT/e$ at room temperature is 0.026 V and the value of the diffusion length has about the same magnitude $10^{-2}$ cm, $(kT/eL_n) \simeq 1$ to 2 V/cm. The value of $\mathscr{E}$ is about $n_p(x = 0)/p_p$ in units of V/cm. For most practical devices where the resistivity of the $p$-type material is between 1 and 10 $\Omega \cdot$ cm ($p_p$ between $10^{16}$ and $10^{15}$ holes/cm³), $n_p(x = 0) \ll p_p$. Consequently the value of $\mathscr{E}$ is small. In our example, $p_p = 10^{16}/\text{cm}^3$ and $n_{p_0} = 10^4/\text{cm}^3$ and with $V_F = 0.52$ V (i.e., $V_F \simeq 20\ kT/e$), $n_p(x = 0) = 4.8 \times 10^8\ n_{p_0} = 4.8 \times 10^{12}/\text{cm}^3$. Then $\mathscr{E}(x = \infty) = 4.8 \times 10^{12}/10^{16} = 4.8 \times 10^{-4}$ V/cm. Thus the voltage drop is negligible in all practical devices.

This result can be stated as follows: There are so many holes in the $p$-type region that the forward current can be carried with a very small electric field in the $p$-region away from the junction. This is equivalent to the general statement that for either forward or reverse biased $p$-$n$ junctions, one can use the approximation that all the applied voltage, $V_F$ or $V_R$, appears across the junction. This approximation does not hold, of course, when the forward applied voltage $V_F$ becomes comparable to the junction potential, $V_0$.

## 4.5 Reverse Bias and the Ideal Diode Equation

Under reverse bias, the edges of the depletion region act as sinks for minority carriers. The depletion region acts as a "collector" of minority carriers and in the example of the asymmetrical diode with a lightly doped $p$-region, we set $n_p(x = 0) = 0$. The reverse current is established by the diffusion of electrons [eq. (4.27)] from the bulk of the $p$-region toward the edge of the depletion region at $x = 0$. The electron–minority carrier distribution is shown in Fig. 4.10 under reverse-bias conditions. Following the procedures developed in Section 4.4.1, the reverse current has the same functional dependence as given in eq. (4.31); that is,

$$J_n(x = 0) = -\frac{eD_n[n_p(x = 0) - n_{po}]}{L_n}$$

and with the boundary condition that $n_p(x = 0) = 0$, the reverse current is given by

$$J_n(x = 0) = \frac{eD_n}{L_n} n_{po} \qquad (4.37)$$

and if the contributions of holes in the $n$-type region are considered, the total reverse current $J_R$ is

$$J_R = e\left(\frac{D_n}{L_n} n_{po} + \frac{D_p}{L_p} p_{no}\right). \qquad (4.38)$$

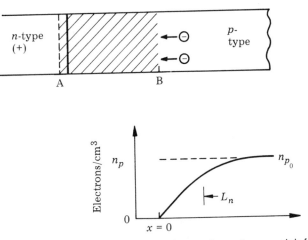

**FIGURE 4.10**   Under reverse bias, the depletion region acts as a sink for minority carriers, so that in the $p$-type region, $n_p = 0$ at $x = 0$, point B. The magnitude of the electron diffusion length $L_n$ is indicated in the plot of $n_p$ versus distance.

Considering the forward current [eq. (4.33)] and the reverse current [eq. (4.37)] together yields

$$J = J_0\left(\exp\frac{eV}{kT} - 1\right)$$ (4.39a)

where

$$J_0 = -e\left(\frac{D_n}{L_n}n_{p_0} + \frac{D_p}{L_p}p_{n_0}\right)$$ (4.39b)

and $V$ is the applied voltage (positive for forward bias) and the minus sign in eq. (4.39b) is used because the current convention of positive charge flowing toward positive values of $x$. The minus sign is usually omitted.

Equation (4.39a) is called the ideal diode equation, as it represents the case where current flow is determined *only* by minority-carrier diffusion in the $p$-type and $n$-type bulk regions adjacent to the edges of the depletion region. Under forward bias minority carriers are injected or "emitted" into the bulk regions, and under reverse bias the minority carriers are collected from the bulk regions. This ideal equation neglects the electric fields (which are very small except under high forward bias) in the bulk regions. The ideal diode equation also neglects the contribution from electrons and holes that are generated within the depletion region and are swept across the high-field region. This space-charge-generation current can be the dominant contribution to the reverse current and leads to a voltage-dependent reverse current, as discussed in the next section.

## 4.6 Space-Charge Generation

In the depletion (space-charge) region, the concentrations of electrons and holes are below thermal equilibrium values. Consequently, the balance between recombination and generation discussed in Chapter 3 is upset. In bulk material at thermal equilibrium the generation rate $G$ equals the recombination rate $R$,

$$G = R.$$ (4.40)

In the depletion region we assume that the carriers are swept out (Fig. 4.11) by the electric field before recombining, so that $R = 0$. The generation rate can be written, following the discussion in Section 3.8, as

$$G = \frac{n_i}{2\tau_0}$$ (4.41)

where the average lifetime $\tau_0$ is approximated by

$$\tau_0 = \frac{\tau_n + \tau_p}{2}.$$ (4.42)

For the asymmetrical junction where the width of the depletion region, $d$, is closely

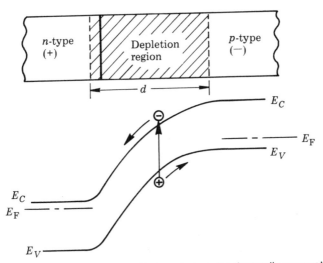

**FIGURE 4.11**  Under reverse bias, electron–hole pairs thermally generated within the depletion region are swept across the region by the electric field.

approximated by the width of the depletion region in the $p$-type region, $d_p$, the generation current $J_G$ is given by

$$J_G = e \frac{n_i}{2\tau_0} d_p = e \frac{n_i}{2\tau_0} \left( \frac{2\epsilon V}{eN_A} \right)^{1/2} = \frac{n_i}{\tau_0} \left( \frac{\epsilon eV}{2N_A} \right)^{1/2} \tag{4.43}$$

If we choose $d_p = 10^{-4}$ cm, $n_i = 10^{10}/\text{cm}^3$, and $\tau_0 = 10^{-5}$ s, then $J_G = 8 \times 10^{-9}$ A/cm². In reverse bias, the generation current can dominate over the diffusion current and leads to a current proportional to the width of the depletion layer, $J_G \propto d_p \propto V_R^{1/2}$ where $V \simeq V_R$ for $V_R \gg V_0$.

## 4.7 Transit and Charge Storage Times

The frequency responses of $p$-$n$ junctions is determined by the transit time, the time required for a carrier to traverse the depletion region, and the storage time, the time required to change the amount of charge injected (or stored) in the bulk material adjacent to the depletion region. The transit time $T_R$ is generally negligible compared to the charge storage time, which is proportional to the minority-carrier lifetimes, $\tau_n$ and $\tau_p$.

The transit time to traverse the width $d$ of the depletion region is determined by the drift velocity $v_d$:

$$T_R = \frac{d}{v_d} = \frac{d}{\mu \mathscr{E}(x)}. \tag{4.44}$$

If we consider our example of the asymmetrical $p$-$n$ junction and use the approximation that $\mathscr{E}(x) = \mathscr{E}_{\text{max}}/2$ [eq. (4.7)] and $\mathscr{E}_{\text{max}} = 2V/d$ with $d =$

$(2\epsilon V/eN_A)^{1/2}$, then for holes traversing the depletion region in the p-type material,

$$T_R = \frac{2\epsilon}{e\mu_p N_A} = 2\epsilon\rho = 2\tau_{DR}. \tag{4.45}$$

In this case the transit time is independent of the width of the depletion layer and applied voltage and depends only on the dielectric relaxation time $\tau_{DR} = \epsilon\rho = 10^{-12}\rho$, where $\rho$ is the resistivity. Another limiting case is where the carrier reaches a saturated drift velocity, $v_s$, where $v_s \cong 10^7$ cm/s. In this case

$$T_R = \frac{d}{v_s} \tag{4.46}$$

where for $d$ of the order of 1 $\mu$m ($10^{-4}$ cm), the transit time is of the order of $10 \times 10^{-12}$ s (10 ps).

Under forward bias, carriers are injected into the bulk material. For the asymmetrical junction, electrons are injected into the p-type material, where they diffuse and recombine. The distribution of injected electrons is given by eq. (4.30).

$$n_p(x) = n_p(x = 0)e^{-x/L_n}$$

where $L_n = (D_n\tau_n)^{1/2}$, the electron diffusion length. The total amount of stored charge, the number of electrons stored per unit area in the p-type region (shaded area in Fig. 4.12), is

$$Q_n = e\int_0^\infty n_p(x = 0)e^{-x/L_n}\, dx$$
$$= -en_p(x = 0)L_n. \tag{4.47}$$

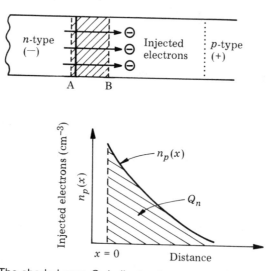

**FIGURE 4.12** The shaded area $Q_n$ indicates the amount of charge of electrons injected into the p-type material under forward bias.

The stored charge is related to the forward current $J_n = -eD_n n_p(x = 0)/L_n$ [eq. (4.31)] by $n_p(x = 0) = -J_n L_n/eD_n$, so that

$$Q_n = L_n\left(\frac{J_n L_n}{D_n}\right) = J_n \tau_n. \tag{4.48}$$

The relation [eq. (4.48)] is a statement that the current $J_n$ is determined by the requirement that the population of injected electrons in the $p$-type region is resupplied every $\tau_n$ seconds, so that $J_n = Q_n/\tau_n$. Consequently, the minority-carrier lifetime is the characteristic time governing the forward-bias transient response. The lifetimes can be reduced by adding recombination centers in order to reduce switching speeds.

## 4.8 Schottky Barrier (or Surface Barrier) Diode

If one deposits a layer of metal on $n$-type silicon or gallium arsenide and measures the current–voltage characteristics, the results are similar to those of $p$-$n$ junctions (Fig. 4.13). There is a forward-bias direction with the $n$-type semiconductor negative where the current $J = J_s \exp(eV/nkT)$ and for the other voltage polarity a reverse, voltage-independent current. The factor $n$ is called the ideality factor and typically has values between 1.0 and 1.2 for well-behaved barriers. As in the case of the $p$-$n$ junction, the currents are orders of magnitude below those that would be found if ohmic contacts were placed on the semiconductor. As we will show, the behavior of the Schottky barrier—or metal–semiconductor—diode follows that of the $p$-$n$ junction except for the crucial difference that under forward bias, minority carriers are not injected into the semiconductor.

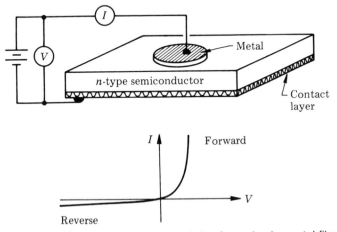

**FIGURE 4.13** Schottky barrier $I$–$V$ characteristics for a circular metal film deposited on an $n$-type semiconductor.

The rectifying behavior of the metal–semiconductor contact was known long before the discovery of the *p-n* junction. Research workers in the 1930s investigated the structure in detail. The current–voltage behavior is due to the presence of a potential barrier at the metal/*n*-type semiconductor interface, as shown in Fig. 4.14. Here we have represented the energy diagram of the metal as a horizontal line at the Fermi energy, $E_F$, with shading to denote the states filled with electrons. In the semiconductor at a distance away from the interface, the position of the conduction-band edge $E_C$ relative to $E_F$ is set by the electron concentration $n$ and the density of states $N_C$ through the relation [eq. (3.8)] $n = N_C \exp(-(E_C - E_F)/kT)$. In thermal equilibrium, the Fermi levels in both metal and semiconductor are aligned as shown in the figure.

At the metal–semiconductor interface, there is an energy barrier $e\phi_B$, where $\phi_B$ is the barrier height potential in volts for electrons in the metal to enter the semiconductor. The magnitude of $e\phi_B$ depends on the metal and semiconductor but typically is about two-thirds of the energy gap $E_G$ of *n*-type semiconductors. In this discussion we assume that the Fermi level is fixed, often called "pinned," at a constant energy level at the surface of the semiconductor. Although such states are not allowed in the bulk of the semiconductor crystal, as discussed in Chapter 12, they are allowed at the surface, where the crystal lattice terminates.

The barrier height $\phi_B$ at a metal-semiconductor interface is less than the work function which is the energy required to remove an electron from the surface of the metal into vacuum. The work function $W$ of a metal or semiconductor is defined as the difference in potential energy of an electron between the Fermi level and the vacuum level where an electron has zero kinetic energy. The work function is equal to the threshold energy for photoelectric emission from samples at absolute zero. The work functions for common metals such as Al, Ni and Cu lie in the range between four and five electron volts (Cs has a low work function of about

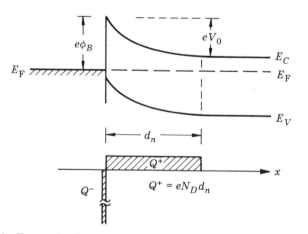

**FIGURE 4.14**   Energy level and charge distribution of a Schottky barrier diode under equilibrium conditions with $e\phi_B$ the metal–semiconductor barrier height and $eV_0$ the potential barrier in the *n*-type material.

2.1 eV) and Si and Ge have work functions of about five eV. In semiconductors, the electron affinity, $\chi$, is defined as the energy required to remove an electron from the conduction band edge $E_c$ to the vacuum level.

In the simplest approximation, the barrier height between a metal and semiconductor in contact is just the difference between the work function and electron affinity, $e\phi_B = W - \chi$ (see also Fig. 14.6). In practice, it is set by the position of the Fermi-level at the semiconductor surface.

## 4.8.1 Depletion Region and Capacitance

To establish an internal electric field, there must be a space-charge region. This is established by a sheet of charge per unit area $Q^-$ at the interface that is balanced by the fixed positive charge $Q^+$ of the donors in the $n$-type semiconductor,

$$Q^- = Q^+ = eN_D d_n \tag{4.49}$$

where $N_D$ is the concentration of donors and $d_n$ is the width of the depletion region in the $n$-type material. This is the same depletion approximation as that used in the discussion of the $p$-$n$ junction, where we assumed that all the mobile charge carriers are swept out (depleted) from the space-charge region. The maximum electric field $\mathscr{E}_{max}$ is located at the metal–semiconductor interface [as eq. (4.6)]:

$$\mathscr{E}_{max} = -\frac{eN_D d_n}{\epsilon} \tag{4.50}$$

where $\epsilon = \epsilon_r \epsilon_0$ is the permittivity of the semiconductor with $\epsilon_r$ = dielectric constant and $\epsilon_0$ = permittivity of free space, $\epsilon \simeq 10^{-12}$ F/cm for silicon. The negative sign reflects the sign convention with the electric field directed toward the interface.

From the semiconductor side, $eV_0$ is the potential barrier that electrons surmount in order to flow into metal. Using the same procedure as for the $p$-$n$ junction [eq. (4.8)] yields

$$V_0 = \frac{eN_D d_n^2}{2\epsilon}, \tag{4.51}$$

a value which could also be determined from the relationship indicated in Fig. 4.14 that

$$e\phi_B = eV_0 + (E_C - E_F). \tag{4.52}$$

The metal–semiconductor potential barrier $\phi_B$ can be determined from measurement of the capacitance as a function of reverse-bias voltage $V_R$ (positive voltage applied to the $n$-type semiconductor). As in the case of the abrupt, asymmetrical junction [eq. (4.15)], the small-signal capacitance $C$ for a metal–semiconductor junction of area $A$ is given by

$$C = \frac{A\epsilon}{d_n} = \frac{A\epsilon}{[2\epsilon(V_0 + V_R)/eN_D]^{1/2}} \tag{4.53}$$

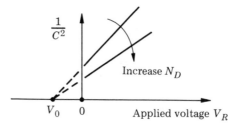

**FIGURE 4.15** The $1/C^2$ versus applied voltage relations for a Schottky barrier diode, showing the change in slope for increasing concentrations of donors, $N_D$.

where $V_0 + V_R$ is now the total potential in the $n$-type semiconductor. In general, the parameter $1/C^2$ is plotted

$$\frac{1}{C^2} = \frac{1}{A^2}\frac{2(V_0 + V_R)}{eN_D\epsilon}, \tag{4.54}$$

so that the intercept on the voltage axis gives $V_0$, as shown in Fig. 4.15. The slope of the $1/C^2$ versus $V_R$ line gives the value of the number of ionized donors and hence the value of the electron concentration $n$, $n = N_D$, where we assume that all the donors are ionized in the depletion region. From the value of $n$ and knowledge of the density of states $N_C$ in the conduction band, the value of $E_C - E_F$ is found which when added to $V_0$ gives the value of $\phi_B$.

## 4.8.2 Current–Voltage Characteristics

We have tacitly assumed that the carrier concentration $n$ is much less than the density of states $N_C$ and that Maxwell–Boltzmann statistics applied to the description of the occupation of states in the conduction band of the semiconductor. For the metal, at energies many $kT$ above the Fermi energy $E_F$ we can also use Maxwell–Boltzmann statistics. The number of electrons in the metal that have an energy greater than the barrier height $e\phi_B$ will be proportional to $\exp(-e\phi_B/kT)$. In the semiconductor, the number of electrons with energy greater than $eV_0$ will be proportional to $\exp(-eV_0/kT)$. Under thermal equilibrium conditions, no applied voltage, the number of electrons able to surmount the barrier is equal from both directions. That is, the current density $J_m$ of electrons in the metal surmounting the barrier and flowing into the semiconductor is equal to the current density $J_{sc}$ in the opposite direction of electrons flowing into the metal; $J_m = J_{sc}$. Under forward bias, with negative voltage $V_F$ applied to the semiconductor as shown in Fig. 4.16, the potential barrier in the semiconductor is reduced from $V_0$ to $(V_0 - V_F)$, just as in the case of the $p$-$n$ junction. Consequently, the number of electrons that can flow into the metal is now *increased* by a factor $\exp(eV_F/kT)$. Then $J_{sc} \gg J_m$ and the current density $J$ is given by

$$J = J_s \exp\left(\frac{eV_F}{kT}\right)$$

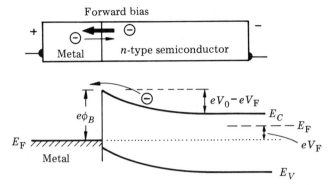

**FIGURE 4.16**  Energy levels versus distance for a Schottky barrier diode under forward bias, showing electrons injected into the metal.

where $J_s$ is a constant. Under reverse applied voltage $V_R$, the barrier in the semiconductor is increased by $V_R$ while the barrier height $e\phi_B$ for electrons in the metal remains unchanged [neglecting any small barrier changes due to image forces; Yang (1978)]. Then $J_{sc} \ll J_m$. Consequently, the current–voltage characteristics can be written

$$J = J_s \left[ \exp\left(\frac{eV}{nkT}\right) - 1 \right] \qquad (4.55)$$

where $J_s = AT^2 \exp(-e\phi_B/kT)$ is derived from thermionic emission theory with $A$ the Richardson constant ($A = 4\pi emk^2/h^3 \simeq 120 \text{ A/cm}^2$) and $n$ the ideality factor, $n \simeq 1$.

We have used $J_s$ as the constant rather than $J_0$ used for $p$-$n$ junction to emphasize the difference in the underlying physical concepts. Although in *both* cases the exponential dependence on applied voltage is due to changes in the potential barriers, the constant $J_0$ for the $p$-$n$ junction is determined by minority-carrier diffusion and recombination, and the constant $J_s$ for the Schottky barrier is determined by thermionic emission.

Although the forward-bias behavior of the Schottky barrier differs from that of the $p$-$n$ junction in terms of injection of minority carriers, both structures can be utilized under reverse bias in field-effect structures in which the junction depletion width modulates the width of the channel region. Similarly, both act as "collectors" of minority carriers created in the bulk material adjacent to the depletion region.

## 4.9 Solar Cells and Light-Emitting Diodes

The current–voltage characteristics of $p$-$n$ junctions are governed by the behavior of minority carriers and the presence of a potential barrier and depletion region between the $n$- and $p$-type regions. Under forward bias, the potential barrier is decreased and carrier injection occurs. Under reverse bias, charge collection occurs as the electric field in the depletion region acts as a sink for minority carriers which arrive at the edges of the depletion region. The solar cell provides an example of the collection process and the light-emitting diode illustrates the injection process.

Silicon solar cells are made by forming a thin, heavily doped, $n$-type region on a thick, lightly doped $p$-type substrate, as illustrated in Fig. 4.17. Visible light, solar irradiation, is incident on the top, $n$-type region, and the photons penetrate into the Si, forming hole–electron pairs when the photon energy exceeds the band-gap energy, $E_G$. As an approximation, we consider that the generation rate $G_n$ of electron–hole pairs created per second is uniform throughout the silicon. The

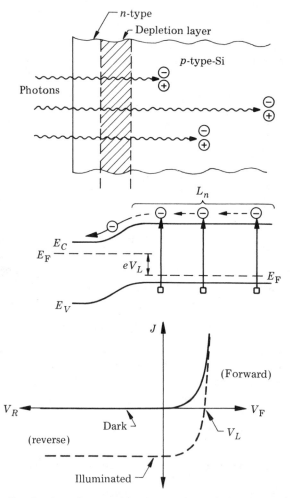

**FIGURE 4.17** Illumination of an $n$-$p$ junction produces hole–electron pairs which diffuse to the depletion region. The minority carriers (electrons in $p$-type material) are swept across the depletion region and increase the reverse current.

$n$-type region and the depletion region are thin compared to the $p$-type substrate, so that we are concerned primarily with electron–hole generation within the $p$-type region. Under normal conditions of illumination, the hole concentration due to the thermally ionized acceptors is far in excess of the photon-generated holes, so that we are concerned only with electrons in the $p$-type region, the minority carriers.

In the fabrication of solar cells from single-crystal Si, great care is taken to maintain high values of the minority-carrier lifetime, $\tau_n$, so that long diffusion lengths, $L_n = (D_n \tau_n)^{1/2}$, in excess of 100 $\mu$m can be obtained. The number of electrons generated per unit area per second in the $p$-type material within a diffusion length of the depletion region is $G_n L_n$. All of these carriers can diffuse to the edge of the depletion region, which collects them by sweeping them across the depletion region. The current density $J_G$ due to these optically generated carriers is

$$J_G = eG_n L_n. \tag{4.56}$$

Under reverse-bias conditions (positive applied voltage on $n$-type), this current density appears as an increase in current, shown in Fig. 4.17, as the electrons are swept from the $p$-type to the $n$-type region. The current for illumination conditions is voltage independent because $L_n \gg d$; the width, $d$ of the depletion region is the only reverse voltage-dependent parameter.

Under forward-bias conditions, illumination causes a decrease in the forward current. The total current density can be written

$$J = J_0 \left[ \exp\left(\frac{eV}{kT}\right) - 1 \right] - J_G \tag{4.57}$$

where the first term is the usual diode equation [eq. (4.39a)] and the second term is due to illumination.

Under open-circuit conditions where no current is measured, $J = 0$, a potential $V_L$ appears across the $n$-$p$ junction (the $n$-region is negative with respect to the $p$-region). To counteract the flow of optically generated carriers, the junction potential is adjusted to lower the barrier so that an equal and opposite flow of electrons from $n$-type to $p$-type is established. The open-circuit voltage (neglecting the dark reverse current) is

$$V_L = \frac{kT}{e} \ln \frac{J_G}{J_0}. \tag{4.58}$$

Solar cells can be made from silicon, gallium arsenide, or other semiconductors. For efficient light-emitting diodes, however, only special semiconductor materials are used. These materials, such as gallium arsenide, have a direct gap (Chapter 12) so that recombination of electrons and holes occurs with the emission of a photon whose energy is close to that of the band gap. Light-emitting diodes are operated by forward biasing the $n$-$p$ junction. The electrons injected from the $n$-type region into the $p$-type region will recombine with the holes leading to photon emission. This light emission at band-gap energies is direct evidence of carrier injection in forward-bias $p$-$n$ junctions. The properties of light-emitting diodes are discussed in Chapter 13.

## GENERAL REFERENCES

BALIGA, B. J., *Modern Power Devices,* Wiley, New York, 1987.

FRASER, D. A., *The Physics of Semiconductor Devices,* 3rd ed. Clarendon Press, Oxford, 1983.

MULLER, R. S., and T. I. KAMINS, *Device Electronics for Integrated Circuits,* Wiley, New York, 1977.

NEUDECK, G. W., *The PN Junction Diode,* Addison-Wesley, Reading, Mass., 1983.

SOLYMAR, L., and D. WALSH, *Lectures on the Electrical Properties of Materials,* 3rd ed. Oxford University Press, Oxford, 1984.

STREETMAN, B. G., *Solid State Electronic Devices,* 2nd ed. Prentice-Hall, Englewood Cliffs, N.J., 1980.

YANG, E. S., *Fundamentals of Semiconductor Devices,* McGraw-Hill, New York, 1978.

## PROBLEMS

**4.1** Consider an idealized semiconductor p-n junction with equal densities of states, mobilities, and lifetimes, and with the p-type region a factor of 100 more heavily doped than the n-type region.

    **(a)** Under forward bias, which region has positive voltage applied, and what is the direction of conventional current flow?

    **(b)** Which is the predominant carrier for forward-bias current flow, electrons or holes, near the junction region and far away from the junction in the n-type material and in the p-type material?

    **(c)** Write a general expression for the current density in forward and reverse bias for the predominant current carrier in terms of mobilities and lifetimes in the diffusion approximation.

    **(d)** Does the forward-bias current increase, decrease, or remain unchanged (within 20%) if the n-type doping is increased by a factor of 10, if the carrier lifetimes are increased a factor of 10, and if the band gap $E_G$ is increased by 10%?

    **(e)** Under light illumination of the junction, do the forward- and reverse-bias currents increase, decrease, or remain unchanged?

    **(f)** Under 10 V reverse bias, does the junction capacitance increase, decrease, or remain unchanged (within 20%) if the doping concentration in the p-type region is increased by a factor of 5, if the doping concentration in the n-type region is increased by a factor of 5, if the area of the junction is doubled, and if the carrier mobilities are doubled?

**4.2.** Using the properties of Si at 300 K listed in Table 2.4 (mobilities), Table 2.5 (dielectric constant), and Table 3.3 ($N_C$, $N_V$, and $E_G$) for an abrupt step junction with $p_p = 10^{18}/\text{cm}^3$ and $n_n = 10^{16}/\text{cm}^3$, calculate the following:

(a) The resistivities of the $n$-type and $p$-type material in $\Omega$-cm.

(b) The positions of the Fermi level in the $n$-type and $p$-type material.

(c) The potential $V_0$ across the $p$-$n$ junction using $E_G$ and $E_F$.

(d) The hole concentration $p_n$ in the $n$-type material using values of $p_p$ and $V_0$.

(e) The width of the depletion region.

**4.3.** For a Si $p$-$n$ junction with $p_p = 10^{18}/\text{cm}^3$, $n_n = 10^{16}/\text{cm}^3$, $n_i = 10^{10}/\text{cm}^3$, minority-carrier lifetimes $= 10^{-6}$ s, mobilities $= 10^3 \text{ cm}^2/\text{V} \cdot \text{s}$, and junction area $= 10^{-4} \text{ cm}^2$. Calculate the following for 0.5 V forward bias at 300 K:

(a) The concentration at the edge of the depletion region of holes injected into the $n$-type region and of electrons injected into the $p$-type region.

(b) The minority-carrier diffusion length.

(c) The current in amperes.

**4.4.** For the same conditions as Problem 4.3 but with 2.5 V reverse bias, calculate the following for $n_n = N_D$ and $p_p = N_A$:

(a) The width of the depletion region, the maximum value of the electric field, and the capacitance.

(b) For carrier diffusion only, the reverse current in amperes.

(c) For space-charge generation only, the reverse current.

**4.5.** For the same conditions as Problems 4.3 and 4.4 but with the $n$-type and $p$-type regions each 2 mm long, calculate the following:

(a) The electric field in the $n$-type and $p$-type regions near the contacts (far away from the junction) under forward bias.

(b) The voltage in the $n$- and $p$-type region by assuming that the length from contact to depletion region is 2 mm.

(c) The current in the $n$-type material if the $p$-type region were removed and replaced with an ohmic contact to the $n$-type region with $V = 0.5$ V.

**4.6.** For reverse bias on the step junction $p$-$n$ junction in Problem 4.4, calculate the transit time of a carrier (assume an average electric field). Compare your answer with that of the dielectric relaxation time, $\tau_{DR} = \rho\epsilon$.

**4.7.** You place a Schottky barrier with area $= 10^{-4} \text{ cm}^2$ on $n$-type Si where the barrier height $\phi_B = 0.7$ V and there are $10^{16}$ donors/cm$^3$ in the $n$-type region. Calculate the following using table values where necessary:

(a) The potential $V_0$ in the $n$-type region.

(b) The width of the depletion region and capacitance under 5 V reverse bias.

(c) The reverse current density assuming Richardson's constant $A = 4\pi emk^2/h^3$, where $k$ is Boltzmann's constant.

**4.8.** Starting conditions: Abrupt $p$-$n$ junction in Si ($E_G = 1.1$ eV) at 300 K with $N_A = p_p = 10^{17}/\text{cm}^3$, $N_D = n_n = 10^{15}/\text{cm}^3$, carrier lifetimes are equal ($10^{-7}$ s), and carrier mobilities are equal ($10^3 \text{ cm}^2/\text{V} \cdot \text{s}$). Assume majority-carrier concentrations ($n_n$, $p_p$); and lifetimes, mobilities, and energy gap are temperature independent; and that the current is determined by injection and diffusion of minority carriers. The magnitude of the p-n junction current increases, decreases, or remains unchanged (change less than 20%).

(a) Under forward-bias voltage of 0.2 V
   (1) if forward bias increases from 0.2 to 0.3 V
   (2) if irradiated with light $(E > E_G)$
   (3) if carrier lifetime decreases to $10^{-8}$ s.

(b) Under reverse bias voltage of 2.0 V
   (1) if reverse bias increases from 2.0 to 3.0 V
   (2) if irradiated with light $(E > E_G)$
   (3) if carrier lifetime decreases to $10^{-8}$ s.

**4.9.** You have n-type direct-gap InGaAs that you can make ohmic contacts on and can make Schottky barriers on by depositing a silver layer. You observe rectifying behavior with a forward current that increases with voltage as exponential $(eV/kT)$.

(a) Under forward bias, would you make the n-type material positive or negative with respect to the metal?

(b) If you removed the Schottky barrier and replaced it with an ohmic contact, would you get more, less, or about the same current for the same applied voltage as in part (a)?

(c) Which of the following devices could you make with this direct-gap material using Schottky barriers and ohmic contacts: light-emitting diodes, solar cells, bipolar transistors, or field-effect transistors?

**4.10.** You make a solar cell by diffusing a thin p-region on an n-type semiconductor substrate and illuminate in sunlight.

(a) You measure the voltage between the p-type and n-type region. Is the polarity of the voltage positive or negative of the p-region (with respect to the n-region)? Is the magnitude of the voltage less than, equal to, or greater than the band gap $E_G$?

(b) You have the choice of irradiating the solar cell with γ-rays to decrease carrier lifetime or heating in an $H_2$ atmosphere to increase carrier lifetime. Do you choose to (increase or decrease) carrier lifetime to increase the solar cell efficiency?

**4.11.** You have an abrupt (step) p-n junction with $10^{18}$ donors (electrons)/cm³ and $10^{16}$ acceptors (holes)/cm³ with a junction potential of 0.6 V $(V_0)$ at 300 K.

(a) With no applied voltage, what is the electron concentration in the p-type material?

(b) The electron mobility is 1000 cm²/V · s and the lifetime is $10^{-6}$ s; under forward bias if the injected electron concentration at the edge of the depletion region is $10^{14}$/cm³, what is the electron current density?

(c) If the area of the junction is $10^{-2}$ cm² and the dielectric constant $\epsilon$ is $10^{-12}$ F/cm, what is the capacitance of the p-n junction with 10 V reverse bias?

# CHAPTER 5

# Bipolar and Field-Effect Transistors

## 5.1 Introduction

Key components of the integrated circuit are three-terminal devices: bipolar junction transistors and field-effect transistors. In these devices a small signal applied to one of the terminals can modulate large changes in current through the other two terminals.

The *bipolar junction transistor* (BJT), is composed of two *p-n* junctions in series. For the *pnp* transistor shown in Fig. 5.1, the thin (less than 1 μm wide) *n*-type region is the common element—the base—between the forward-biased *p-n* junction emitter and the reverse-biased *n-p* junction collector. In a properly designed *pnp* bipolar transistor, the emitter current $I_E$ is primarily holes injected into the base and the collector current $I_C$ is primarily the collected current of holes from the base so that $I_C \simeq I_E$. The base current $I_B$ is typically two orders of magnitude below $I_C$ or $I_E$. A small modulation of the emitter–base voltage, $V_{EB}$, makes a large change in the forward bias current, $I_E$, because the emitter current is proportional to $\exp{(eV_{EB}/kT)}$, as given in Chapter 4. Consequently, a small change in base current $I_B$ leads to a large change in collector current $I_C$. The collector is an *n-p* junction under reverse bias, so the collector current is nearly independent of the voltage $V_{BC}$ between base and collector.

In this chapter we discuss homojunction transistors made of one semiconductor, silicon or gallium arsenide. Chapter 13 covers *heterojunction bipolar transistors* (HBT) where the emitter and base are made of semiconductors with different

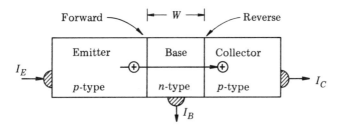

**FIGURE 5.1**  Schematic of a *pnp* bipolar transistor with base width *W* showing holes injected into the base from the forward-bias emitter and collected by the reverse-bias base–collector junction.

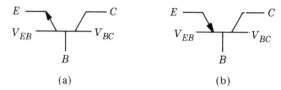

(a)                              (b)

**FIGURE 5.2** Circuit symbols for (a) *npn* and (b) *pnp* bipolar transistors.

energy gaps. The operation of the *pnp* HBT is the same as that discussed above for the *pnp* transistor.

The current gain $\beta$ is defined by

$$\beta = \frac{I_C}{I_B}. \tag{5.1}$$

Note however, that there is no true "gain" or increase of current because current is conserved with $I_E = I_B + I_C$.

The conventional circuit symbols for the *pnp* and *npn* transistors are shown in Fig. 5.2. The arrows on the emitter–base connection indicate the flow of conventional current (positive charge flow). The notations $E$, $B$, and $C$ as well as $V_{EB}$ and $V_{BC}$ are not shown in circuit diagrams.

The field-effect transistor is composed of a source (a source of charge carriers), a gate (a control for the amount of charge carriers that flow from the source), and a drain (a collecting place or drain for the charge carriers that pass the gate). A *junction field-effect transistor* (JFET) is shown in Fig. 5.3. There is an *n*-type *source* region that provides the electrons, an *n*-type *channel* region (fewer dopants than in the source or drain) for the flow of electrons between source and drain, and an *n*-type *drain* region. The *p*-type layer on top of the channel region is the gate electrode. Under operational conditions the gate is reverse biased and changes in the gate voltage $V_G$ ($V_G$ is negative with respect to the *n*-channel) produce a change in the width of the depletion region in the channel and hence a change in the effective thickness of the channel region. With a positive voltage on the drain terminal (drain voltage $V_D$ positive), a current $I_{SD}$ from source to drain will flow through the channel. The source and drain regions have a larger donor concentration (i.e., are more heavily doped) than the channel region and hence provide very low resistance ohmic contacts to the channel. The resistance $R$ of the channel region is

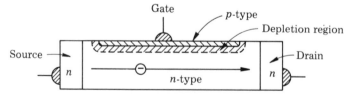

**FIGURE 5.3** Schematic of an *n*-channel junction field-effect transistor (JFET) with *p*-type diffused gate, showing electrons moving from source to drain in a channel whose thickness is controlled by the width of the depletion region.

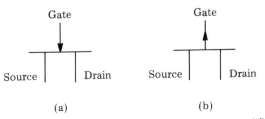

**FIGURE 5.4**  Circuit symbols for (a) *n*-channel and (b) *p*-channel field-effect transistors.

inversely proportional to the thickness of the channel, $R \propto t^{-1}$, so that an increase in gate voltage $V_G$ will increase the width of the depletion region, thus decreasing the thickness of the channel region and hence decreasing the current. The reverse-bias gate current $I_G$ is much less than the source–drain current $I_{SD}$, so that a small modulation signal on the gate produces a large change in the source–drain current.

Field-effect transistors can be made with electrons flowing through *n*-channels or with holes flowing through *p*-type channels (with *p*-type source and drain regions). These structures are referred to as *n*-channel and *p*-channel field-effect transistors, respectively. The circuit symbols are shown in Fig. 5.4 with arrows to the gate. The notations source, gate, and drain are not shown in circuit diagrams.

The gates can be made with diffused junctions, a *p*-type diffused region on an *n*-type channel for the example in Fig. 5.3, with a metal gate—a metal–semiconductor or Schottky barrier—such as used in GaAs integrated circuits, or with an insulating or oxide region between the gate metal and the channel. Acronyms abound (Table 5.1): The *field-effect transistor* is a FET and the *metal–semiconductor field-effect transistor* is a MESFET. The *junction FET* is a JFET, while a *metal–insulator–semiconductor* FET is a MISFET or more commonly an IGFET, for *insulated-gate FET*. In silicon structures where the insulating layer is an oxide layer (generally $SiO_2$) under the gate metal, the *metal–oxide-semicon-*

**Table 5.1**  Acronyms Commonly Used in Transistor Technology[a]

| BJT | Bipolar junction transistor |
| --- | --- |
| CMOS | Complementary MOS |
| CCD | Charge-coupled device |
| FET | Field-effect transistor |
| HBT | Heterojunction bipolar transistor |
| IGFET | Insulated-gate FET |
| JFET | Junction FET |
| MESFET | Metal–semiconductor FET |
| MISFET | Metal–insulator gate FET |
| MOSFET | Metal–oxide semiconductor FET |
| poly Si | Polycrystalline silicon |

[a]See also Table 13.1 for acronyms used in heterojunction transistor terminology and Table 15.1 for those used in packaging.

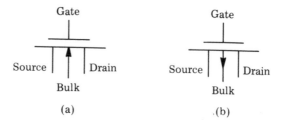

**FIGURE 5.5** Circuit symbols for (a) *n*-channel and (b) *p*-channel insulated-gate field-effect transistors (IGFETs).

ductor (or *s*ilicon) structure is a MOSFET. When there are *n*-channel and *p*-channel MOSFETs that are paired for logic circuits, the *c*omplementary MOS transistor circuit is referred to as CMOS. The gates in IGFETs are sometimes fabricated with heavily doped *poly*crystalline *s*ilicon, leading to poly Si gates or polygates. The metal–oxide-semiconductor structure is a capacitor and integrated-circuit elements can be built with closely space arrays of MOS capacitors. *C*harge-*c*oupled *d*evices (CCDs) allow the charge stored in one capacitor to be transferred to an adjacent device by changes in the voltages of the adjacent gates.

The circuit symbols for the IGFET or MOSFET are shown in Fig. 5.5. There is a difference between these symbols and those for the JFET (Fig. 5.4). The two structures have both source and drain regions and gates that control the magnitude of the source–drain currents. As we will see in later sections of this chapter, the channels are formed differently: the JFET (or MESFET) has a channel built into the structure, while the IGFET has a channel induced by the field applied from the gate to the semiconductor. In Fig. 5.5 the arrow represents the direction of *p*-type to *n*-type between the bulk and the induced channel at the surface. The labels are not used in circuit diagrams.

## 5.2 Bipolar Transistor

Schematic diagrams of the cross section and dopant distributions in an *npn* bipolar junction transistor in silicon are shown in Fig. 5.6a and b. The transistor is made by diffusing donors at high concentrations, $N_D$, into the Si to form the emitter and diffusing a lower concentration $N_A$ of acceptors deeper into the *n*-type Si to form the *p*-type base. The fabrication procedure is to form the base first and then the emitter. The width $W$ of the base region is typically 1 $\mu$m ($10^{-4}$ cm) or less and is tightly controlled in the diffusion process, as this width is one of the critical parameters in transistor operation. In this chapter we approximate the diffusion profiles of donors $N_D$ and acceptors $N_A$ as step functions (Fig. 5.6c) as we did in Chapter 4. These ideal step distributions are useful in formulating the operation of the transistor but do ignore some of the secondary effects associated with the graded dopant profiles at the emitter–base and base–collector junctions. We also use the block diagram shown in Fig. 5.6d for clarity.

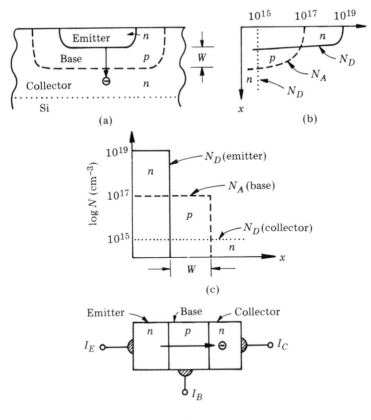

**FIGURE 5.6** An *npn* bipolar transistor with base width *W*: (a) cross section with electrons injected from emitter to base; (b) dopant concentration profile; (c) idealized dopant profile with abrupt or step changes in donor and acceptor concentrations; (d) schematic of the step junction *npn* transistor in block form.

The *pnp* transistor has basically the same structure as that of the *npn* transistor but is fabricated by diffusing a large concentration of acceptors to form the *p*-type emitter and diffusing a lower concentration of donors deeper to form an *n*-type base on the *p*-type substrate material that forms the collector. Again, the base region is usually formed first. We illustrate the operation of the bipolar transistor with a *pnp* transistor shown in Fig. 5.7a with the emitter grounded.

The concept of operation is that the emitter–base *p-n* junction is forward biased. With the emitter containing orders of magnitude more acceptors than the concentration of donors in the base, most of the current is given by holes injected into the base region, as discussed in Chapter 4. If the width *W* of the base is sufficiently small so that very little recombination occurs, almost all the injected holes diffuse to the reverse-biased collector. An energy-band diagram of the *pnp* transistor is shown in Fig. 5.7b for zero bias and in Fig. 5.7c for operating conditions.

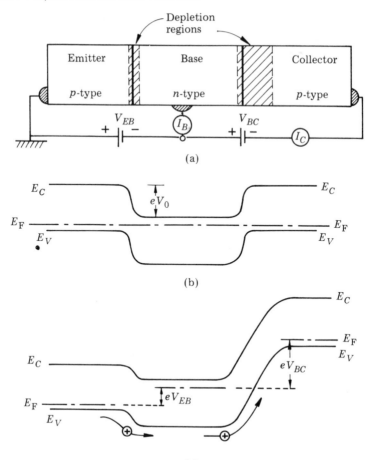

**FIGURE 5.7** A *pnp* bipolar transistor with forward-bias emitter-to-base voltage $V_{EB}$ and reverse-bias base-to-collector voltage $V_{BC}$: (a) schematic with shaded depletion regions; (b) energy levels at equilibrium with junction potential $V_0$; (c) energy levels under operating bias voltage showing holes injected into and collected from the base.

There are three key features to the *pnp* (also the *npn*) transistor.

1. The emitter current is determined by the injection of majority carriers from the emitter into the base. In practice this is achieved by making the dopant concentration in the emitter at least two orders of magnitude greater than the dopant concentration in the base. In heterojunction bipolar transistors this is achieved by use of epitaxial semiconductors where the band gap of the emitter is greater than that of the base (Chapter 13).

2. The carriers injected into the base do not recombine in the base, but diffuse across the base to the reverse-biased collector. In practice this is achieved by making the base width $W$ much less than the diffusion length $L$ of minority

carriers in the base. In general, $W \leq 0.01\ L$, where $L = (D\tau)^{1/2}$ and $\tau$ is the minority-carrier lifetime (Chapter 3).

3. The current across the reverse-biased collector is determined by the collection of carriers injected from the emitter that arrive at the boundary of the base–collector depletion region. In practice this is achieved by making a high-quality base–collector junction which has a very small reverse-bias leakage current. The dopant concentration in the collector is less than that in the base, so that the depletion region extends primarily into the collector under reverse bias.

## 5.3 Emitter–Base Current

The dopant concentration distribution and distribution of injected carriers in the base is shown in Fig. 5.8 for the condition that the base width $W$ is much less than the hole diffusion length $L_p$, and that the acceptor concentration $N_A$ in the emitter is a factor of 100 greater than the donor concentration $N_D$ in the base. In Fig. 5.8 we define the edge of the depletion layer between emitter and base as $x = 0$ and the edge of the depletion layer between base and collector as $x = W$. Following the treatment in Chapter 4, the current from emitter to base (for a heavily doped emitter) is determined by the injection of holes. Further, the total emitter current $I_E$ is given by the diffusion current of holes at $x = 0$. That is,

$$I_E = I_p(\text{diffusion})|_{x=0} \tag{5.2}$$

which can be written as

$$I_E = I_p(\text{diffusion})\Big|_{x=0} = -eD_p \frac{dp}{dx}\Big|_{x=0} A \tag{5.3}$$

where $A$ is the area in cm$^2$ and $D_p$ is the diffusion constant for holes.

To find the gradient $dp/dx$ of injected holes, we rely on the expressions derived in Chapter 4. The concentration of holes injected at $x = 0$ is given by

$$p_n(x = 0) = p_{n_0} \exp\left(\frac{eV_{\text{app}}}{kT}\right) \tag{5.4}$$

where $p_{n_0}$ is the minority-carrier (or hole) distribution in thermal equilibrium in the $n$-type base at a point well away from the depletion region:

$$p_{n_0} = p_p \exp\left(-\frac{eV_0}{kT}\right) \tag{5.5}$$

where $eV_0$ is the barrier height between the emitter and base and $p_p$ is the hole concentration in the emitter region. From the relations given in Chapter 4,

$$p_p = N_A \quad \text{and} \quad p_{n_0} = \frac{n_i^2}{N_D} \tag{5.6}$$

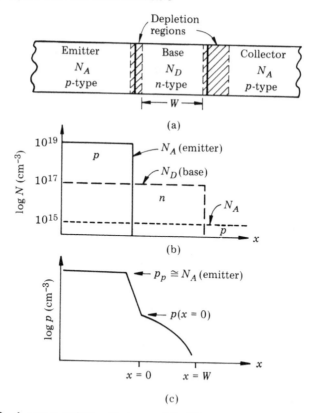

**FIGURE 5.8** A *pnp* transistor with base width *W* between depletion regions: (a) schematic; (b) step concentration profile of acceptors and donors; (c) log concentration of holes in the emitter, $p_p = N_A$, and concentration $p(x = 0)$ of holes injected into the base at the edge of the depletion region.

where $N_A$ and $N_D$ are the acceptor and donor concentrations in the emitter and base, respectively, and $n_i$ is the intrinsic carrier concentration ($n_i \simeq 10^{10}/\text{cm}^3$ in silicon at room temperature).

To this point we have only assumed that the emitter has a much larger dopant concentration than the base, so that $p_p \gg n_n$. Now we invoke the fact that the base width is much less than the hole diffusion length $L_p$, as shown in Fig. 5.9. Under these conditions the hole concentration can be expressed as

$$p_n(x) = p_n(x = 0)\left(1 - \frac{x}{W}\right) \tag{5.7}$$

where the hole concentration at $x = W$ can be approximated by zero. This condition follows because every hole that diffuses to the boundary of the depletion layer between the base and collector (or $x = W$) will be swept across the depletion region so that $p_n(x = W)$ is forced to zero. We use a linear approximation because $W \ll L_p$ and thus neglect hole recombination in the base.

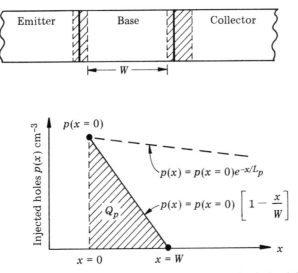

**FIGURE 5.9** A *pnp* transistor with base width $W$ showing the holes injected into the base at the edge of the depletion region, $x = 0$, for a uniform hole gradient with $p(W) = 0$ showing the injected hole charge $Q_p$ as shaded. The dashed line is the hole concentration for the base width $W \gg L_p$, the hole diffusion length.

The dashed line in Fig. 5.9 illustrates the condition that $W \gg L_p$, so that

$$p_n(x) = p_n(x = 0) \exp\left(-\frac{x}{L_p}\right) \qquad (5.8)$$

or

$$p_n(x) \simeq p_n(x = 0)\left(1 - \frac{x}{L_p}\right) \qquad (5.9)$$

where $L_p$ is the minority-carrier (hole) diffusion length. This is the "wide-base" or standard *p-n* junction condition (see Section 4.41 or 3.9).

The hole gradient $dp/dx$ is thus

$$\frac{dp}{dx} = -\frac{p_n(x = 0)}{W} \qquad \text{(transistor)} \qquad (5.10)$$

or

$$\frac{dp}{dx} = -\frac{p_n(x = 0)}{L_p} \qquad \text{(\textit{p-n} junction).} \qquad (5.11)$$

From eq. (5.3), the emitter current $I_E$ can be written as

$$I_E = \frac{eD_p}{W} p_n(x = 0) A \qquad \text{(transistor)} \qquad (5.12)$$

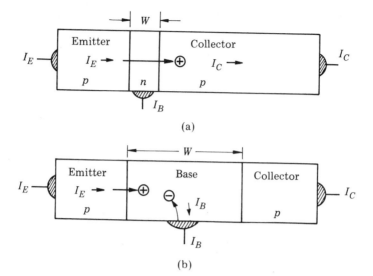

**FIGURE 5.10**   Two *pnp* transistors of base width $W$ and hole diffusion length $L$: (a) narrow base with $W << L$, where the emitter current $I_E \simeq I_C$, the collector current; (b) wide base, $W > L$, where $I_E \simeq I_B$, the base current.

or

$$I_E = \frac{eD_p}{L_p} p_n(x = 0) A \qquad (p\text{-}n \text{ junction}). \qquad (5.13)$$

Although these expressions are similar, there is a vast difference in their implications to the operation of a transistor as shown in Fig. 5.10. In the narrow-base transistor (Fig. 5.10a), all the holes injected from the emitter diffuse across the base and reach the collector. Thus we have

Narrow-base transistor:

$$I_E \simeq I_C \qquad \text{and} \qquad I_B \simeq 0. \qquad (5.14)$$

In the wide-base transistor (Fig. 5.10b) all the holes injected from the emitter diffuse in the base region and recombine there. None of the injected holes reach the collector:

Wide-base transistor:

$$I_E \simeq I_B \qquad \text{and} \qquad I_C \simeq 0. \qquad (5.15)$$

## 5.4 Contributions to the Base Current

In the ideal narrow-base transistor of eq. (5.14), the base current $I_B$ is nearly zero. In practical cases the value of $I_B$ is $1/100$ to $1/1000$ of the value of $I_C$, so it is reasonable to consider that $I_B \simeq 0$. As shown in Fig. 5.11, there are three contributions to the base current.

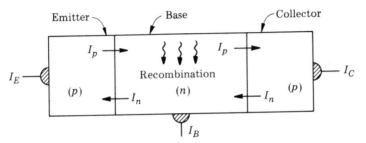

**FIGURE 5.11**  A *pnp* transistor showing the various contributions to the base current: recombination in the base, electrons injected from base to emitter, and electrons from collector to base.

1. We have assumed that in the forward-biased *p-n* junction, all the current is due to holes from the emitter, $I_E \simeq I_p$. However, electrons are also injected from the base into the emitter. These electrons also give rise to a current $I_n$. From the relations developed in Chapter 4 we can write

$$\frac{I_p}{I_n} \simeq \frac{p_p}{n_n} \tag{5.16}$$

where $p_p$ is the hole concentration in the emitter and $n_n$ is the electron concentration in the base. The injected electrons contribute to the emitter current,

$$I_E = I_p + I_n \tag{5.17}$$

but not to the collector current,

$$I_C \simeq I_p \ (E). \tag{5.18}$$

Therefore,

$$I_n \text{ is one component of } I_B. \tag{5.19}$$

It is desirable to minimize the value of $I_n$ by making $p_p \gg n_n$. In heterojunction bipolar transistors $I_n$ is minimized by making the emitter energy gap greater than that of the base. The dopant concentration in the base can then be greater than that in the emitter.

2. As the width of the base region is increased, an increasing number of the injected holes will recombine with the electrons (majority carriers) present in the base. Every time a hole recombines with an electron (on the average), an electron must be supplied from the contact to the base region. In this case

$$e \cdot R_b \text{ is one component of } I_B \tag{5.20}$$

where $R_b$ is the total number of recombination events occurring between injected holes and electrons in the base. We can make a close approximation to the current $I_B$ due to recombination in the base by considering the amount of charge $Q_p$ per unit area stored in the base and the hole lifetime $\tau_p$. From the triangular distribution of charge shown in Fig. 5.9,

$$Q_p = \frac{eWp_n(x = 0)}{2}. \tag{5.21}$$

This stored charge must be replaced due to recombination every $\tau_p$ seconds (see eqn. 4.48) so that for a transistor of area $A$,

$$I_B = \frac{Q_p A}{\tau_p} = \frac{eWp_n(x = 0)}{2\tau_p} A. \tag{5.22}$$

The emitter current $I_E = [eD_p p_n(x = 0)A]/W$ from eq. (5.12). The ratio of emitter-to-base currents (under the condition that recombination is the sole contribution to $I_B$) is

$$\frac{I_E}{I_B} = \frac{2D_p \tau_p}{W^2} = \frac{2L_p^2}{W^2} \tag{5.23}$$

where the hole diffusion length $L_p = (D_p\tau_p)^{1/2}$. A typical value of $L_p$ is $10^{-2}$ cm for $\tau_p = 10^{-5}$ s; so that for a base width of 1 $\mu$m ($10^{-4}$ cm), $I_E/I_B = 2 \times 10^4$. This ratio [eq. (5.23)] indicates the importance of minimizing the base width.

3. We have assumed that the $p$-$n$ junction characteristics of the collector are perfect and the reverse-bias current $I_R$ is zero. However, there is a finite leakage current (often at the surface) or reverse current $I_n(C)$ of electrons from the collector to the base.

$$I_n(C) \text{ contributes to the base current.} \tag{5.24}$$

The total contribution of all three to the base current is then

$$I_B = I_n + eR_b + I_n(C). \tag{5.25}$$

Again, $I_B$ is very small in practical devices compared to $I_E$ or $I_C$.

## 5.5 Base Transit Time

In a *pnp* transistor, the time $T_R$ required for holes to diffuse across the base from the emitter to the collector can determine the maximum frequency of operation of the device. Since the base width $W$ is much less than the hole diffusion length, the transit time $T_R$ is much less than the hole lifetime. That is, many holes enter and leave the base region during the $\tau_p$ seconds required for an electron–hole recombination event. If we use the same hole storage $Q_p$ argument as in Section 5.4 and assume that

$$I_E = \frac{Q_p}{T_R} A \tag{5.26}$$

we have from eqs. (5.22) and (5.23),

$$T_R = \frac{W^2}{2D_p}. \tag{5.27}$$

For a width of $10^{-4}$ cm and $D = 10$ cm$^2$/s, $T_R = 0.5 \times 10^{-9}$ s, which sets an upper frequency limit $1/T_R$ of about 2 GHz for a base of this width.

## **5.6** Operating Characteristics of the *pnp* Transistor

Figure 5.12 shows schematics of the voltages and currents and the $I$–$V$ characteristics of a typical device. The three main features of the device are:

**1.** The collector current will be nearly equal to the emitter current.
**2.** The emitter–base voltage controls both currents.
**3.** Both currents are nearly independent of the collector–base voltage.

A figure of merit is the dc current gain $\beta$ [eq. (5.1)]:

$$\beta = \frac{I_C}{I_B} > 100.$$

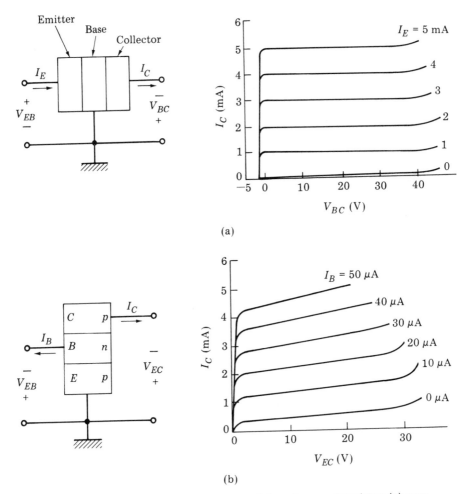

(a)

(b)

**FIGURE 5.12**  Operating $I$ versus $V$ characteristics of a *pnp* transistor: (a) common-base circuit; (b) common-emitter circuit. [From Neudeck (1983).]

If one utilizes the circuit shown in Fig. 5.12b and measures only $I_B$ and $I_C$, there would be an apparent gain $\beta$ in current. However, current is not created by magic and the concept of current "gain" can be a misleading physical concept.

## 5.7 JFETs and MESFETs

In this section we consider the operation of junction field-effect transistors (JFETs) and metal–semiconductor FETs (MESFETs), also called Schottky barrier FETs (Schottky FETs). As indicated in Fig. 5.13, the JFET on an $n$-type semiconductor operates the same way as the MESFET. In both cases, the operation of the FETs is based on change in the resistance of the $n$-type semiconductor by decreasing the cross-sectional area due to increasing the width of the depletion layer under the gate. In the MESFET the reverse-biased gate is a metal–semiconductor barrier (Schottky barrier) and in the JFET, the gate is established by a reverse-biased $p$-$n$ junction. In both cases, the depletion layer serves to isolate the gate contact from the source–drain current $I_{SD}$; that is, only a small current flows in the gate circuit compared to $I_{SD}$. Further, the potential is constant across the gate region in both FETs. This constant potential on the gate means that because of the potential difference between source and drain, the width of the depletion region varies along the FET.

The junction FET is more commonly used in Si-device applications due to the ease of fabrication of $p$-$n$ junctions and the reproducibility of the junction $I$–$V$ characteristics. In this section we concentrate on the JFET. However, MESFETs

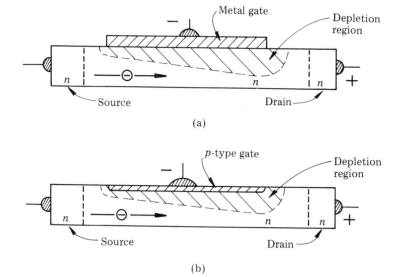

(a)

(b)

**FIGURE 5.13** Cross sections of $n$-channel field-effect transistors (FETs) with depletion region shaded: (a) metal–gate, metal–semiconductor MESFET or Schottky barrier FET; (b) $p$-type gate JFET.

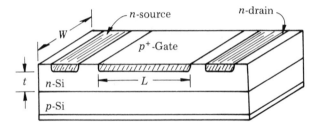

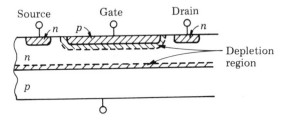

**FIGURE 5.14** Three-dimensional sketch and cross section of an *n*-channel JFET formed on an *n*-type epitaxial layer on a *p*-type substrate, showing the channel width *W* and length *L*.

are used in GaAs integrated circuits where one wants to avoid a thermal-diffusion step to form a junction in the fabrication process.

The junction FET is generally made (Fig. 5.14) on a thin, 3- to 5-µm-thick (3 to $5 \times 10^{-4}$ cm) *n*-type epitaxial layer (single-crystal layer) grown on a *p*-type Si substrate. This *n*-type layer forms the channel and has a thickness *t*. The source and drain contacts are made to the *n*-type Si channel by diffusing donors to form heavily doped *n*-type regions. The gate is formed by a diffusion of acceptors to form a heavily doped *p*-type layer extending a length *L*. The FET has a width *W* (extending into the page). As in Section 5.6, we use block diagrams to illustrate device operation.

We will treat *n*-channel JFET operation by forming source and drain regions on an *n*-type channel layer on a *p*-type substrate. The thickness *t* of the channel extends from the bottom of the diffused *p-n* junction gate to the *p*-type substrate.

In Fig. 5.15 we show the situation where the gate is grounded ($V_G = 0$) and there is a small potential $V_D$ between source (grounded) and drain. Hence electrons will flow from source to drain (or conventional current from drain to source). The electrons are constrained to flow in the channel region between the two depletion regions. The resistance of the channel is given by

$$R = \frac{\rho L}{A} \tag{5.28}$$

or

$$R = \frac{L}{(eN_D\mu)W(t - 2d)} \tag{5.29}$$

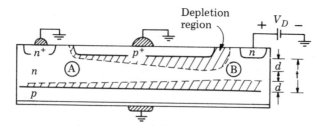

**FIGURE 5.15**   Cross section of an *n*-channel JFET showing the asymmetrical width *d* of the depletion region with the width near the gate–drain contact, B, wider than that near the source–gate contact, A.

where $\rho$ = resistivity of *n*-type Si = $(eN_D\mu)^{-1}$
  $L$ = length of the channel under the gate
  $W$ = width of the channel (see Fig. 5.14)
  $t$ = thickness of the epitaxial layer between diffused gate and substrate
  $d$ = thickness of the depletion regions (and we assume that *d* is the same for the gate and substrate junctions).

As given by eq. (4.13), the width *d* of the depletion region is given by

$$d = \left[\frac{2\epsilon(V_0 + V)}{eN_D}\right]^{1/2}$$ (5.30)

where *V* is the voltage difference between the gate and the semiconductor and $V_0$ is the junction potential. In Fig. 5.15, $V = V_G = 0$ at point A and $V = V_D$ at point B. For $V_D \ll V_0$ (where $V_0 \approx 0.7$ V), the depletion width does not change appreciably with changes in $V_D$, and the source-to-drain current $I_{SD}$ increases with $V_D$ (or is ohmic in behavior). This is often called the linear region of the *I–V* characteristics of a JFET, in which the channel acts as a resistor.

  For increased values of $V_D$ (i.e., $V_D \gtrsim V_0$), the widths of the depletion regions increase [eq. (5.30)], thus decreasing the cross-sectional area *A* of the channel. Hence the value *R* of the channel resistance increases.

  In the JFET, the voltage along the channel will increase from the source to the

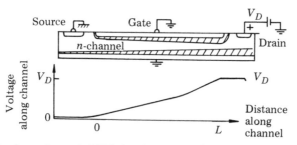

**FIGURE 5.16**   An *n*-channel JFET showing the nonlinear voltage along the channel.

drain. In the present case with the gate grounded ($V_G = 0$), the voltage increases from zero at the source to $V_D$ at the drain. This situation is shown schematically in Fig. 5.16. The diffused gate region and the substrate have the same potential along the length $L$ (in this case at zero potential). Consequently, the potential difference between gate and $n$-type channel varies from $V_0$ at $x = 0$ to $V_0 + V_D$ at $x = L$. The greatest potential difference occurs at the edge of the gate near the drain ($x = L$).

With further increases in the drain voltage $V_D$, the two depletion regions nearly touch and "pinch off" the channel near $x = L$. As shown in Fig. 5.17, there is no further increase in $I_{SD}$ with an increase in $V_D$; this regime of operation of the JFET is called the "saturation" region and it sets in when $V_D > V_D(\text{sat})$. The value of $V_D(\text{sat})$ can be found from eq. (5.30) by setting $V = V_D(\text{sat})$ and $d = t/2$:

$$V_D(\text{sat}) = \frac{eN_D t^2}{8\epsilon} - V_0 \qquad (V_G = 0). \qquad (5.31)$$

The $I_{SD}$ versus $V_D$ characteristics of a JFET with $V_G = 0$ are shown in the upper curve of Fig. 5.17.

As shown in Fig. 5.17, for an $n$-channel FET as one increases the gate voltage (makes $V_G$ negative) to increase the reverse bias on the gate, the source-to-drain current $I_{SD}$ decreases. This is due to the increase in the depletion widths. The onset of the saturation region occurs at lower values of drain voltage and in analogy to the derivation of eq. (5.31), $V_D(\text{sat})$ is given by

$$V_D(\text{sat}) = \frac{eN_D t^2}{8\epsilon} - V_G - V_0. \qquad (5.32)$$

The value of $I_D(\text{sat})$ can be crudely estimated from

$$I_D(\text{sat}) = \frac{V_D(\text{sat})}{R(\text{sat})} \qquad (5.33)$$

by use of eq. (5.29) to obtain the resistance $R$ of the channel at $V = V_D(\text{sat})$.

The transit time $T_R$ of the electrons from source to drain is determined by the drift velocity $v_d = \mu_n \mathscr{E}$, where $\mu_n$ is the electron mobility and $\mathscr{E}$ is the electric

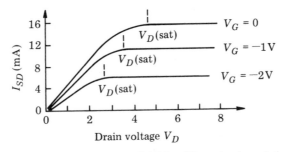

**FIGURE 5.17**  *I–V* characteristics of a JFET for different gate voltages. The dashed lines indicate the drain voltage at the onset of saturation.

field. For the case of an applied voltage $V_D$ between source and drain and a uniform voltage gradient along the channel length $L$, $\mathscr{E} = V_D/L$, so that

$$v_d = \frac{\mu V_D}{L} \tag{5.34}$$

and

$$T_R = \frac{L}{v_d} = \frac{L^2}{\mu V_D}. \tag{5.35}$$

For high electric fields ($\mathscr{E} \gtrsim 10^3$ V/cm), the drift velocity becomes independent of $\mathscr{E}$ and has a saturated drift velocity $v_s \simeq 10^7$ cm/s. The transit time is then

$$T_R = \frac{L}{v_s} \tag{5.36}$$

and would have a value of $10^{-10}$ s for $L = 10^{-3}$ cm.

In summary, there are three features of the $I$–$V$ characteristics of a JFET (or a MESFET).

1. The current ($I_{SD}$) is ohmic for low values of voltage ($V_D$) between source and drain. The FET acts as a resistor. This is the linear range of operation.
2. With an increase in $V_D(V_D > V_0)$, the channel resistance increases and above a certain voltage $V_D(\text{sat})$, the current saturates and becomes nearly independent of $V_D$.
3. With an increase in gate voltage, $V_G$, the saturation current decreases. The value of $V_D(\text{sat})$ decreases as $V_G$ increases.

## 5.8 Insulated-Gate FET and Metal–Oxide–Semiconductor FET

The metal–oxide–semiconductor FET is characterized by having an oxide layer ($SiO_2$) between the gate contact and the semiconductor. This oxide layer has a very high resistivity ($\rho > 10^{16} \ \Omega \cdot$ cm) and prevents current flow from the semiconductor to the gate just as the depletion layer does in the JFET. Because of the importance of the insulating layer, the MOSFET is also called an *insulator–gate* FET (IGFET) or a *metal–insulator–semiconductor* FET (MISFET). An *n*-channel MOS (Fig. 5.18) is fabricated by diffusing donors into the source and drain regions (forming *n*-regions) on *p*-type Si and growing an oxide layer that acts as the gate oxide.

The *n*-channel is formed by applying sufficient positive voltage to the gate so that there is an accumulation of electrons in the region under the oxide. These electrons form the *n*-channel and conduct current between the *n*-type source and drain regions.

Let us consider the simplest case with a metal–oxide layer on a *p*-type semiconductor where there are no surface states at the oxide–semiconductor interface nor charge in the oxide. As indicated in Fig. 5.19a, there is no bending of the

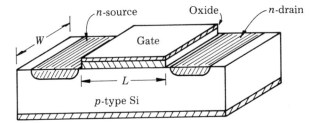

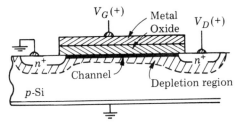

**FIGURE 5.18** Three-dimensional sketch and cross section of an *n*-channel MOSFET on *p*-type Si showing the channel width *W* and length *L*.

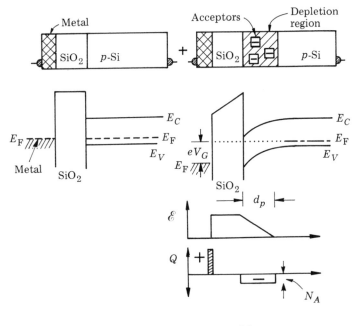

**FIGURE 5.19** An *n*-channel MOSFET on *p*-type Si, showing energy levels, electric field, and charge distribution for (a) equilibrium, flat band and (b) positive voltage on the gate for $V_G$ less than that required for a channel.

conduction and valence bands in the semiconductor, and the Fermi level in the metal and $p$-type semiconductor line up. With a small positive bias applied to the metal, $V_G$, part of the gate voltage, is in the oxide layer $V_{ox}$ and part in the semiconductor, $V_S$, as given by

$$V_G = V_{ox} + V_S \tag{5.37}$$

and the electric field in the oxide layer is given by

$$\mathscr{E}_{ox} = \frac{V_{ox}}{t} \tag{5.38}$$

where $t$ is the thickness of the oxide layer.

The presence of a voltage $V_S$ in the semiconductor leads to the establishment of depletion region as shown in Fig. 5.19b. The electric field $\mathscr{E}_s$ at the semiconductor surface is given (as in the Schottky barrier diode) from Gauss's law by

$$\mathscr{E}_s = -\frac{Q_s}{\epsilon_r \epsilon_0} = -\frac{Q_s}{\epsilon} \tag{5.39}$$

where $\epsilon$ is the permittivity of Si ($\epsilon \approx 10^{-12}$ F/cm) and $Q_s$ is the number of negatively charged acceptors per unit area in the depletion layer. That is,

$$Q_s = -eN_A d_p \tag{5.40}$$

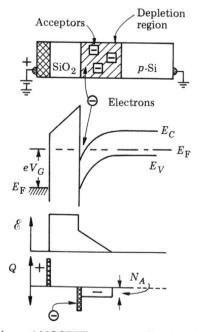

**FIGURE 5.20** An $n$-channel MOSFET on $p$-type Si with sufficient positive voltage on the gate to form an $n$-channel, showing energy levels, electric field, and charge distribution.

where $N_A$ is the number of acceptors/$cm^3$ and $d_p$ is the width of the depletion layer in the $p$-type Si. The amount $Q_M$ of positive charge on the metal plate is equal to $Q_s$ (with no trapped charge in the oxide):

$$Q_M = Q_s. \tag{5.41}$$

With an increase in gate voltage $V_G$, more voltage appears across the semiconductor, and the bands bend further until some free electrons begin to accumulate at the semiconductor surface. The gate voltage at which this occurs is called the "turn-on" or "threshold" voltage $V_T$. The band structure and charge distribution for this case are shown in Fig. 5.20. With a further increase in gate voltage even more electrons accumulate. If we make the approximation that all the gate voltage above $V_T$ is used to accumulate more electrons and hence the depletion layer does not increase, we can determine the number of free electrons based on the concept of charge on the plates of a capacitor. The thickness of the capacitor is given by $t$, the oxide thickness, so that the value $C_{ox}$ of the capacitance per unit area is

$$C_{ox} = \frac{\epsilon_{ox}}{t} \tag{5.42}$$

where $\epsilon_{ox}$ is the permittivity of $SiO_2$ ($\epsilon_{ox} \approx \epsilon_{Si}/3 \approx 0.34 \times 10^{-12}$ F/cm). From the general relation that $VC = Q$, the value of $Q_n$, the charge density per unit area of electrons, is given by

$$Q_n = C_{ox}(V_G - V_T). \tag{5.43}$$

The capacitance voltage characteristic of a MOS capacitor is shown in Fig. 5.21. For negative applied voltage, the capacitance is determined by the oxide thickness [eq. (5.42)]. For positive voltages, the capacitance decreases as the depletion layer forms and increases. At large negative voltages above $V_T$, the capacitance becomes voltage independent, with a value of $C_{ox}C_d/(C_{ox} + C_d)$, where $C_d$ is the capacitance associated with the depletion layer. The exact shape of the $C$–$V$ characteristics depends on the measurement frequency, the number of interface charges, and the presence of mobile or fixed charges in the oxide layer.

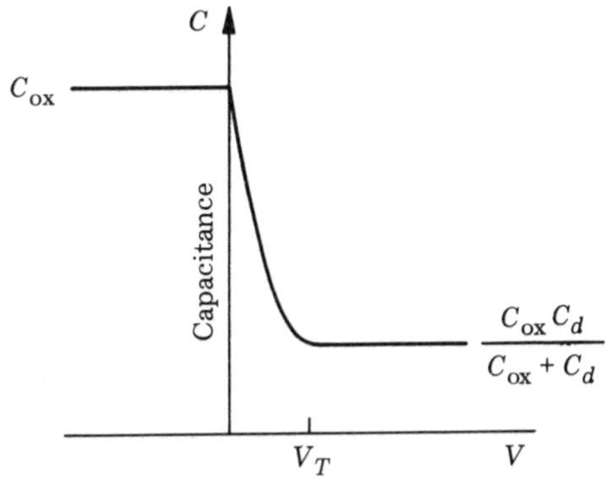

**FIGURE 5.21**  Capacitance–voltage characteristics of a MOS capacitor (at high frequencies).

# 5.9 Threshold Voltage and Oxide Charge

In the operation of a MOSFET, the threshold voltage $V_T$ required to "turn on" the channel conductance is an important parameter. In real devices there is both charge within the oxide layer and charge at the oxide–silicon interface. In addition, we must consider the influence on the potential caused by the metal contact on the $SiO_2/Si$ structure just as in the case of the Schottky barrier diode discussed in Section 4.8. The threshold voltage will be the sum of all these contributions (Streetman, 1980):

$$V_T = V_{ms} + V_{ox} + V_d + V_i \tag{5.44}$$

where the various contributions are due to work function differences at the metal–semiconductor interface ($V_{ms}$), the charge in the oxide and at the interface ($V_{ox}$), the charge in the depletion region ($V_d$), and the voltage required to invert the surface ($V_i$).

Let us consider each of these contributions for an $n$-channel MOS formed on a $p$-type substrate where $N_A = 10^{15}$ acceptors/cm$^3$. We choose an oxide thickness of 34 nm ($3.4 \times 10^{-6}$ cm) to give a convenient value of the capacitance based on eq. (5.42):

$$C_{ox} = \frac{\epsilon_{ox}}{t} = \frac{3.4 \times 10^{-13} \text{ F/cm}}{3.4 \times 10^{-6} \text{ cm}} = 10^{-7} \text{ F/cm}^2$$

where

$$\epsilon_{ox} = 3.4 \times 10^{-13} \text{ F/cm}.$$

**Inversion Voltage $V_i$**

The surface of $p$-type Si will be inverted (i.e., have an $n$-type channel) whenever the Fermi level at the surface crosses the point where $n = p = n_i$). If we choose the inversion voltage as that condition where the surface is as strongly $n$-type as the substrate is $p$-type (the strong inversion condition), then from the relations in Chapter 4,

$$V_i = \frac{2kT}{e} \ln \frac{N_A}{n_i} \tag{5.45}$$

where $n_i$ is the intrinsic carrier concentration. If we chose $n_i = 10^{10}$/cm$^3$ for Si and $kT/e$ as 0.026 V ($T = 300$ K), then for $N_A = 10^{15}$/cm$^3$,

$$V_i = 0.6 \text{ V}.$$

This voltage is positive for an $n$-channel and represents just the onset of strong inversion.

### Depletion Region Voltage $V_d$

If all the voltage $V_i$ appears across the width of the depletion region, then from eq. (5.30), the width $d_p$ of the depletion region is

$$d_p = \left(\frac{2\epsilon_{Si}V_i}{eN_A}\right)^{1/2}$$

$$= \left(\frac{2 \times 10^{-12} \times 0.6}{1.6 \times 10^{-19} \times 10^{15}}\right)^{1/2} = 0.87 \times 10^{-4} \text{ cm} = 0.87 \text{ μm}.$$

The numbers of ionized acceptors/cm$^2$ in the depletion region $N_A^- d_p$ is

$$N_A^- d_p = N_A d_p = 0.87 \times 10^{11} \text{ acceptors/cm}^2$$

and the charge $Q_d$ is

$$Q_d = eN_A^- = 1.6 \times 10^{-19} \times 0.87 \times 10^{11} = 1.4 \times 10^{-8} \text{ C/cm}^2$$

a negative quantity. Since the voltage needed to deplete this charge is applied across the oxide, we express the depletion voltage $V_d$ as

$$V_d = \frac{Q_d}{C_{ox}} = \frac{1.4 \times 10^{-8} \text{ C/cm}^2}{10^{-7} \text{ F/cm}^2} = 1.4 \times 10^{-1} \text{ V}.$$

### Voltage $V_{ox}$ Due to Oxide Charge $Q_{ox}$

The effective charge $Q_{ox}$ in the insulator is due to plus charges at the oxide-semiconductor interface, often designated $Q_{ss}$, and charges introduced during the growth of the oxide. For carefully treated oxides, $Q_{ss} \approx 1.5 \times 10^{10}$ charges/cm$^2$ on (100) surfaces and about $5 \times 10^{10}$/cm$^2$ on (111) surfaces. For these reasons MOS devices are almost always made on (100)-oriented silicon. We will choose $6 \times 10^{10}$/cm$^2$ as the effective number of oxide charges/cm$^2$, so that the effective oxide charge $Q_{ox}$ is $Q_{ox} = 6 \times 10^{10}$/cm$^2$ $\times 1.6 \times 10^{-19} = 10^{-8}$ C/cm$^2$.

The negative oxide voltage $V_{ox}$ is then

$$V_{ox} = \frac{Q_{ox}}{C_{ox}} = \frac{10^{-8} \text{ C/cm}^2}{10^{-7} \text{ F/cm}^2} = 10^{-1} \text{ V}.$$

### Metal–Semiconductor Voltage

The work-function difference, $V_{ms}$, between metal and Si depends on the metal and the semiconductor dopant concentration (see section 4.8). For Al on Si doped to $10^{15}$/cm$^3$ an estimated value of $V_{ms}$ is about $-0.3$ volts ($n$-type) and about $-0.9$ volts ($p$-type) assuming the Al work function $W_{Al} = 4.3$ eV and an electron affinity $\chi$ of 4.3 eV.

The threshold voltage (eqn. 5.44) for this MOSFET is, therefore,

$$V_T = -0.9 - 0.1 + 0.14 + 0.6 = -0.26 \text{ volt}.$$

For an $n$-channel device ($p$-substrate), one normally would expect a positive threshold voltage to invert the surface to strongly $n$-type. For this particular example, the threshold voltage is $-0.26$ volt; this means that a channel exists at $V_G = 0$ due to the values of $V_{ms}$ and $V_{ox}$. A voltage more negative than $-0.26$ volt is required to turn the channel off. This device is called a depletion-mode device since it is normally "on." If the substrate doping concentration ($N_A$) is increased, $V_T$ can become positive due to $V_i$ and $V_d$ becoming more positive.

When $V_T$ for an $n$-channel device is greater than 0 volt ($V_T \geq 0$), the device is normally "off." A voltage more positive than the threshold voltage is required to form the channel; this device is called an enhancement-mode device (normally off). For the case of a $p$-channel device with an Al gate, the threshold voltage is always negative because $V_{ms}$, $V_{ox}$, $V_d$ and $V_i$ are all negative. The $p$-channel device is always an enhancement-mode device.

The important aspect of this discussion is that all the charges are of the magnitude of about $10^{11}/\text{cm}^2$, equivalent to one ten-thousandth of a monolayer. This means that extraordinary precautions must be taken to maintain the cleanliness of the environment. Further, one can manipulate the threshold voltage by use of ion implantation techniques (Chapter 8).

## 5.10 Operating Characteristics of $n$- and $p$-Channel MOSFETs

The voltage polarities and a cross section of an $n$-channel MOSFET are shown in Fig. 5.22a and the $I$–$V$ characteristics in Fig. 5.22b. For gate voltages $V_G$ less than the turn-off voltage $V_T$, there is no $n$-channel connection between the source and drain. Consequently, the source–drain current is zero ($I_{SD} = 0$) irrespective of the drain voltage $V_D$. (This is not exactly true because there is always a small leakage current across a reverse-biased $p$-$n$ junction.) For high values of the gate voltage ($V_G > V_T$) and low values of drain voltage ($V_D < V_T$) the $n$-type channel between source and drain behaves like a resistor. The value of a resistor with $Q_n$ electrons per unit area having a mobility $\mu_n$ is given by

$$R = -\frac{L}{W\mu_n Q_n e} \tag{5.46}$$

where the minus sign arises because $Q_n$ is negative and $L$ and $W$ are the channel length and width, respectively.

The conductance $g_d$ of the $n$-channel is

$$g_d = \frac{dI_D}{dV_D}\bigg|_{V_G = \text{constant}} \tag{5.47}$$

which in the linear regime where $I_D = V_D/R$ is

$$g_d = \frac{1}{R} = e\mu_n Q_n \frac{W}{L} \tag{5.48}$$

where $\mu_n$ is the electron mobility in the channel.

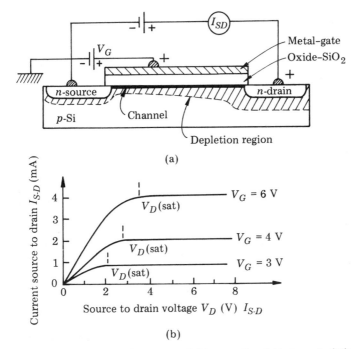

FIGURE 5.22 (a) Schematic diagram and (b) operating I–V characteristics of an n-channel MOSFET for different gate voltages. The dashed lines show the drain voltage at the onset of saturation.

From eq. (5.43), $Q_n = C_{ox}(V_G - V_T)$ and

$$g_d = \frac{e\mu_n W C_{ox}(V_G - V_T)}{L}. \tag{5.49}$$

If one assumes a value of $Q_n$ of $6 \times 10^{11}$ electrons/cm² or $10^{-7}$ C/cm² and a mobility of 1000 cm²/V · s, then

$$g_d = \frac{10^{-4}W}{L}$$

in units of (ohms)$^{-1}$ or siemens. For a gate length $L$ of 1 μm and a gate width $W$ of 10 μm, $g_d = 10^{-3}$ S or 1 mS.

The mutual transconductance $g_m$ is defined as

$$g_m = \frac{dI_D}{dV_G}\bigg|_{V_D = \text{constant}} \tag{5.50}$$

and in the region (the "saturation" region) where the drain current $I_D$ remains nearly constant for increased values of $V_D$, the expression for $g_m$ is approximately that given by eq. (5.49).

Again we are faced with a situation (as in the JFET) where the potential varies along the channel between source and drain. Since both $V_G$ and $V_D$ are positive with respect to the source (and $p$-type semiconductor), as $V_D$ increases the potential difference between the gate to the $n$-type channel decreases. The effect is most pronounced near the drain. If $V_D$ becomes sufficiently large so that the gate-to-channel potential drops below $V_T$, the turn-on voltage, the $n$-channel nearly disappears and the current saturates as shown in Fig. 5.22.

In summary, there are three features of the $I$–$V$ characteristics of an $n$-channel MOSFET:

1. The current ($I_{SD}$) is ohmic (linear region) for low values of source-to-drain voltages ($V_D$).

2. With an increase in drain voltage $V_D$, the channel resistance increases and above a certain voltage $V_D$(sat), the current saturates and becomes nearly independent of $V_D$.

3. With an increase in gate voltage, $V_G$, the saturation current *increases*. The value of $V_D$(sat) increases as $V_G$ increases.

There are many features in common in the operation of the MOSFET and JFET, as shown in Table 5.2.

In the previous discussion, we emphasized $n$-channel FETs. A $p$-channel MOS is formed on $n$-type material with $p$-type source and drain regions formed by diffusing acceptors into the $n$-type region. By applying sufficient negative voltage to the gate, there is an accumulation of holes under the oxide layer, thus defining the $p$-channel.

Both $n$-channel and $p$-channel MOSFETs can be formed on $p$-type substrates in which localized $n$-type wells are created by diffusion, as shown in Fig. 5.23. When properly connected in pairs, these $n$- and $p$-channels form the basis of complementary or CMOS circuits.

We have only considered MOSFETs in which a gate voltage $V_G$ greater than the "turn-on" voltage $V_T$ is required for channel conduction. These structures are

**TABLE 5.2** Comparison of $n$-Channel MOSFET and JFET

|  | **MOSFET** | **JFET** |
|---|---|---|
| 1. Fabricated on: | $p$-type | $n$-type |
| 2. Source and drain | $n^+$ | $n^+$ |
| 3. Current carriers | Electrons | Electrons |
| 4. Gate polarity | Positive | Negative |
| 5. Channel | $n$-type | $n$-type |
| 6. Gate isolated by: | Insulator | Depletion region |
| 7. Conduction varied by: | Surface field | Depletion width |
| 8. Conduction parameter varied | Charge in channel | Channel thickness |
| 9. Small values of $V_D$ | Ohmic | Ohmic |
| 10. Large values of $V_D$ | $I_{SD}$ saturates | $I_{SD}$ saturates |
| 11. Increasing values of $V_G$ | $I_{SD}$ increases | $I_{SD}$ decreases |
| 12. Saturation voltage, $V_D$(sat) | Increases with $V_G$ | Decreases with $V_G$ |

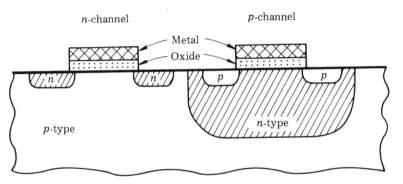

**FIGURE 5.23**  Cross section of a complementary MOS (CMOS) structure with an *n*-type well diffused into the *p*-type substrate.

normally "off" unless a gate voltage is applied and the source–drain current increases. This mode of operation is termed enhancement. By use of ion implantation techniques to introduce dopants into the layer under the oxide, one can adjust the turn-on or threshold voltage $V_T$ and/or can form a channel layer with zero gate voltage. With a channel present at $V_G = 0$, the structures are normally "on" and by proper voltage polarity on the gate the source–drain current can be decreased—the "depletion" mode of operation.

The MOSFET is an extremely flexible circuit element because one can have *n*- or *p*-channel devices with currents that increase (enhancement) or decrease (depletion) with gate voltage. The capacitance of a MOS structure can also be used for charge storage. The stored charge can be shifted from one capacitor to an adjacent element by control of the gate voltages.

## GENERAL REFERENCES

BALIGA, B. J., *Modern Power Devices*, Wiley, New York, 1987.

COLCLASER, R. A., *Microelectronics: Processing and Device Design*, Wiley, New York, 1980.

GIBBONS, J. F., *Semiconductor Electronics*, McGraw-Hill, New York, 1966.

GRAY, P. R., and R. G. MEYER, *Analysis and Design of Analog Integrated Circuits*, 2nd ed. Wiley, New York, 1984.

MULLER, R. S., and T. I. KAMINS, *Device Electronics for Integrated Circuits*, Wiley, New York, 1977.

NEUDECK, G. W., *The Bipolar Junction Transistor*, Addison-Wesley, Reading, Mass., 1983.

NICOLLIAN, E. H., and J. R. BREWS, *MOS Physics and Technology*, Wiley, New York, 1982.

PIERRET, R. F., *Field Effect Devices*, Addison-Wesley, Reading, Mass., 1983.

STREETMAN, B. G., *Solid State Electronic Devices*, 2nd ed., Prentice-Hall, Englewood Cliffs, N.J., 1980.

SZE, S. M., *Physics of Semiconductor Devices,* Wiley, New York, 1981.

YANG, E. S., *Fundamentals of Semiconductor Devices,* McGraw-Hill, New York, 1978.

## PROBLEMS

**5.1**   A *pnp* transistor with a base width of 1 μm operates with an emitter-to-collector current of 0.3 A for an emitter-to-base voltage of 0.3 V and a base-to-collector voltage of 3.0 V.

   **(a)** With respect to the base, what are the voltage polarities (positive or negative) of the emitter and collector?

   **(b)** For a current gain = 100, what is the base current?

   **(c)** Changing only one parameter at a time, does the emitter-to-collector current increase, decrease, or remain unchanged (changes by more than 20%) if:

   **(1)** the base-to-emitter voltage increases from 0.3 V to 0.4 V?

   **(2)** the base-to-collector voltage increases from 3.0 V to 4.0 V?

   **(3)** the minority-carrier lifetime in the base is decreased from $10^{-4}$ s to $10^{-9}$ s?

   **(4)** the base width is increased to 100 μm?

   **(5)** the hole concentration in the emitter is doubled?

   **(6)** the hole concentration in the collector is doubled?

**5.2.**   Consider a Si *pnp* transistor with a $10^{-4}$-cm-wide base operated at 300K with the following parameters: carrier lifetimes = $10^{-4}$ s and mobilities = 1000 cm²/s. Emitter: $10^{19}$ holes/cm³; base: $10^{17}$ electrons/cm³, collector, $10^{15}$ holes/cm³, $n_i = 10^{10}$/cm³.

   **(a)** With a forward bias of 0.26 V, what is the concentration of holes injected into the base at the edge of the emitter–base junction?

   **(b)** If the applied voltage is doubled, what is the hole concentration?

   **(c)** If the injected hole concentration is $10^{13}$/cm³, what is the current density of holes across the base?

   **(d)** If the base in part (c) was $10^{-1}$ cm long what would be the hole current density in the base region?

**5.3.**   Assume that you want to make an ideal *npn* bipolar transistor with a current gain of 100 and an emitter current of 0.1 mA.

   **(a)** Would the collector current be greater than, equal to (within 20%), or less than the emitter current?

   **(b)** Would the dopant concentration in the base region compared to the dopant concentration in emitter be a factor of $10^2$ greater, a factor of 10 greater, about equal to, a factor of 10 smaller, or a factor of $10^2$ smaller?

   **(c)** Would the width of the base compared to the carrier diffusion length be a factor of $10^2$ greater, about equal to, or a factor of $10^2$ smaller?

   **(d)** Would the collector current increase, decrease, or remain unchanged

   **(1)** if the emitter–base voltage is increased by 20%?

   **(2)** if the collector–base voltage is increased by 20%?

**5.4.** You want to design an *npn* transistor with a gain of approximately 100 from a semiconductor with an electron mobility of 1000 $cm^2/V \cdot s$ and a lifetime of $4 \times 10^{-6}$ s. The base region has a concentration of $10^{17}$ acceptors/$cm^3$.

(a) With respect to the base, what applied voltage polarity (positive, negative) and what voltage (0.5 V or 5 V) would you use for the emitter and collector?

(b) What dopant concentration ($10^{19}/cm^3$, $10^{17}/cm^3$, $10^{15}/cm^3$) would you use for the emitter and collector regions?

(c) What base width would you choose ($10^{-8}$ cm, $10^{-6}$ cm, $10^{-4}$ cm, $10^{-2}$ cm)?

(d) Keeping all applied voltages the same, would the collector current (increase, decrease, or remain unchanged within $\pm 20\%$) if

(1) the base width is decreased by a factor of 2?

(2) the emitter width is decreased by a factor of 2?

(3) the emitter area is decreased by a factor of 2?

**5.5.** Using the properties of Si at 300 K listed in Table 2.4 (mobilities), Table 2.5 (dielectric constant), and Table 3.3 ($N_C, N_V, E_G$) for a *npn* abrupt junction transistor with $n_n = 5 \times 10^{18}/cm^3$, $p_p$ (base) $= 5 \times 10^{16}/cm^3$, $n_n = 10^{15}/cm^3$, and base width $= 10^{-4}$ cm:

(a) Draw the band diagram under equilibrium conditions.

(b) Calculate the position of the Fermi levels in the emitter, base, and collector.

(c) Calculate $V_0$ for the emitter–base and base–collector junctions.

(d) Calculate the width of the depletion regions in the emitter–base and base–collector junctions.

(e) What are the minority–carrier concentrations in the three regions?

**5.6.** Using the parameters in Problem 5.5 with an electron lifetime of $10^{-5}$ s in the base and a hole lifetime of $10^{-7}$ s in the emitter, with $10^{13}$ electrons/$cm^3$ injected into the base at the edge of the depletion region (assume carrier diffusion):

(a) What is the electron current density in the base?

(b) What would be the electron current density in a 1-mm-wide base?

(c) What is the magnitude of the emitter–base applied voltage?

(d) What is the concentration of holes injected from the base into the emitter?

(e) What is the base-to-emitter hole current density? Compare this value to the electron current density injected in the base.

**5.7.** Using the parameters in Problem 5.5, with carrier lifetimes of $10^{-5}$ s in the base and collector and with 5 V reverse bias on base-to-collector junction:

(a) In the carrier diffusion approximation, what is the reverse current density?

(b) Calculate the width of the depletion region.

(c) What is the space-charge-generated reverse current?

(d) Compare these currents with the emitter–base current with $10^{13}$ electrons/$cm^3$ injected into the base.

**5.8.** For a $p$-channel junction FET on an epitaxial layer of $p$-type Si:

    **(a)** Dopants are diffused into the material to make the source, drain, and gate junction. Name the dopants (donors or acceptors) for the source, drain, and gate.

    **(b)** The applied voltages are drain voltage = 5 V and gate voltage = 2 V. What are the voltage polarities (positive, negative) with respect to the source for the drain and gate?

    **(c)** Is the width of the depletion region under the gate junction near the drain less than, greater than, or equal to that near the source?

**5.9.** You make a JFET on an $n$-type epitaxial layer of Si (1 μm thick) by diffusing a $p$-type gate junction and making diffused source and drain regions.

    **(a)** Would the source and drain also be $p$-type? (Yes, no, doesn't matter)

    **(b)** Would this be ,an $n$-channel or a $p$-channel FET?

    **(c)** Would the source–drain current increase, decrease, or remain unchanged if you increase the gate voltage by 50%?

    **(d)** If you increased the thickness of the epitaxial layer would the gate voltage be different to change the source–drain current by 50%? (More, less, doesn't matter)

    **(e)** Assuming that you are in the linear carrier velocity versus electric field regime ($v = \mu E$), would you get a faster FET by increasing or decreasing the channel length $L$ (distance between source and drain) and by increasing or decreasing source–drain voltage $V_{S\text{-}D}$?

**5.10.** To build an $n$-channel MOSFET:

    **(a)** Are the source and drain diffused with donors or acceptors, and is the drain voltage positive or negative?

    **(b)** Is the MOSFET made on $n$-type, intrinsic, or $p$-type material, and is the gate a Schottky barrier, $n$-$p$ junction, or metal/oxide barrier?

    **(c)** Under an enhancement mode (source–drain current increases with gate voltage $V_G$) is the gate positive or negative, and under a depletion mode (current decreases with $V_G$), is the gate positive or negative?

**5.11.** You want to build an $n$-channel MOSFET that can operate in the enhancement or depletion mode but want to change from the conventional use of $p$-type substrate and $n$-type source and drain.

    **(a)** Could you use an $n$-type layer on the $p$-substrate for the channel?

    **(b)** Could you make the source or the drain $p$-type?

    **(c)** Could you bias the gate so you could get large gate currents for a special circuit application?

    **(d)** Could you use the conventional structure but instead of a $p$-type substrate use a $p$-type epitaxial layer on an $n$-type substrate or on a heavily doped $p^+$-substrate?

**5.12.** Consider a Schottky barrier on $p$-type Si with an acceptor concentration = $10^{17}/\text{cm}^3$ and the dielectric constant = $10^{-12}$ F/cm.

(a) What voltage polarity (positive or negative) would be applied to the semiconductor in the forward bias, and would electrons or holes be injected into the metal?

(b) Neglecting the barrier height value, what is the width of the depletion region in centimeters with 10 V reverse bias?

(c) A Schottky barrier field-effect transistor is made on p-type silicon.

   (1) The source and drain contact regions would be heavily doped n- or p-type Si?

   (2) Would the applied potential to the drain be positive or negative with respect to the source?

   (3) Would the applied potential to the gate be positive or negative with respect to the source?

# 6

# Crystallography and Crystalline Defects

## 6.1 Introduction

Integrated circuits are fabricated on single-crystal wafers sliced from semiconductor ingots. The chemical properties (such as etch rates) and physical properties (such as cleavage planes) of the semiconductors depend on its crystallographic directions and planes. Wafers are commonly identified by their surface orientation and reference flats cut on the edges. Figure 6.1 shows a Si wafer with a reference flat. The flat of this material is used to denote the $<110>$ direction or the cleavage planes. If scribe lines are drawn on wafers parallel to the reference flats with a pointed tool, the wafer may be cleaved easily along the scribe lines.

To improve the quality of the material on which the devices are built, an epitaxial growth process is often used. This process is a technique to grow a thin crystalline layer on a crystalline substrate (Fig. 6.2) so that the grown layer bears a certain crystallographic relationship with respect to the underlying substrate. Epitaxy not only offers the possibility of reducing defect density and unwanted impurities in the crystalline layers, but also has the advantage of providing layers of different doping concentrations and carrier types than those in the substrate. In Si technology, the epitaxial layer is Si; this type of epitaxy is termed homoepitaxy. In GaAs and other III–V compound technology, heteroepitaxy is often practiced, such as growing $Al_xGa_{1-x}As$ epitaxial layers on GaAs, as discussed in Chapters 13 and 14. The emphasis in this chapter is on the crystal lattice and the defects present in semiconductors. These defects play a major role in semiconductor technology.

## 6.2 Crystallography

Semiconductor wafers, even without any epitaxial layers, contain imperfections. Defects in a crystalline material can be categorized as (1) point defects, (2) line defects, (3) planar defects, and (4) volume defects. These defects will be characterized in the following sections. Although defects are generally undesirable and often unavoidable in a crystal, in some cases they can be put to use to improve the device performance, such as in defect "gettering" of unwanted impurities. To characterize crystalline defects, a certain amount of crystallography is necessary.

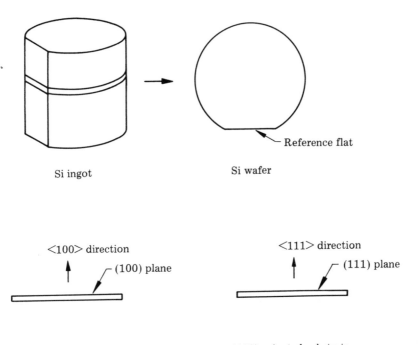

**FIGURE 6.1** Schematics showing a Si wafer cut from a Si ingot. The reference flat denotes the <110> direction. For (100)-oriented substrates, the <110> direction is the line of intersection between a (111) plane and the surface (100) plane. There are two orthogonal sets of <110> directions on a (100) plane. For (111)-oriented substrates, the <110> direction is the line of intersection between another (111) plane and the surface (111) plane. There are three sets of <110> directions on a (111) surface making 60° angles with each other.

A crystal is a solid composed of atoms arranged in a pattern periodic in space. It is, however, sometimes more convenient to consider a set of points to define crystals. Such a set of points, called lattice points, have a fixed relation in space, meaning that the lattice when viewed in a particular direction from one lattice point would appear exactly the same when viewed in the same direction from *any other* lattice point. This set of points is defined such that each point in this array has identical surroundings. The lattice points usually coincide with atom positions. However, in complicated crystals, the individual atoms may not have identical surroundings, whereas the lattices describing the complicated crystals will. Because

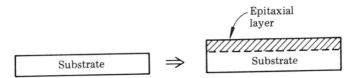

**FIGURE 6.2** Growth of epitaxial layers.

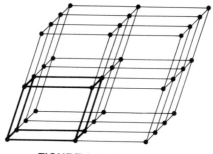

**FIGURE 6.3** Point lattice.

of this property of a lattice point, the space containing this set of points can be divided by three sets of planes into a set of cells each identical in size, shape, and orientation; such a cell is called a unit cell. Figure 6.3 shows a point lattice; the heavy dark lines outline a unit cell. This cell repeats itself in space and can be described by three unit vectors **a, b,** and **c,** called crystallographic axes, that are related to each other in terms of their lengths $a$, $b$, and $c$ and the angles $\alpha$, $\beta$, and $\gamma$ (see Fig. 6.4). Any direction in the cell can be described as a linear combination of the three axes:

$$\mathbf{r} = n_1\mathbf{a} + n_2\mathbf{b} + n_3\mathbf{c}. \tag{6.1}$$

In addition,

$$\mathbf{r} = \mathbf{r} + P\mathbf{a} + Q\mathbf{b} + R\mathbf{c} \tag{6.2}$$

where $P$, $Q$, and $R$ are integers. This means that the unit cell may be translated in space via a linear combination of the unit vectors without changing the surroundings (i.e., the unit cell repeats itself in space).

In dividing the space by three sets of planes, various shapes and sizes of unit cells can be produced. As it turns out, only seven different kinds of cells are necessary to describe all possible point lattices. These seven kinds of cells define seven crystal systems, shown in Fig. 6.4. Each corner of the unit cell of these seven systems has a lattice point, but not in the interior of the cells or on the cell faces. It is possible to place more points either in the center of the unit cell or on the cell faces without violating the general definition of a point lattice that each has identical surroundings for certain crystal systems. Based on this arrangement of points, a total of 14 so-called Bravais lattices can be produced for the seven crystal systems. Figure 6.5 shows four of these lattices: the cubic and hexagonal Bravais lattices. The number of lattice points in a unit cell, $N_u$, is given by

$$N_u = N_i + \frac{N_f}{2} + \frac{N_c}{8} \tag{6.3}$$

where $N_i$ is the number of points in the interior, $N_f$ is the number of points on faces (each $N_f$ is shared by two cells), and $N_c$ is the number of points on corners (each $N_c$ point is shared by eight cells). When $N_u = 1$, the cell is called a primitive

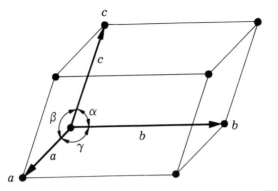

| System | Axial lengths and angles | Bravais lattice |
|---|---|---|
| Cubic | Three equal axes at right angles<br>$a = b = c, \alpha = \beta = \gamma = 90°$ | Simple<br>Body-centered<br>Face-centered |
| Tetragonal | Three axes at right angles, two equal<br>$a = b \neq c, \alpha = \beta = \gamma = 90°$ | Simple<br>Body-centered |
| Orthorhombic | Three unequal axes at right angles<br>$a \neq b \neq c, \alpha = \beta = \gamma = 90°$ | Simple<br>Body-centered<br>Base-centered<br>Face-centered |
| Rhombohedral* | Three equal axes, equally inclined<br>$a = b = c, \alpha = \beta = \gamma \neq 90°$ | Simple |
| Hexagonal | Two equal coplanar axes at 120°,<br>third axis at right angles<br>$a = b \neq c, \alpha = \beta = 90°, \gamma = 120°$ | Simple |
| Monoclinic | Three unequal axes,<br>one pair not at right angles<br>$a \neq b \neq c, \alpha = \gamma = 90° \neq \beta$ | Simple<br>Base-centered |
| Triclinic | Three unequal axes, unequally inclined<br>and none at right angles<br>$a \neq b \neq c, \alpha \neq \beta \neq \gamma \neq 90°$ | Simple |

*Also called trigonal.

**FIGURE 6.4** Unit cell and the seven crystal systems.

cell. A unit cell is not necessarily a primitive cell; however, any nonprimitive unit cell can be "referred" to a primitive cell, as shown in Fig. 6.6.

In considering the Bravais lattice of the face-centered cubic lattice shown in Fig. 6.6, it is customary to use the conventional unit cell of the FCC cell rather than the primitive cell. Since it is a cubic system ($a = b = c, \alpha = \beta = \gamma = 90°$) the length $a$ is called the lattice parameter. The number of atoms per unit cell with $N_i = 0$, $N_f = 6$, and $N_c = 8$ is

$$N_u = \frac{6}{2} + \frac{8}{8} = 4 \text{ atoms/unit cell.} \tag{6.4a}$$

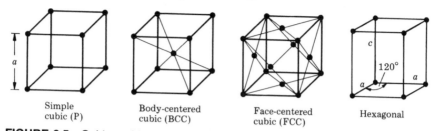

**FIGURE 6.5** Cubic and hexagonal Bravais lattices. The basal plane of a hexagonal lattice is the plane defined by the two *a* axes.

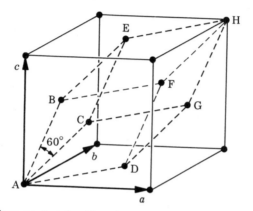

**FIGURE 6.6** Face-centered cubic point lattice referred to a rhombohedral primitive cell. The FCC unit cell is shown by the solid lines. The rhombohedral primitive cell is outlined by the dashed lines. Points A and H are on the corners of the FCC cell as well as of the rhombohedral cell; points B, C, D, E, F, and G are on the corners of the rhombohedral cell but on the faces of the FCC cell. [After Cullity (1956).]

A number of commonly used metals in IC technology belong to the FCC crystal structure, such as Al ($a$ = 0.405 nm) and Au ($a$ = 0.408 nm). Many semiconductors have a diamond cubic structure, which is not one of the Bravais lattices. The diamond structure can be considered to be two interpenetrating FCC lattices with two atoms associated with one lattice point. The diamond structure is therefore based on the a face-centered cubic Bravais lattice with eight atoms per unit cell, as shown in Fig. 6.7a.

The crystal structure of Si is diamond cubic with a lattice parameter $a$ of 0.543 nm at room temperature. Other group IVB elements, such as Ge and Sn (gray tin), also have the diamond structure. These elements have four outer-shell electrons, readily available to share electrons with four neighboring atoms, resulting in highly directional covalent bonds (Fig. 6.7b). The number of atoms/unit cells of the diamond lattice is given by $N_i$ = 4, $N_f$ = 6, and $N_c$ = 8, so that

$$N_u = 4 + \frac{6}{2} + \frac{8}{8} = 8 \text{ atoms/unit cell.} \tag{6.4b}$$

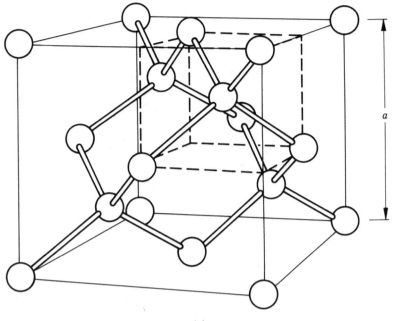

(a)

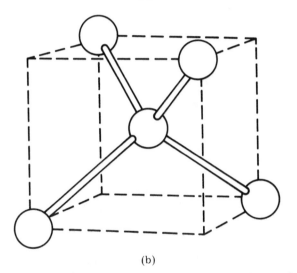

(b)

**FIGURE 6.7** (a) Diamond lattice. The atoms are connected by covalent bonds. The cube outlined by the dashed lines shows one tetrahedral unit. (b) Tetrahedral unit of the diamond lattice.

The number of atoms/cm³, $N$, is given by

$$N = \frac{\text{atoms/unit cell}}{a^3}.$$  (6.4c)

To determine the number of atoms/cm$^2$ on a crystal surface, take the number of surface atoms/unit cell for a given orientation and divide by the surface area.

$$N'_s = \frac{\text{atoms}}{\text{cm}^2} = \frac{\text{surface atoms/unit cell}}{\text{surface cell area}}. \qquad (6.4d)$$

For example, the 100 surface of the Si lattice, Fig. 6.7, has an area $a^2$ and 2 atoms/unit cell (4 corner atoms shared with 4 adjacent cells so each contributes $\frac{1}{4}$ and one center atom). Then $N'_s(100) = 2/a^2$. The 111 surface of the Si lattice is a triangle of length $a\sqrt{2}$ (area $= \sqrt{3}a^2/2$) with 3 corner atoms (each shared by 6 adjacent cells) and 3 side atoms (each shared by 2 adjacent cells). The number of surface atoms/cell is $(\frac{3}{6} + \frac{3}{2}) = 2$ atoms/unit cell ($N'_s(111) = 2.3/a^2$). The 110 surface has the highest number of atoms/cm$^2$ with 4 atoms/cell and a surface cell area of $\sqrt{2}a^2$ ($N'_s(110) = 2.8/a^2$).

Since group IVB elements have a diamond lattice, it is conceivable that a compound between groups III and V elements or between groups II and VI elements may also have a crystal structure similar to the diamond lattice. As it turns out, III–V compounds such as GaAs and AlSb and II–VI compounds such as ZnS do indeed have a crystal structure very similar to the diamond lattice. This structure is called the zinc blende structure, shown in Fig. 6.8. In the zinc blende structure, the interior atoms of the cubic unit cell are different from those at the corners; other than that, the zinc blende structure is identical in atom position and stacking sequence to the diamond lattice. From Fig. 6.8 it can be seen that one of the FCC sublattices of zinc blende consists of atoms of one element (e.g., Ga), and the other FCC sublattice consists of atoms of a different element (e.g., As).

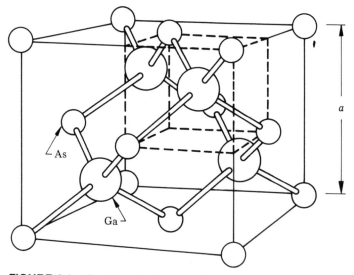

**FIGURE 6.8**   Zinc blende structure, showing Ga and As atoms.

## 6.3 Directions and Planes

Given a unit cell, the position of any lattice point in a cell may be described in terms of its coordinates; if the vector for the origin of the unit cell to the given point is $x\mathbf{a} + y\mathbf{b} + z\mathbf{c}$, where $x$, $y$, and $z$ are fractions, then the coordinates of the given point are $x$, $y$, and $z$. The direction of any line in a lattice may be described by drawing a line through the origin parallel to the given line and then assigning the coordinates of any point on the line. Let's say that the line goes through the origin and the point with coordinates $u$, $v$, $w$, where these numbers are not necessarily integers. The direction $[uvw]$, written in brackets, are the indices of the direction of the line. Since this line also goes through $2u$, $2v$, $2w$ and $3u$, $3v$, $3w$, and so on, it is customary to convert $u$, $v$, $w$ to a set of smallest integers by multiplication.

**EXAMPLE 1**

Find the coordinates of points A, B, C, and D and the directions of the lines going through them in Fig. 6.9.

**Solution:** We can define point A as the origin of the lattice; therefore, the coordinates of point A is 000. Point B has the coordinates of 1 $\frac{1}{2}$ 0, since point B can be brought from point A, the origin, by a vector $1\mathbf{a} + \frac{1}{2}\mathbf{b} + 0\mathbf{c}$. The direction of the line AB is [1 $\frac{1}{2}$ 0], or more conventionally, [210]. The coordinates of C are 100, and the direction of line AC is [100], which is also the vector $\mathbf{a}$, the crystallographic axis. The coordinates of D are $\frac{1}{3}$ 1 1, relative to point A. The direction of the line CD, however, is not [133]. Had a line been

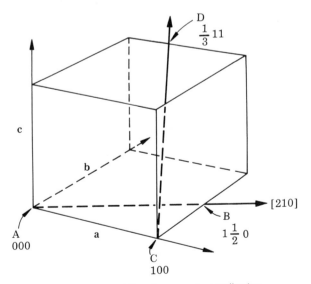

**FIGURE 6.9** Directions and coordinates

**TABLE 6.1** Conventions Used to Indicate Directions and Planes in Crystallographic Systems

A. Directions: Line from origin to point at $u$, $v$, $w$
   1. Specific directions are given in brackets, $[uvw]$.
   2. Indices $uvw$ are the set of smallest integers.
      $[\frac{1}{2} \frac{1}{2} 1]$ goes to $[112]$.
   3. Negative indices are written with a bar, $[\bar{u}vw]$.
   4. Directions related by symmetry are given by $<uvw>$.
      $[111]$, $[1\bar{1}1]$, and $[\bar{1}11]$ are all represented by $<111>$.

B. Planes: Plane that intercepts axes at $1/h$, $1/k$, $1/l$
   1. Orientation is given by parentheses, $(hkl)$.
   2. $hkl$ are Miller indices.
   3. Negative indices are written with a bar, $(\bar{h}kl)$.
   4. Planes related by symmetry are given by $\{hkl\}$.
      $(100)$, $(010)$, and $(\bar{1}00)$ are planes of the form $\{100\}$.

C. In cubic systems: bcc, fcc, diamond
   1. Direction $[hkl]$ is perpendicular to plane $(hkl)$.
   2. Interplanar spacing: $d_{hkl} = a/\sqrt{h^2 + k^2 + l^2}$.

---

drawn between A and D, the direction of the line AD would be $[133]$. The line CD is drawn between C and D; the direction indices must therefore be modified accordingly. This can be done by drawing a line from A parallel to the line CD. An equivalent operation would be to shift the origin from A to C. The coordinates of D relative to C are $-\frac{2}{3}$ 1 1; this can be seen by traveling from C to D via the three crystallographic axes. Note that the direction of the line CA is $[\bar{1} 0 0]$. The direction of the line CD is therefore $[-\frac{2}{3} 1 1]$ or $[\bar{2}33]$. The bar on top of the number in brackets means that it is negative.

---

The orientation of planes in a lattice can also be defined by a set of numbers, called the Miller indices. We can define the Miller indices of a plane as the reciprocals of the fractional intercepts that the plane makes with the crystallographic axes. Table 6.1 gives conventions used to indicate directions and planes used in crystallographic systems. In these conventions, if the plane is parallel to the axis, the intercept is taken to be at infinity, $\infty$; the reciprocal of $\infty$ is $1/\infty = 0$.

## EXAMPLE 2

Find the Miller indices of the plane shown in Fig. 6.10.

**Solution:** To find the Miller indices, the intercepts need to be defined.

|  | a | b | c |
|---|---|---|---|
| The intercepts | 3 | 2 | 2 |
| The reciprocals | $\frac{1}{3}$ | $\frac{1}{2}$ | $\frac{1}{2}$ |

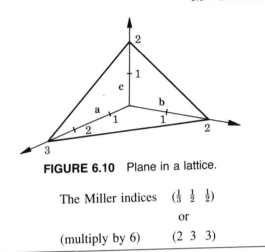

**FIGURE 6.10** Plane in a lattice.

The Miller indices $(\frac{1}{3} \ \frac{1}{2} \ \frac{1}{2})$

or

(multiply by 6)  (2 3 3)

Figure 6.11 shows the Miller indices of some lattice planes. The planes are conventionally written as $(hkl)$ and directions as $[uvw]$. For the cubic system, the direction $[hkl]$ is perpendicular to the plane $(hkl)$. The plane $(hkl)$ is parallel to the $(\overline{hkl})$ plane, which is on the opposite side of the origin (see Table 6.1).

The dot product of $(hkl) \cdot [uvw]$ is the sum: $hu + kv + lw$. If the dot product of $(hkl) \cdot [uvw] = 0$, the direction $[uvw]$ lies on the plane $(hkl)$.

**EXAMPLE 3**

Determine if the plane (111) contains the following directions: [100], [$\overline{2}$11], and [$\overline{1}$10].

**Solution:** If the plane contains a direction, that direction should lie on the given plane. To check this, we take the dot products:

$(111) \cdot [100] = 1$  [100] does not lie on (111)

$(111) \cdot [\overline{2}11] = 0$  [$\overline{2}$11] lies on (111)

$(111) \cdot [\overline{1}10] = 0$  [$\overline{1}$10] lies on (111)

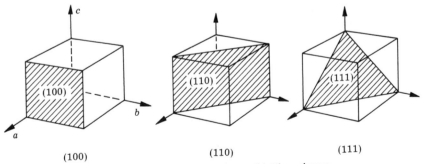

**FIGURE 6.11** Miller indices of lattice planes.

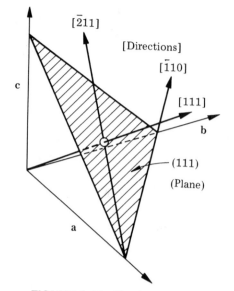

**FIGURE 6.12** The (111) plane.

## EXAMPLE 4

Show that the [111] direction is perpendicular to the (111) plane (Fig. 6.12).

**Solution:** From Example 3 we know that [$\bar{2}$11] and [$\bar{1}$10] lie on (111). Since both of these two vectors are perpendicular to the [111] vector (the dot products are zero), the [111] direction must be perpendicular to the (111) plane.

Table 6.2 gives distances between atoms and planes in the cubic system and Fig. 6.13 shows the lattice configuration for the zinc blende structure containing A and B atoms. With the knowledge of crystallography summarized above and in

**TABLE 6.2** Distances Between Atoms and Planes in the Cubic System with Lattice Parameter $a$, Atomic Density $N$, and Number $N_u$ of Atoms per Unit Cell

| | Simple | bcc | fcc | Diamond |
|---|---|---|---|---|
| **A. Distance between Atoms along Crystal Axis Directions** | | | | |
| $<100>$ | $a$ | $a$ | $a$ | $a$ |
| $<110>$ | $\sqrt{2}a$ | $\sqrt{2}a$ | $a/\sqrt{2}$ | $a/\sqrt{2}$ |
| $<111>$ | $\sqrt{3}a$ | $\sqrt{3}a/2$ | $\sqrt{3}a$ | $\sqrt{3}a/4$ and $3\sqrt{3}a/4$ |
| **B. Interplanar Spacings along Planar Directions** | | | | |
| $\{100\}$ | $a$ | $a/2$ | $a/2$ | $a/4$ |
| $\{110\}$ | $a/\sqrt{2}$ | $a/\sqrt{2}$ | $a/2\sqrt{2}$ | $a/2\sqrt{2}$ |
| $\{111\}$ | $a/\sqrt{3}$ | $a/2\sqrt{3}$ | $a/\sqrt{3}$ | $a/4\sqrt{3}$ and $3a/4\sqrt{3}$ |

$N = N_u/a^3$ with $N_u$ = 1, 2, 4, and 8 for simple cubic, bcc, fcc, and diamond lattices, respectively.

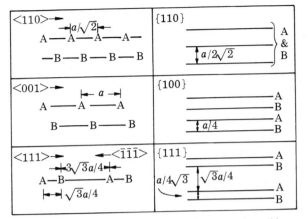

**FIGURE 6.13**   Lattice configuration for the zinc blende lattice with *a* as the lattice constant with A and B denoting Ga and As atoms in GaAs.

Fig. 6.8, it can be seen that the zinc blende unit cell (using GaAs as an example) can be generated by constructing a Ga FCC sublattice with the origin assigned at 000, followed by constructing an As FCC sublattice with an origin at $\frac{1}{4}\frac{1}{4}\frac{1}{4}$ of the Ga sublattice. It can also be seen in Fig. 6.13 that the atomic layer sequence viewed along the [111] direction is different from that viewed along the [$\bar{1}\bar{1}\bar{1}$] direction. This variation in sequences causes significantly different surface properties between the (111) surface and the ($\bar{1}\bar{1}\bar{1}$) surfaces, referred to as A and B faces. The (111) Ga face (A face) has gallium atoms with no free electrons. The ($\bar{1}\bar{1}\bar{1}$) As face [sometimes it is just called the (111) As face without the bars, the B face] consists of arsenic atoms with two free electrons (Ghandhi, 1983). The (111) As face is thus electronically more active than the (111) Ga face. Etching of the (111) As face occurs very rapidly, with a resultant smooth polish. The (111) Ga face etches very slowly, so that all imperfections become delineated. The (111) As face oxidizes more readily and has a higher As vapor pressure ($T < 770°C$) than the (111) Ga face.

## 6.4 Crystalline Defects

As mentioned before, defects in a crystalline material may be categorized as (1) point defects, (2) line defects, (3) planar defects, and (4) volume defects. Table 6.3 lists some of the examples of these four types of defects, and Fig. 6.14 shows schematically some of the point defects in a two-dimensional simple cubic lattice.

### 6.4.1 Point Defects

A point defect is a deviation in the periodicity of the lattice arising from a single point. Other defects, such as dislocations and stacking faults, extend over many lattice sites. Point defects may exist as a result of thermal equilibrium, whereas

**TABLE 6.3** Crystalline Defects in Semiconductors

| Defect Type | Examples |
|---|---|
| Point | Vacancies, interstitials, impurity atoms, antisite defects, extended point defects [not as localized as a simple point defect (e.g., an interstitial atom can form a defect complex with one or more neighboring atoms)] |
| Line | Dislocations |
| Plane | Stacking faults, twins, and grain boundaries |
| Volume | Precipitates and voids |

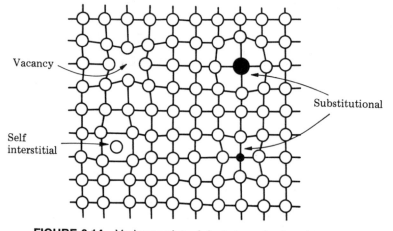

**FIGURE 6.14** Various points defects in a simple cubic lattice.

other types of defects form due to nonequilibrium conditions such as occur during crystal growth or during thermal or mechanical processing of the material. Point defects may be categorized as (1) native defects such as a vacancy, and (2) impurity-related defects due to the introduction of an impurity atom into the lattice. For semiconductors, point defects not only cause structural disturbances, but also often introduce electronic states in the band gap. If an attractive potential exists between a native defect and an impurity atom, they may interact and form a defect complex, such as a vacancy–impurity pair.

## 6.4.2 Native Defects and Shallow Dopants

For Si, there are three types of native defects: the vacancy, the interstitial, and the interstitialcy. The vacancy is an empty lattice site, referred to as $V$ in the literature. Depending on the configuration of the unsatisfied bonds due to the missing atom, a vacancy in Si can be either neutral or negatively or positively charged. A vacancy is also referred to as a Schottky defect. A Si atom residing in the intersticies of the Si lattice is defined as a self-interstitial. A Frenkel pair is a vacancy–interstitial pair formed when an atom is displaced from a lattice site to an interstitial site. An interstitialcy consists of two atoms in nonsubstitutional positions configured about a single lattice site. Because of the similarity between an interstitial and an interstitialcy, a distinction is generally not made; both are identified as $I$ in the literature. When an impurity such as a shallow dopant (e.g., As in Si) occupies a lattice site, it is known as a substitutional defect. Such a defect when surrounded only by Si atoms on their normal sites is identified as $A$. When a vacancy $V$ forms next to $A$, it is known as an impurity vacancy pair, usually referred to as an $AV$ defect. If one of the atoms in the interstitialcy defect is a dopant atom, the defect is referred to as an $AI$ defect. If the impurity atom occupies an interstitial site, it is designated as $A_i$; if the impurity atom occupies a substitutional site, it is designated as $A_s$.

For GaAs and other III–V compounds and alloys, it should be noted that defects may occupy lattice sites either in the Ga (group III) sublattice or in the As (group V) sublattice or both. For example, an empty As lattice site is referred to as the $V_{As}$ defect. A Si atom occupying a Ga site is referred to as the $Si_{Ga}$ defect. Since Si is a group IV element substituting for Ga, a group III element, this $Si_{Ga}$ defect is an $n$-type dopant. A Si atom occupying an As site is referred to as a $Si_{As}$ defect, which is a $p$-type dopant for GaAs. For Si atoms diffused into GaAs, or implanted Si atoms after thermal activation, the Si atoms tend to occupy the sites in the Ga sublattice preferentially. As a result of the concentration of $Si_{Ga}$ being greater than the concentration of $Si_{As}$, Si is generally an $n$-type dopant for GaAs and its alloys. It has been observed that Ge behaves similarly to Si, but carbon is found to be an acceptor in GaAs. When the concentration of $Si_{As}$ becomes comparable to that of $Si_{Ga}$, the semiconductor is said to have been compensated and the electrical conductivity of the semiconductor decreases. It is also possible for a Ga atom to be located on an As site, $Ga_{As}$, or an As atom on a Ga site, $As_{Ga}$. These defects are called antisite defects. It has been suggested that the commonly observed deep-level donor center, EL2, in GaAs is due to the antisite defect $As_{Ga}$ (Lagowski et al., 1982; Kirkpatrick et al., 1985). The other antisite defect, $Ga_{As}$, is believed to be a double acceptor (i.e., can trap two holes).

## 6.4.3 Thermodynamics of Point Defects

Native point defects exist under thermodynamic equilibrium conditions due to the increase in configurational entropy of the crystal which their presence creates. This can be seen by considering the configurational entropy of mixing between vacancies and atoms in a lattice starting from the Boltzmann expression of entropy:

$$S = k \ln W, \tag{6.5}$$

where $S$ is the entropy, $W$ the number of ways for arranging the defects in a crystal, and $k$ the Boltzmann factor. Assuming that there are $n$ vacancies in a crystal of $N$ lattice sites, we obtain

$$W = \frac{N(N - 1) \cdots (N - n + 1)}{n!} \tag{6.6}$$

$$= \frac{N!}{(N - n)!\, n!}.$$

Using the Stirling's approximation of $\ln x! \simeq x \ln x; x \to$ large

$$S = k \left[ N \ln N - (N - n) \ln (N - n) - n \ln n \right]. \tag{6.7}$$

The free energy of a system is

$$G = H - TS \tag{6.8}$$

where $G$ is the Gibbs free energy, $H$ the enthalpy, $T$ the absolute temperature, and $S$ the entropy. The change of the free energy of the lattice due to the presence of the vacancies is

$$\Delta G = nH_v^f - T(S + nS_v^f) \tag{6.9}$$

where $H_v^f$ is the enthalpy of formation of a vacancy and $S_v^f$ is the vibrational entropy, which is generally small compared to $S$ [eq. (6.5)]. Under thermodynamic equilibrium,

$$\frac{\partial \Delta G}{\partial n} = 0. \tag{6.10}$$

From eq. (6.9) we have

$$\frac{\partial \Delta G}{\partial n} = H_v^f - TS_v^f - T\frac{\partial S}{\partial n} = 0. \tag{6.11}$$

From eq. (6.7) we have

$$\frac{\partial S}{\partial n} = k \ln \frac{N - n}{n}.$$

Substituting into eq. (6.11) yields

$$H_v^f - TS_v^f - kT \ln \frac{N - n}{n} = 0$$

or

$$\frac{n}{N - n} = \exp\left(\frac{S_v^f}{k}\right) \exp\left(-\frac{H_v^f}{kT}\right). \tag{6.12}$$

Since $N \gg n$, eq. (6.12) may be written as

$$\frac{n}{N} = \exp\left(\frac{S_v^f}{k}\right) \exp\left(-\frac{H_v^f}{kT}\right). \tag{6.13}$$

The derivation above is for uncharged vacancies (neutral vacancies); therefore, the neutral vacancy concentration $[V^x]$ is

$$[V^x] = N \exp\left(\frac{S_v^f}{k}\right) \exp\left(\frac{-H_v^f}{kT}\right). \tag{6.14}$$

We use the bracket notation to indicate concentrations. For metals, $H_v^f$ is between 0.5 and 1 eV. The rule of thumb is: $H_v^f \simeq 10kT_m$ ($T_m$ = melting point in K). For

Si, $H_v^f \simeq 2.4$ eV and for Ge, $H_v^f \simeq 2.2$ eV. The term exp $(S_v^f/k)$ is of the order of 1. [$S_v^f \simeq 1.1 \, k$ for Si, so that exp (1.1) is a factor of 3.] The equilibrium concentration of neutral vacancies in silicon is independent of the position of the Fermi level, and with $N = 5.0 \times 10^{22}/\text{cm}^3$ (the concentration of Si lattice sites),

$$[V^x] = 5.0 \times 10^{22} \times 3 \times e^{-2.4/kT} = 1.5 \times 10^{23} e^{-2.4/kT} \qquad (6.14a)$$
$$= 1.3 \times 10^{13}/\text{cm}^3 \text{ at 1200 K.}$$

There is a major difference between defects (vacancies and interstitials) in semiconductors and those in metals; that is, defects in a semiconductor can be charged electrically, whereas defects in a metal are considered neutral. Since defects can be charged (or ionized), the concentration of these defects becomes a function of the Fermi level position in the semiconductor. Let us use the charged states of vacancies in Si as an example. It is generally accepted that the single vacancy in Si can have four charge states (Van Vechten, 1980): $V^+$, $V^x$, $V^-$, and $V^=$, where $+$ refers to a donor state, $x$ a neutral species, and $-$ an acceptor state. Figure 6.15 shows the energy levels of the charged vacancies in the Si band gap at 0 and 1400 K, $V^+$ state has an energy level $E^+$, $V^-$ has an energy level $E^-$, and so on. These energy levels are estimated values but are sufficiently accurate for demonstration purposes.

The concentration of charged vacancies is governed by the Fermi–Dirac statistics (see Chapter 12) and is given by (using $V^-$ as an example)

$$[V^-] = \frac{[V_T]}{1 + e^{E^- - E_F)/kT}} \qquad (6.15)$$

where $E_F$ is the Fermi level, $E^-$ is the energy level of the acceptor state vacancy, and $[V_T]$ is the total vacancy concentration:

$$[V_T] = [V^x] + [V^-] + [V^=] + [V^+]. \qquad (6.16)$$

After some algebraic manipulation of eqs. (6.15) and (6.16) (see the problems) we find that

$$[V^-] = [V^x] \exp [(E_F - E^-)/kT] \qquad (6.17)$$

and similarly,

$$[V^+] = [V^x] \exp [(E^+ - E_F)/kT]. \qquad (6.18)$$

Upon examining eq. (6.17), we find that the concentration $[V^-]$ is large compared to $[V^x]$ only when $(E_F - E^-) \gg kT$ and positive. This means that $[V^-]$ is significant when the Si sample is strongly $n$-type (i.e., $E_F$ is located far above $E^-$). As the Si sample becomes less $n$-type and $E_F$ goes below $E^-$, $[V^-]$ decreases accordingly. For $p$-type samples where $E_F$ is far below $E^-$, $[V^-]$ is negligible small compared to $[V^x]$. Figure 6.16 shows the graphical representation of $[V^-]$, $[V^=]$, and $[V^+]$ as a function of the Fermi level $E_F$ for two temperatures. According to eq. (6.18) and Fig. 6.16, the concentration $[V^+]$ is large compared to $[V^x]$ when $E_F$ is located far below $E^+$ (i.e., strongly $p$-type). $[V^+]$ becomes negligibly small for $n$-type Si.

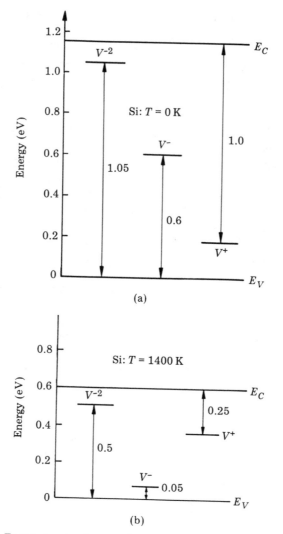

**FIGURE 6.15** Energy levels of vacancies of Si at (a) 0 K and (b) 1400 K. [After Van Vechten (1980).]

## EXAMPLE 5

Calculate the concentration $[V^-]$ at 1400 K in an $n$-type Si doped with a donor density of $2.8 \times 10^{20}/\text{cm}^3$.

**Solution:** We start by calculating the Fermi level of this sample at 1400 K. From eq. (3.10) we have

$$E_F = E_C + kT \ln \frac{n}{N_C}.$$

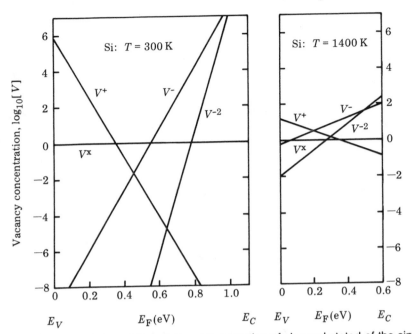

**FIGURE 6.16**  Variation of the relative concentration of charged stated of the single Si vacancy with position of $E_F$ at two temperatures. [After Van Vechten (1980).]

The value of $n$ is taken to be the donor density ($2.8 \times 10^{20}/\text{cm}^3$). The value of $N_C$ at 1400 K can be calculated using eq. (3.6) and Table 3.3:

$$N_C(1400 \text{ K}) = \frac{N_C (300 \text{ K})(1400)^{3/2}}{(300)^{3/2}} = 2.8 \times 10^{20} \text{ cm}^{-3}$$

We see that $N_C$ is just equal to $N_D$. We may then take $E_F$ to be located at $E_C$ as suggested by eq. (3.10).

At 1400 K, the band gap of Si is reduced from 1.15 eV at 300 K to 0.6 eV at 1400 K (Van Vechten, 1980). The energy level of $E^-$ tracks the band gap and is located at 0.05 eV $+ E_V$ at 1400 K (Fig. 6.15). The concentration $[V^-]$ is given by eq. (6.17):

$$[V^-] = [V^x] \exp\left(\frac{E_F - E^-}{kT}\right) = [V^x] \exp\left(\frac{E_C - 0.05 - E_V}{kT}\right)$$

$$= [V^x] \exp\left(\frac{E_G - 0.05}{kT}\right) = [V^x] \exp\left(\frac{0.55}{0.12}\right)$$

$$= 98[V^x].$$

Therefore, $[V^-]$ is about 100 times higher than $[V^x]$ at 1400 K, in agreement with the result shown in Fig. 6.16. The equilibrium concentration of neutral vacancy $[V^x]$ at 1400 K is

$$[V^x] = 1.5 \times 10^{23} \exp(-2.4/kT) = 3.4 \times 10^{14}/\text{cm}^3.$$

Therefore,

$$[V^-] = 3.3 \times 10^{16}/\text{cm}^3.$$

It should be noted that the creation of ionized vacancies does not diminish the equilibrium concentration of neutral vacancies $[V^x]$. More neutrals are simply created to maintain their concentration at the level indicated by eq. (6.14), and the total concentration of vacancies, $[V_T]$, increases accordingly.

Generalizing from the vacancy concentrations [eqs. (6.14) to (6.18)], the equilibrium concentrations of any charged native defect, $X$, may be related to that of the neutral state:

$$\frac{[C_x^-]}{[C_x^0]} = \exp\left(-\frac{E_x^- - E_F}{kT}\right) \tag{6.19}$$

$$\frac{[C_x^+]}{[C_x^0]} = \exp\left(-\frac{E_F - E_x^+}{kT}\right) \tag{6.20}$$

where $[C_x^0]$ is the concentration of the neutral defects, $[C_x^-]$ and $[C_x^+]$ are the concentrations of the charged defects with energy level $E_x^-$ and $E_x^+$, respectively, and $E_F$ the Fermi level. Equations (6.19) and (6.20) are general for dilute doping concentrations. If the charged defects are vacancies, eqs. (6.19) and (6.20) reduce to eqs. (6.17) and (6.18). When the doping concentration increases to high concentrations such that dopant–dopant and dopant–defect interactions cannot be ignored, these equations may no longer hold. Point defects play very significant roles in solid-state diffusional processes and the issues are addressed in Chapter 7.

## 6.4.4 Deep Level Centers

Shallow dopant impurities have small ionization energies (such as As and P in Si with ionization energies of ~0.04 eV). There are chemical impurities and charged point defects that form deep energy states in semiconductors. The energy levels of these centers in the band gap are usually far away from the band edges; thus they are called deep level centers. Typical deep level impurities are oxygen and metallic elements in Si and in GaAs. A deep level impurity may have several energy levels, with each energy level being either an acceptor state or a donor state. For example, Au and Si has an acceptor state at 0.54 eV from the conduction band edge, $E_C$, and a donor state 0.35 eV from the valence band edge, $E_V$ (Fig. 6.17). Other than chemical impurities, charged point defects such as $V^-$ in Si, and $V_{Ga}$ and antisite defects in GaAs, also form deep level states. It should be noted that there are many deep levels centers in III–V compounds and alloys whose physical origins have not been clearly identified. A typical example is the DX center in AlGaAs. These centers are found to be deep donors of unknown origin—therefore, DX centers.

A deep center may act either as a trap or as a recombination center, depending on the impurity, temperature, and other doping conditions. Consider a minority

Au levels in Si

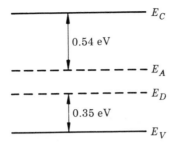

**FIGURE 6.17**   Au in Si has an acceptor state, $E_A$ ($\approx E_C - 0.54$ eV) and a donor state, $E_D$ ($\approx E_V + 0.35$ eV). For strongly p-type Si, the Fermi level $E_F$ is below $E_D$; the donor states are ionized (the electron in the state is given away) and become positively charged. The acceptor state at $E_A$ is not ionized (the state has not received an electron because $E_F$ is far away below $E_A$). For strongly n-type Si, $E_F$ is far above $E_A$; the acceptor states are ionized (the state has received an electron) and become negatively charged. The Au donor states are not ionized in strongly n-type Si.

carrier (holes in an n-type semiconductor, for example) captured at an impurity center. The minority carrier stays at the center for a period of time and is then ejected thermally into the band from which it came; this impurity center is known as a trap. Before ejection, if a majority carrier is captured, recombination of the carrier pair takes place and the center is therefore a recombination center.

For deep centers to be effective recombination centers, they should be (1) singly charged and attractive to the minority carriers and (2) neutral after the capture of a minority carrier. Capture of majority carriers by neutral centers is highly probable because of the large number of majority carriers. For the impurity centers to be effective traps, they should be (1) doubly charged and attractive to the minority carriers and (2) repulsive to the majority carriers, after capturing one minority carrier. Before capturing of majority carrier (not a very probable event because of the small capture cross section), the minority carrier is ejected. The capture efficiency of carriers by these impurities depends on the capture cross section, which is a measure of the probability of the capturing event. The cross section of attractive centers is $\approx 10^{-12}$ to $10^{-15}$ cm$^2$, the cross section of neutral centers is $\approx 10^{-15}$ to $10^{-17}$ cm$^2$, and the cross section of repulsive centers is $\approx 10^{-22}$ cm$^2$.

As mentioned previously, Au can be either a deep acceptor or a deep donor in Si (Au can be neutral as well). These impurities are called amphoteric dopants. For example, substitutional In is a deep acceptor in Si, interstitial In in Si is a deep donor. Si in GaAs is a donor when occupying a vacant Ga site, and an acceptor when occupying a vacant As site. These elements can be either donors or acceptors, depending on their lattice location. Let's consider a Si wafer containing Au. In strongly p-type Si, Au donors are ionized and become positively charged, thus attractive to the minority carriers, that is, the electrons (Fig. 6.17). After the capture of an electron, the Au donor state becomes neutral and is now effective in capturing the majority carriers, the holes. The donor states of Au are, therefore, recombination centers in $p$-type Si. If more shallow dopants (phosphorus, for example) are introduced into this sample to make it less $p$-type, the Fermi level will rise. As $E_F$ crosses $E_D$, the Au donor states are not likely to be ionized and cease to be electron traps. Further increase in shallow $n$-type dopants causes $E_F$ to rise above $E_A$. The Au acceptor states are now ionized and become negatively charged, thus effective in capturing holes. After capturing the minority carriers, the centers become neutral and ready for the capture of the majority carriers, the electrons. In strongly $n$-type Si, the Au acceptor states are effective recombination centers.

Since deep level impurities can be effective recombination centers, the presence of these defects may reduce the minority-carrier lifetime (see Section 3.8). A long minority-carrier lifetime is usually beneficial to device operations. It is sometimes necessary to reduce the amount of deep impurities by "gettering" processes.

The detection and characterization of deep level defects are difficult even for high-resolution electron microscopy (HREM). An experimental technique known as deep level transient spectroscopy (DLTS) has been developed to characterize these defects (Lang, 1974). One convenient way to perform DLTS is to measure the capacitance change of a sample when a pulse of charge is injected into the sample, normally under a reverse bias. From the change of the capacitance as a function of temperature, the defect density and defect energy level may be deduced.

## 6.5 Line Defects

Line defects in a crystalline material are known as dislocations. In contrast to point defects, dislocations are formed due to nonequilibrium conditions such as thermal and mechanical processing and epitaxy. Under equilibrium conditions, there is no requirement for the presence of dislocations or any other defects (except native point defects) in the crystal. For this reason it is possible to grow dislocation free single crystals of semiconductors. For example, Si wafers contain typically less than 100 dislocations/cm$^2$, compared to $10^8$ dislocations/cm$^2$ or more in the metal interconnect lines on a chip. Chemical etching of the Si wafer reveals the dislocations at the surface in the form of small pits (Jenkins, 1977). The etch pit density (EPD) is used as a means to indicate wafer quality, with values less than $10^2$/cm$^2$ indicative of high quality. Defect etches are also used in GaAs (Stirland, 1988).

A dislocation may be created in the following manner in a simple cubic lattice (see Fig. 6.18): Make a cut on the crystal along any plane. Let the crystal on one side of cut shift by a vector parallel to the cut surface relative to the other half (along line AA in Fig. 6.18a). Then join the atoms on either side of the cut. A *screw* dislocation is thereby created (Fig. 6.18a). If the shift is perpendicular to the cut, an *edge* dislocation is created (Fig. 6.18b). If the shift is neither parallel nor perpendicular to the cut, the dislocation is called a mixed dislocation. An edge dislocation may be viewed also as having an extra half-plane inserted into the crystal (see Fig. 6.19). In addition to the edge, screw, and mixed dislocations, any dislocation may be either a perfect dislocation or an imperfect (or partial) dislocation. For a perfect dislocation the shift or translational vector connects identical

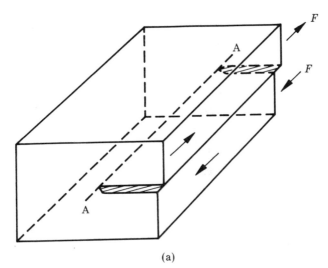

(a)

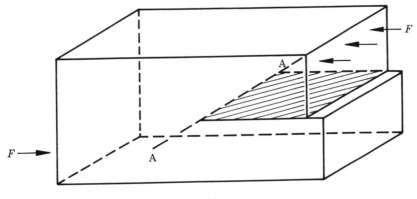

(b)

**FIGURE 6.18** (a) Parallel shift to line AA—screw dislocation; (b) perpendicular shift to line AA—edge dislocation.

lattice points. For an imperfect (or partial) dislocation, the translational vector does not end on an identical lattice point.

The translational or shift vector of a dislocation is commonly known as a Burgers vector, **b.** The Burgers vector of an edge dislocation is readily determined by drawing a Burgers circuit around an edge dislocation. The Burgers vector **b** is perpendicular to the dislocation line for an edge dislocation (Fig. 6.20). For a screw dislocation, the Burgers vector **b** is parallel to the dislocation line.

Several properties of a dislocation should be noted (Read, 1953):

1. For a given dislocation, there is only one Burgers vector, no matter what shape the dislocation line has.

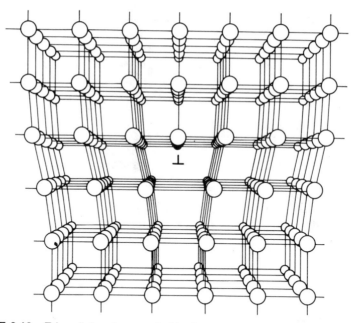

**FIGURE 6.19** Edge dislocation created by inserting an extra half-plane of atoms.

2. A dislocation must end on itself, thus forming a loop, or on other dislocations, thus forming a network, or on surfaces such as an external surface or a grain boundary.
3. In general, dislocations in real crystals form three-dimensional networks. The sum of the Burgers vectors at the node or point of junction of the dislocation is zero.
4. The slip (or glide) plane of a dislocation contains the dislocation line and its Burgers vector.

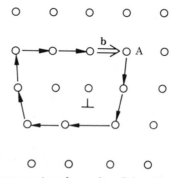

**FIGURE 6.20** The Burgers vector of an edge dislocation is determined by drawing a Burgers circuit around a dislocation. Look along the dislocation (the symbol ⊥ represents a dislocation going into the plane of the page) and draw a circuit (starting from atom A) in the clockwise direction. The Burgers vector **b** is the vector to close the circuit.

**5.** When a dislocation slips on its slip plane, the atoms move in the direction of the Burgers vector, but the dislocation line moves normal to itself on the slip plane.

Dislocations in diamond cubic and zinc blende lattice usually lie along $<110>$ directions due to the directionality of the bonds [the directionality is sometimes referred to as the Peierls force in a crystal; Hirth and Lothe (1968)]. The Burgers vector is of the $(a/2) <110>$ type (i.e., a Burgers vector with magnitude of $a/2$ directed along the $<110>$ direction). The dislocations in these crystals are therefore usually 60° mixed dislocations or screw dislocations. It should be pointed out that for III–V compounds the unbounded atoms at the dislocation may be on either the III or V element.

## 6.5.1 Force on Dislocations

Consider a crystal containing a single edge dislocation (Fig. 6.21). The crystal has a volume of $D \times l \times H$. In order for the edge dislocation to move, a shear stress must be applied to the crystal. Stresses are defined as force per unit area (N/m²; 1 N/m² = 10 dyn/cm²). Shear stresses are due to forces applied tangentially to the surfaces. Normal stresses, on the other hand, are due to forces applied perpendicular to the surfaces. In Fig. 6.21 the shear forces are applied to the top and bottom surface of the crystal, resulting in a shear stress $\tau$ in the directions indicated by the arrows in the figure. The surfaces on which the shear forces are applied have an area of $D \times l$.

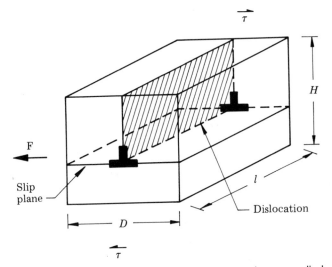

**FIGURE 6.21**  Edge dislocation contained in a crystal under an applied shear stress $\tau$. The crystal has dimensions $D \times l \times H$. The dislocation with length $l$ slips from the right-hand side of the crystal to the left-hand side of the crystal. This motion of the dislocation causes the crystal to slip by a distance $b$.

The work done by an applied shear stress $\tau$ which causes the crystal to slip by a distance $b$ (the Burgers vector) is

$$W_\tau = \tau l D b \tag{6.21}$$

where $W_\tau$ is the work done by the applied shear stress, $l$ the length of the edge dislocation, $b$ the Burgers vector, and $D$ the width of the crystal. The upper half of crystal slips by a distance $b$ relative to the lower half of the crystal when the edge dislocation moves from the right-hand side of the crystal to the left-hand side of the crystal as a result of the applied shear stress.

When a shear stress is applied to the crystal and causes the dislocation to move, there must be an equivalent force acting on the dislocation to cause the dislocation motion. We label as $F_d$ this equivalent force on the dislocation due to the applied stress. The unit of $F_d$ is force per unit length of dislocation (N/m; this unit is not to be confused with that of stress, N/m$^2$). We can understand the meaning of a force on dislocation by noting that $F_d$ is a direct result of an applied shear stress which causes the dislocation to move. Since the dislocation moves due to $F_d$ across the slip plane, the work done by the dislocation, $W_d$, is

$$W_d = F_d l D. \tag{6.22}$$

To check the consistency of units, we note that work done is in units of N-m and $l$ and $D$ are in units of meters, conforming that $F_d$ has the units of N/m.

The force $F_d$ on the dislocation causes the dislocation to travel from the right-hand side to the left-hand side of the crystal. Since this movement of the dislocation across the crystal causes the crystal to slip by an amount $b$, the work done by dislocation is equal to the work done by the shear stress on the crystal:

$$W_d = W_\tau. \tag{6.23}$$

Therefore,

$$F_d = \tau b. \tag{6.24}$$

The force on the dislocation results from the applied shear stress on the crystal, and the resulting force acts normal to the dislocation. For a screw dislocation, $F_d$ is also given by $F_d = \tau b$ and acts normal to the dislocation.

## 6.5.2 Stress Field Around Dislocations

Dislocations are essentially due to atoms out of normal registry. At the core of a dislocation, the atomic arrangement may be quite disruptive. At some distance from the core of the dislocation, the atoms tend to return to registry. As a result of the gradual decay of the disruptive atomic arrangement away far the core, the lattice around a dislocation is strained more severely near the core and less so away from the core. The elastic stress associated with the strain therefore follows the same trend. As in Chapter 1, the strain is $dl/l$.

Consider a screw dislocation lying along an arbitrary direction, $z$ (Fig. 6.22). We assume that the lattice is isotropic where the elastic constants are independent

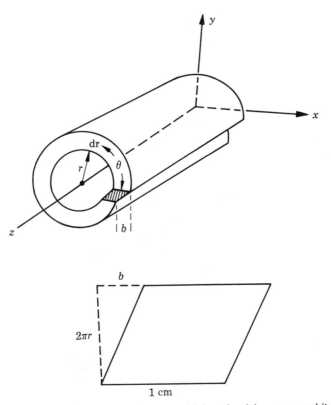

**FIGURE 6.22** Displacement around a screw dislocation lying on an arbitrary $z$ axis. The lower drawing shows the "unfolded" material around the dislocation.

of the orientation of the crystal axis. If we "unfold" a shell of material with thickness $dr$ at a distance $r$ away from the screw dislocation, it is clear that a shear strain is associated with the material surrounding the dislocation due to the Burgers vector. The shear strain is given by

$$\varepsilon_{\theta z} = \frac{b}{2\pi r} \qquad r > b \tag{6.25}$$

where we use the subscript $\theta$ to indicate cylindrical coordinates and $z$ to indicate the direction of displacement.

The shear stress is given by Hooke's law,

$$\sigma_{\theta z} = G\varepsilon_{\theta z} = \frac{Gb}{2\pi r} \tag{6.26}$$

where $G$ is the shear modulus of the material. In isotropic materials, $G = \dfrac{Y}{2(1 + \mu)}$ where $Y$ is Young's modulus and $\mu$ is Poisson's ratio. Transforming the stress field into the Cartesian coordinate system ($x$, $y$, $z$ system), we have

$$\sigma_{zx} = -\frac{Gb}{2\pi}\frac{y}{x^2 + y^2}$$

$$\sigma_{zy} = \frac{Gb}{2\pi}\frac{x}{x^2 + y^2}. \tag{6.27}$$

From these equations we note that the elastic stresses around a dislocation decrease as a function of $1/r$. Near the core of the dislocation where $r \simeq b$, the strains are no longer elastic and these equations do not apply.

## 6.5.3 Energy of Dislocations

The strain energy of the lattice due to the presence of a dislocation is known as the energy of the dislocation. The total strain energy $\Gamma$ is the area of the elastic stress–strain relationship, which obeys Hooke's law in this case, so that $\Gamma = \frac{1}{2}$ stress $\times$ strain. The differential strain energy of the material containing a screw dislocation $\Gamma_s$ at a distance $r$ from the dislocation core is

$$\frac{d\Gamma_s}{dV} = \tfrac{1}{2}\sigma_{\theta z}\varepsilon_{\theta z}. \tag{6.28}$$

From eqs. (6.25) and (6.26), eq. (6.28) can be written

$$\frac{d\Gamma_s}{dV} = \frac{1}{2}\frac{Gb^2}{4\pi^2 r^2} \tag{6.29}$$

where $\Gamma_s$ is the total energy of the screw dislocation per unit length of the screw dislocation and $V$ is the volume of the material.

$$d\Gamma_s = \frac{1}{2}\frac{Gb^2}{4\pi^2 r^2}\,dV = \frac{1}{2}\frac{Gb^2}{4\pi^2 r^2}\,2\pi r\,dr$$

$$= \frac{Gb^2}{4\pi}\frac{dr}{r} \tag{6.30}$$

$$\Gamma_s = \int_{r_c}^{r}\frac{Gb^2}{4\pi}\frac{dr}{r} + U_c \tag{6.31}$$

where $r_c$ is the radius of the dislocation core ($r_c \simeq b$) and $U_c$ is the energy of the dislocation core where elasticity theory does not hold. Upon integration, we have

$$\Gamma_s = \frac{Gb^2}{4\pi}\ln\frac{r}{b} + U_c. \tag{6.32}$$

$U_c$ is estimated to be between $Gb^2/12$ and $Gb^2/5$ (liquid-like atomic arrangement; thus $U_c$ is estimated by the heat of fusion). Assuming that $U_c \simeq 2(Gb^2/4\pi)$, twice the lower limit, we obtain

$$\Gamma_s = \frac{Gb^2}{4\pi}\left(\ln\frac{r}{b} + 2\right). \tag{6.33}$$

We note that the strain energy increases as $\ln r$. For a well-annealed metallic crystal where the dislocation density $\rho$ is about $10^8/cm^2$, the average distance $L$ between two dislocations is $10^{-4}$ cm $\left(L \simeq \dfrac{1}{\sqrt{\rho}}\right)$. If these two dislocations are of the opposite signs, their stress fields approximately cancel each other at middistance (i.e., $r = L$). Using $b = 10^{-8}$ cm, this approximation leads to

$$\Gamma_s = \frac{Gb^2}{4\pi}\left(\ln \frac{10^{-4}\ cm}{10^{-8}\ cm} + 2\right) \tag{6.34}$$

$$\simeq \frac{Gb^2}{4\pi}(12) \simeq Gb^2.$$

For an edge dislocation, the dislocation energy is modified by Poisson's ratio $\mu$ (Poisson's ratio = lateral strain/longitudinal strain). The energy for an edge dislocation $\Gamma_e$ is

$$\Gamma_e = \frac{\Gamma_s}{1 - \mu} \tag{6.35}$$

where $\mu$ is Poisson's ratio. In general, the strain energy

$$\Gamma = \alpha Gb^2 \tag{6.36}$$

where $\alpha$ is between 0.5 and 1. For semiconductors where the dislocation density may be lower than $100/cm^2$, and $L \simeq 0.1$ cm,

$$\Gamma \simeq \frac{Gb^2}{4\pi}\left(\ln \frac{0.1\ cm}{10^{-8}\ cm} + 2\right) \tag{6.37}$$

$$\simeq 1.5Gb^2.$$

It should be noted that $\Gamma$ is not a very sensitive function of the dislocation density due to the logarithmic dependence of $r$ (or the size of the crystal). For Si, $G = 7.5 \times 10^{10}$ N/m$^2$, $b = (a/2)$ [110], and $a = 0.543$ nm, the dislocation energy (in units of eV/length of dislocation) is about 10.4 eV per 0.1 nm. This is a very large amount of energy, and therefore it is very difficult to create dislocations by random thermal fluctuations. Dislocations are created by stresses during thermal and mechanical processing and during crystal growth.

## 6.5.4 Electronic Effects of Dislocations

The single dangling electron at each atomic site along an edge dislocation in Si and Ge is usually considered as an acceptor, attracting an electron to lower its energy form $E_1$ to $E_2$ (Fig. 6.23). Thus the core of the edge dislocation may be negatively charged and repels electrons, thus forming a positively charged space which obstructs the transport of free electrons in the material (Mataré, 1971).

Consider an $n$-type semiconductor with a net donor concentration of ($N_D - N_A$). For a negatively charged dislocation line, a positively charge cylinder of radius $R$ will be formed around the dislocation such that

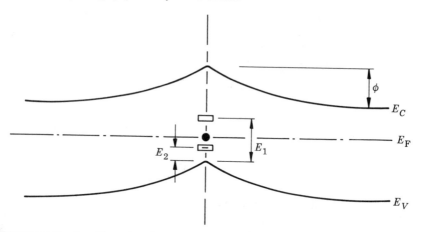

**FIGURE 6.23**  Band bending around an edge dislocation. The dangling electron at the core of the dislocation (running into the page) accepts an electron to lower its energy from $E_1$ (relative to $E_V$) to $E_2$, forming a positively charged space-charge region. The electron barrier $\phi$ is created (band bending) around the dislocation to repel the free electrons in the semiconductor. [After Mataré (1971).]

$$e\pi R^2(N_D - N_A) = \frac{e}{d} = \frac{ef}{c} \qquad (6.38)$$

where $e$ is the electronic charge, $f$ the filling factor (or the probability of an electron occupying the state), $c$ the spacing between the dangling bonds, and $d$ the distance between filled states.

$$f = \frac{c}{d} = \frac{1}{1 + \exp\,[(E_2 - E_F)/kT]}. \qquad (6.39a)$$

The left-hand side of eq. (6.38) is due to the positive charge in the cylinder, and the right-hand side is due to the negative line charge at the dislocation core. The spacing between the dangling bonds, $c$, is

$$c \simeq \frac{a}{\sin \alpha} \qquad (6.39b)$$

where $\alpha$ is the angle between the Burgers vector **b** and the dislocation line and $a$ is the lattice spacing. For a screw dislocation $\alpha = 0$, $c \to \infty$, there are no dangling bonds and a space-charge region does not form. For an edge dislocation, $\alpha = 90°$, $c \simeq a$. Rearranging eq. (6.38) yields

$$R = [d\pi(N_D - N_A)]^{-1/2}$$

$$= \left[\frac{f}{c}\,\frac{1}{\pi(N_D - N_A)}\right]^{1/2} \qquad (6.40)$$

$$= \left\{\frac{[1 + \exp\,((E_2 - E_F)/kT)^{-1}]}{c\pi(N_D - N_A)}\right\}^{1/2}.$$

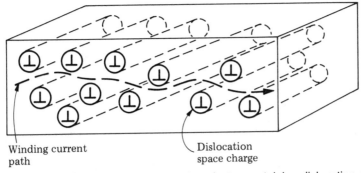

Winding current
path

Dislocation
space charge

**FIGURE 6.24**  Current transport in a semiconductor containing dislocation space-charge cylinders. [After Mataré (1971).]

Assuming that $f \simeq 0.5$ and $c \simeq 10^{-7}$ cm, we have $R \simeq 4$ to 5 μm. With a space-charge region around each dislocation, the transport of carriers normal to the dislocation lines would be affected due to the winding current path, resulting in increased resistivity and reduced carrier mobility. The transport of carriers parallel to the dislocation lines is affected to a lesser amount (see Fig. 6.24).

In addition to the space-charge effects, dislocations may act as generation–recombination centers for carriers. Referring to Fig. 6.23, the dangling bond at the dislocation is negatively charged after accepting an electron to lower its energy. The trapped state ($E_2$) is therefore attractive to the holes. If a hole is captured, the previously captured electron will be annihilated by the captured hole. The dangling bond state goes back up to $E_1$ and the entire process can be repeated (recombination via the dislocation core state). The issue of the recombination–generation processes has been discussed in Chapter 3. The presence of dislocations across a p-n junction causes excessive leakage current, partially due to the recombination process at dislocation core states and partially due to the gettering of metallic impurities along the core, thus forming a conductive path. The electronic behavior of dislocations in GaAs is not firmly established. It is believed that they may be electronically active as recombination centers, similar to those in Si and Ge.

## 6.6  Planar Defects

Planar defects include grain boundaries, stacking faults, and twins. These defects are formed during crystal growth and/or during thermal and mechanical processing of the semiconductor. All three types of planar defects are enclosed by a single dislocation or an array of dislocations separating the faulted area from the normal area or delineating the misorientation between various areas of the semiconductor. In general, these defects are not beneficial to device operations; however, under certain circumstances they may be used to "getter" deep level impurities in the semiconductor.

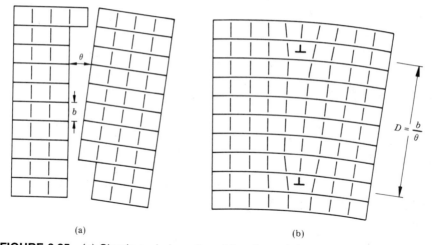

(a)                                                              (b)

**FIGURE 6.25**   (a) Simple grain boundary. The plane of the figure is parallel to a cube face and normal to the axis of relative rotation of the two grains. (a) Two grains have a common cube axis and an angular difference in orientation θ. (b) The two grains are joined to form a bicrystal. The joining requires only elastic strain except where a plane of atoms ends on the boundary in an edge dislocation, denoted by the symbol ⊥. [After Read (1953).]

### 6.6.1  Grain Boundaries

Grain boundaries are not expected to exist in single-crystalline materials; however, in certain devices, such as the polysilicon solar cells, a polycrystalline semiconductor is used. A grain boundary may be viewed as an array of dislocations separating two single grains of crystalline material with a misorientation between them (Fig. 6.25). A tilt boundary is formed when the dislocation array is composed of edge dislocations only; a twist boundary is formed when the array is composed of screw dislocations only (Read, 1953). The misorientation between the adjacent grains increases as the number of dislocations composing the grain boundary increases. Since each dislocation has an energy associated with it [eq. (6.36)], the grain boundary energy should increase with increasing misorientation until the stress fields of the dislocations start to cancel due to overlaps. For a general grain boundary, the structure of the boundary may be modeled as part tilt and part twist.

Electronic states in the semiconductor band gap have been associated with dislocations, resulting in a space region around the dislocation core. It is not surprising that space-charge regions are associated with grain boundaries. The electrical conductivity in a polycrystalline sample is expected in general to be larger in the direction parallel to the grain boundary than that normal to it (Mataré, 1984), similar to what is observed for dislocations (Fig. 6.24).

### 6.6.2  Stacking Faults and Twins

The {111} planes of FCC crystals have a normal stacking sequence of · · · ABC ABC · · · (Fig. 6.26). If an extra {111}-A plane is inserted between a {111}-B and

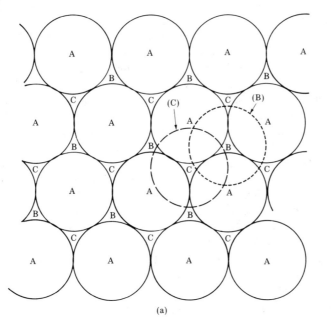

(a)

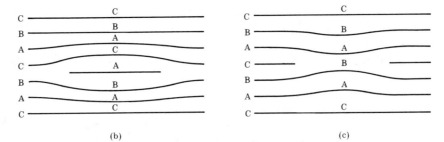

(b)                               (c)

**FIGURE 6.26** Stacking spheres in close packing. (a) When one close-packed layer has been laid down (the A layer), the next layer can go into either of the two sets of hollows (B or C) on the first layer. A, B, and C denote the centers of the spheres in the three possible positions. The normal FCC stacking sequence of the {111} plane is ... ABCABC. ... The normal hexagonal close-packed stacking sequence of the basal plane is ... ABAB. ... (b) Extrinsic stacking fault: a {111}-A plane is inserted between a {111}-B plane and a {111}-C plane. (c) Intrinsic stacking fault: a {111}-C plane is removed between a {111}-B plane and a {111}-A plane.

a {111}-C plane, the stacking sequence is now changed to AB AC (the arrow indicates the inserted plane), the stacking of the first three {111} planes has the sequence ABA, which is not the normal stacking of the FCC crystals but of the hexagonal close-packed (HCP) crystals. An *extrinsic* stacking fault is said to have formed. The faulted area is bounded by a Frank partial dislocation with a Burgers vector of $\mathbf{b} = (a/3) <111>$, a Burgers vector of magnitude $a/3$ lying along the $<111>$ direction. If a {111}-C plane is removed between a {111}-B plane and a {111}-A plane from the ABC sequence, the stacking sequence is changed to

AB $\uparrow$ ABC (the arrow indicates where a plane is removed). An *intrinsic* stacking fault is said to have formed and the boundary of the faulted area is a Shockley partial dislocation with a Burgers vector $\mathbf{b} = (a/6) <112>$.

Semiconductors such as Si and GaAs have the diamond cubic structure, which consists of two interpenetrating FCC lattices. The {111} stacking sequence of the diamond lattice is composed of three double {111} planes with an ABC sequence, similar to that of the FCC lattice. If this ABC sequence is disturbed by inserting or removing a double-{111} plane, a stacking fault is created and bounded by either a Frank or a Shockley partial dislocation (Fig. 6.26).

Stacking faults are created during crystal growth such as epitaxy and during the annealing of ion implanted regions. Extrinsic stacking faults are also formed in Si due to oxidation. Apparently, Si self-interstitials are created during oxidation; these interstitials then coalesce to form extrinsic stacking faults. The presence of HCl (actually chlorine) during oxidation tends to reduce the formation of extrinsic stacking faults. The addition of a chlorine-containing compound during oxidation is a common practice to reduce not only stacking faults but also sodium in the oxide layer. Since stacking faults are bounded by dislocations, the presence of a stacking fault near *p-n* junctions can cause increased leakage current due to recombination and/or gettering of metallic impurities around the fault boundaries.

Twins are related to stacking faults and are also formed during crystal growth. Consider the normal stacking sequence of a FCC lattice of $\cdots$ ABC A $\downarrow$ B $\downarrow$ C $\downarrow$ A $\downarrow$ B $\downarrow$ C $\cdots$. Let us insert a C plane between the second A plane and the B plane, an A plane between the B and C planes, and a B plane between the C and B planes (the arrows indicate where the planes are to be inserted). The

$$\downarrow \quad \downarrow \quad \downarrow \quad \downarrow \quad \downarrow$$

stacking sequence is now changed to ABCA C B A C B A C B A (the arrows indicate the inserted fault planes). The sequence before the twin plane (the second A plane from the left) is ABCABC, whereas the sequence after the twin plane is ACBACB. A twin fault is therefore formed by inserting a fault plane every other plane in the normal FCC stacking sequence ("a twin is a many-faulted thing"). Despite the fact that a twin can be created by inserting many fault planes, the FCC lattice stacking sequence is undisturbed before and after the twin plane. The influence of twins on the electronic device performance has been reported to be less significant than that of the stacking faults.

## 6.7 Volume Defects

Volume defects include voids and local regions of different phases, such as a precipitate or an amorphous phase. In Si, oxygen precipitation is the most important volume defect. Silicon crystals are grown by either the Czochralski technique (Chapter 1) or by the float-zone technique. The typical concentration of oxygen in Czochralski Si crystal is about 10 to 20 ppm (parts per million) or $5 \times 10^{17}$ to $1 \times 10^{18}$ cm$^{-3}$. The float-zone technique introduces less oxygen in Si than does the Czocralski technique. Most of the oxygen in the as-grown crystal is atomically dispersed and occupies interstitial sites. A low-temperature annealing at 300 to

500°C causes the interstitial oxygen to move into substitutional sites and become donors. An increase in donor concentration during the packaging and passivation steps of an MOS (metal–oxide Si) device at ≈450°C causes a drastic change of the threshold voltage of the devices (Wolf and Tauber, 1986). These oxygen donors must be reduced or eliminated by a precipitation step of annealing at 600 to 700°C or at an elevated temperature cycle of about 1100°C. The high-temperature cycle of forming $SiO_2$ precipitates in Si is the basic method used for intrinsic gettering, an issue discussed in the following section.

For precipitation to occur, we need (1) a thermodynamic driving force for the precipitation reaction—this driving force (a deviation from equilibrium) is generally in the form of a supersaturation of the solute atoms or in terms of a chemical reaction such as the formation of $SiO_2$, and (2) atomic mobility—solute atoms must be able to diffuse to the nucleation sites such as dislocations (heterogeneous nucleation) or diffuse together to form nuclei by themselves (homogeneous nucleation).

Other than oxygen precipitates, metallic precipitates of Cu, Co, Ni, and Fe are observed to form and to serve as nucleation sites for stacking faults during epitaxial growth. These impurity precipitates need to be "gettered" for suitable device operation.

Another type of volume defect is due to local amorphous regions encountered in ion-implanted semiconductors, especially in the case of low-dose irradiation where local amorphous regions around the ion tracks do not overlap to form a continuous layer. The amorphous phase can be considered as a structure without long-range order. Figure 6.27 shows the schematic atomic arrangement for an amorphous solid and for a crystalline solid. As one can see, a crystalline solid has long-range atomic order: an amorphous solid has short-range order (the order among the nearest neighbors) but no long-range order (Elliott, 1986). This type of volume defect is readily eliminated by an annealing step, as discussed in Chapter 8.

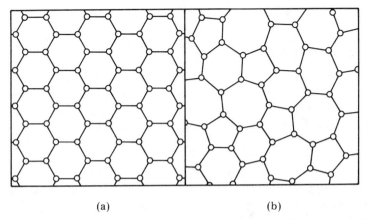

(a)  (b)

**FIGURE 6.27** Schematic atomic arrangement of (a) a crystalline solid and (b) an amorphous solid.

# 6.8 Gettering in Si

As discussed previously, defects and impurities in semiconductors adversely affect the performance of the devices. To improve electronic performance, a processing step known as "gettering" is practiced, especially in Si technology. The basic principle of gettering is utilization of the stress fields of strategically located dislocations to capture impurities at moderate processing temperatures. The dislocations may be introduced by external means on the back side of wafers, known as extrinsic gettering, or from an internal source by creating dislocations away from the active regions of the device, known as intrinsic gettering. For gettering to occur, it is necessary for impurity atoms in the form of a precipitate to (1) dissolve into the host semiconductor, (2) diffuse toward dislocations due to the attractive forces between them, (3) form a precipitate around the dislocation in an inactive region of the semiconductor, and (4) remain without release into the host semiconductor during subsequent processing steps.

## 6.8.1 Extrinsic Gettering

In extrinsic gettering, dislocations are introduced by external means at the back side of wafers. The common methods to introduce dislocations include laser-induced damage, mechanical abrasion, ion implantation, high concentration diffusion, and the deposition of a layer of polysilicon (Wolf and Tauber, 1986). The introduction of dislocations by laser and mechanical abrasion is clearly due to damage. Ion implantation with the proper ions introduces not only dislocations but also planar defects such as stacking faults (Cullis et al., 1978). The diffusion of P from the backside with high surface concentration causes a change in the Si lattice parameter; dislocations are created to accommodate the change (Tseng et al., 1978). These dislocations then act as gettering sites. In the case of deposited polysilicon, the grain boundaries are essentially the active sites for gettering. It should be noted that polysilicon undergoes grain growth at the temperature of gettering, thus causing the gettering efficiency to be reduced.

## 6.8.2 Intrinsic Gettering

Intrinsic gettering is generally used to remove oxygen from active regions of Si. In intrinsic gettering, the dislocations are created by precipitating the dissolved oxygen in the Si using heat treatments only, without any external damage to the backside of the wafers. The advantage of intrinsic gettering is in placing the gettering sites close to the active regions of the device, compared to placing the gettering sites hundreds of micrometers away in extrinsic gettering.

To apply intrinsic gettering to Si, the oxygen concentration should be between 7.5 and $9.5 \times 10^{17}$ cm$^{-3}$. The lower limit ($7.5 \times 10^{17}$ cm$^{-3}$) is somewhat above the threshold concentration (about $6 \times 10^{17}$ cm$^{-3}$) for oxygen to precipitate and the upper limit ($9.5 \times 10^{17}$ cm$^{-3}$) is below the concentration where oxygen precipitation results in such a high density of precipitates that the wafers may warp and may generate gettering sites in the active-device regions (Wolf and Tauber, 1986). The control of the oxygen concentration in the wafers should be rather precise for intrinsic gettering to apply.

Under appropriate heat treatment, the oxygen in the Si precipitates out to form $SiO_2$. There is a large volume change associated with the formation of $SiO_2$ (the volume of $SiO_2$/volume of Si ratio is about 2). To accommodate the compressive stress generated by the presence of $SiO_2$ particles, several stress-releasing mechanisms are involved: (1) formation of dislocation loops, (2) emission of Si self-interstitials, and (3) absorption of vacancies. The Si self-interstitials have the tendency to form extrinsic stacking faults with dislocations as fault boundaries. As a result of the dislocations generated by formation of $SiO_2$ particles, gettering sites become available. The issue now is how to place the $SiO_2$ particles away from the active-device region for gettering purposes. It has been observed that a three-step annealing cycle can achieve intrinsic gettering in a wafer having uniformly distributed oxygen of the appropriate concentration. A three-step high–low–high annealing sequence is described by Wolf and Tauber (1986):

1.  A high-temperature annealing (about 1100°C, for example) to reduce the oxygen concentration near the wafer surface. This step is taken to cause out-diffusion of the oxygen from the surface region, thus reducing the oxygen concentration to about the solubility limit at the processing temperature. The depth of the "denuded" zone should exceed that of the deepest *p-n* junction.
2.  A low-temperature annealing (600 to 800°C, for example) to nucleate $SiO_x$ precipitates. This annealing step for prolonged periods (e.g., 4 to 64 hours) causes the supersaturated interstitial oxygen to diffuse in a local region and nucleate into small clusters.
3.  A high-temperature annealing (900 to 1250°C) for extended periods (4 to 16 hours) to cause growth of $SiO_2$ nuclei. The growth of small clusters to large nuclei (50 to 100 nm) causes the generation of dislocations by forming prismatic dislocation loops and by emitting Si self-interstitials, which coalesce to form extrinsic stacking faults. The gettering sites are thereby created by this three-step annealing sequence.

Although intrinsic gettering is an elegant technique, it does require strict control of process parameters. It is a common practice to implement both extrinsic and intrinsic gettering concurrently.

## 6.8.3 Driving Force for Gettering

Let us consider the driving force of gettering due to the interactions between an impurity and an edge dislocation. Suppose that in a semiconductor there are $n$ substitutional solute atoms with atomic radius $R_1$ ($R_1 \neq R_{host}$). The solute atoms therefore introduce dilatation or contraction into the lattice. Such strain centers interact with the hydrostatic stress fields of dislocations in the crystal. The hydrostatic stress field (either in tension or in compression) is due to the normal stresses associated with a dislocation.

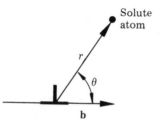

**FIGURE 6.28** Interaction between a solute atom and an edge dislocation. The dislocation runs perpendicular to the plane of paper ($z$ direction). The solute atom is located at $r$ and $\theta$ with respect to the dislocation. The angle $\theta$ is the angle between the direction of the Burgers vector **b** and the direction $r$.

The normal stress field of an edge dislocation in cylindrical coordinate system is given by

$$\sigma_{rr} = \sigma_{\theta\theta} = -\frac{Gb \sin \theta}{2\pi(1 - \mu)r} \tag{6.41}$$

where $G$ is the shear modulus and $\mu$ is Poisson's ratio (see Section 6.5.2). These stresses are either in compression or tension, depending on the angle $\theta$ which is defined by the angle between the direction of the Burgers vector and the direction $r$ (Fig. 6.28). There is a normal stress in the $z$ direction (along the dislocation line) associated with an edge dislocation. This normal stress is

$$\sigma_{zz} = \mu(\sigma_{\theta\theta} + \sigma_{rr}). \tag{6.42}$$

There is also a shear stress field around an edge dislocation just as those around a screw dislocation [eq. (6.26)]. The mean hydrostatic stress field of an edge dislocation may be taken as the average of the three normal stresses associated with an edge dislocation. The mean hydrostatic stress, $P_{edge}$, is

$$P_{edge} = \frac{\sigma_{rr} + \sigma_{\theta\theta} + \sigma_{zz}}{3}$$

$$= \frac{1 + \mu}{3}(\sigma_{rr} + \sigma_{\theta\theta}) \tag{6.43}$$

$$= \frac{-(1 + \mu)Gb \sin \theta}{3\pi(1 - \mu)r}$$

where $\mu$ is Poisson's ratio, $G$ is the shear modulus, and $r$ and $\theta$ are the positions of the solute atom with respect to the dislocation line (Fig. 6.28). Screw dislocations produce only shear stress field, no hydrostatic stress. Substitutional solute atoms do not interact with screw dislocations. Interstitial atoms produce both shear and hydrostatic stress in the lattice; therefore, they interact with both screw and edge dislocations.

The work done, $W$, to increase the size from that of a host atom $R$ to $R_1$ (radius of a solute atom) against the constant mean hydrostatic stress of an edge dislocation is

$$W = P_{edge} \, \Delta V = P_{edge} \, \Delta\left(\frac{4}{3}\pi R^3\right) = P_{edge} \, 4\pi \, R^2 \, \Delta R$$

$$= \frac{-(1 + \mu)Gb \sin \theta}{3\pi(1 - \mu)r} \, 4\pi R^2 \, \frac{R_1 - R}{R}. \tag{6.44}$$

The interaction energy between a solute atom and an edge dislocation, $U$, is just equal to the work done, so that

$$U = \frac{-A \sin \theta}{r} \tag{6.45}$$

where

$$A = \frac{-(1 + \mu)Gb}{3\pi(1 - \mu)} \cdot 4\pi R^2 \, \frac{R_1 - R}{R}.$$

Introducing Cartesian coordinates $r^2 = x^2 + y^2$ and $\sin \theta = y/r$, eq. (6.45) may be written as

$$U = \frac{Ay}{x^2 + y^2}$$

$$x^2 + y^2 - \frac{Ay}{U} = 0 \tag{6.46}$$

$$x^2 + \left[y^2 - \frac{Ay}{U} + \left(\frac{A}{2U}\right)^2\right] - \left(\frac{A}{2U}\right)^2 = 0$$

which can be expressed as

$$x^2 + (y')^2 = \left(\frac{A}{2U}\right)^2 \tag{6.47}$$

where

$$y' = y - \frac{A}{2U}.$$

Each constant interaction energy is represented by a circle, tangent to the line $y = 0$ and centered on the line $x = 0$, as shown in Fig. 6.29. The interaction force to which the solute atoms is subjected acts normal to the equi-energy lines in the direction of the conjugate set of circles, shown as dashed curves. Solute atoms larger than the host will migrate to the lower part of the dislocation in the direction shown by the arrow. Solute atoms smaller than the host will migrate in the opposite direction. The driving force for gettering by dislocation is essentially due to the interaction energy, shown in eq. (6.47).

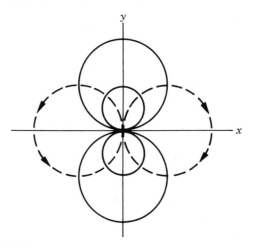

**FIGURE 6.29**   Circles of constant interaction energy.

# REFERENCES

### General References

CASEY, H. C., JR., and M. B. PANISH, *Heterostructure Lasers,* Academic Press, New York, 1978.

CULLITY, B. D., *Elements of X-Ray Diffraction,* Addison-Wesley, Reading, Mass., 1956, Chapter 2.

ELLIOTT, S. R., *Physics of Amorphous Materials,* Wiley, New York, 1986.

HIRTH, J. P., and J. LOTHE, *Theory of Dislocations,* McGraw-Hill, New York, 1968, Chapter 2.

KIRKPATRICK, C. G., R. T. CHEN, D. E. HOLMES, and K. R. ELLIOT, "Growth of Bulk GaAs," in *Gallium Arsenide, Materials, Devices, and Circuits,* ed. M. J. Howes and D. V. Morgan, Wiley, New York, 1985.

MATARÉ, H. F., *Defect Electronics in Semiconductors,* Wiley-Interscience, New York, 1971, Chapters 8 and 10.

READ, W. T., *Dislocations in Crystals,* McGraw-Hill, New York, 1953.

WOLF, S., and R. N. TAUBER, *Silicon Processing for VLSI Era,* Vol. 1, *Processing Technology,* Lattice Press, Sunset Beach, California, 1986, Chapter 2.

### Specific References

CULLIS, A. G., T. E. SEIDEL, and R. L. MEEK, *J. Appl. Phys. 49,* 5188 (1978).

FAHEY, P. M., "Point Defects and Dopant Diffusion in Silicon," Ph.D. thesis, Stanford University, 1985.

GHANDI, S. K., "VLSI Fabrication Principles, Silicon and Gallium Arsenide," *VLSI Fabrication Principles. Silicon and Gallium Arsenide,* Wiley, New York, 1983.

JENKINS, M. W., *J. Electrochem Soc. 124,* 757 (1977).

LAGOWSKI, J., H. C. GATOS, J. M. PARSEY, K. WODA, M. KAMINSKI, and
W. WALAKEIWICZ, *Appl. Phys. Lett. 40*, 342 (1982).

LANG, D. V., *J. Appl. Phys. 45*(7), 3023 (1974).

MATARÉ, H. F., *J. Appl. Phys. 56*(10), 2605 (1984).

STIRLAND, D. J., *Appl. Phys. Lett. 53*, 2432 (1988).

TAN, T. Y., and T. K. TICE, *Philos. Mag. 30*, 615 (1976).

TAN, S. I., B. S. BERRY, and W. F. J. FRANK, in *Ion Implantation in Semiconductors and Other Materials*, ed. B. L. Crowder, Plenum Press, New York, 1973.

TSENG, W. F., T. KOJI, J. W. MAYER, and T. E. SEIDEL, *Appl. Phys. Lett. 33*(5), 442 (1978).

VAN VECHTEN, J. A., "A Simple Man's View of the Thermochemistry of Semiconductors," in *Handbook on Semiconductors*, ed. T. S. Moss, Vol. 3, *Materials, Properties and Preparation*, ed., S. P. Keller, North Holland, Amsterdam, 1980, Chapter 1.

WATKINS, G. D., J. R. TROXELL, and A. P. CHATERJEE, in *Defects and Radiation Effects in Semiconductors 1978*, ed. J. H. Albany, Institute of Physics Conferences Series 46, London, 1979.

## PROBLEMS

**6.1.** Nickel (Ni) is a face-centered cubic (FCC) crystal with an atomic density of $9.14 \times 10^{22}$ atoms/cm$^3$ and an atomic mass of 58.73.
   **(a)** What is the mass density (g/cm$^3$)?
   **(b)** What is the lattice constant, $a$?
   **(c)** If $a = 4.0$ Å, what is the distance between (110) planes?

**6.2.** Silver (Ag) is face-centered cubic with a lattice parameter $a = 4.09$ Å.
   **(a)** What is the spacing between (123) planes?
   **(b)** What is the atomic density of Ag (atoms/cm$^3$)?
   **(c)** How many Ag atoms/cm$^2$ are in a layer $10^{-4}$ cm thick?

**6.3.** Consider a simple cubic crystal and planes with (100), (111), and (110) Miller indices and with an atomic density of $4 \times 10^{22}$ atoms/cm$^3$.
   **(a)** Which planes would have the greatest spacing between them?
   **(b)** Which planes would have the largest number of atoms/cm$^2$ on them?
   **(c)** What is the lattice constant $a$ (Å)?

**6.4.** You have two cubic materials with the same lattice parameter $a = 4.0$ Å. Material A is a simple cubic with atomic mass of 40. Material B is a face-centered cubic with atomic mass of 20.
   **(a)** Which has the greater atomic density (number of atoms/cm$^3$)?
   **(b)** Which has the greater mass density (g/cm$^3$)?
   **(c)** Which material has the greater spacing between (100) planes?

**6.5.** You are given a sphere with a 1-cm radius of the simple cubic material "Cornellian" with an atomic mass of 30. You weigh it and find that it weighs 6.0 g.
(a) How many atoms are there in the sphere?
(b) What is the lattice parameter, $a$, of the material?
(c) You have a 1-cm$^3$ cube of Cornellian and are told that the lattice parameter is 4.0 Å.
  (1) What is the spacing between (321) planes?
  (2) How many atoms would be on the outer layer of atoms on the cube?

**6.6.** Explain why a dislocation must terminate on itself or on a surface or volume fault.

**6.7.** Show that the diamond cubic lattice has a $\cdots$ ABC $\cdots$ sequence.

**6.8.** Explain why summation of the Burgers vector is zero at a dislocation node.

**6.9.** Based on eqs. (6.17) and (6.18), show that

$$\frac{[V^=]}{[V^x]} = \exp\left(-\frac{E_x^= + E_x^- - 2E_F}{kT}\right)$$

and

$$\frac{[V^{++}]}{[V^x]} = \exp\left(-\frac{2E_F - E_x^{++} - E_x^-}{kT}\right).$$

**6.10.** Assuming there are only $V^x$ and $V^+$ in a semiconductor, show that $[V^+] = [V^x]e^{(E^+ - E_F)/kT}$. (*Hint:* One may start with $[V_T] = [V^+] + [V^x]$ and recognize that $[V^+]$ is the concentration of positively charged vacancies.)

**6.11.** The normal stress field around an edge dislocation is given by eq. (6.41). In the Cartesian coordinate system:

$$\sigma_{xx} = -\frac{Gb}{2\pi(1-\mu)}\frac{y(3x^2 + y^2)}{(x^2 + y^2)^2}$$

$$\sigma_{yy} = \frac{Gb}{2\pi(1-\mu)}\frac{y(x^2 - y^2)}{(x^2 + y^2)^2}$$

$$\sigma_{zz} = \mu(\sigma_{xx} + \sigma_{yy})$$

$$\sigma_{xy} = \frac{Gb}{2\pi(1-\mu)}\frac{x(x^2 - y^2)}{(x^2 + y^2)^2}.$$

Work out the sign and direction of the shear stress $\sigma_{xy}$ and the normal stresses $\sigma_{xx}$ and $\sigma_{yy}$ for the cases where $y < x$ and $y > x$.

# 7

# Diffusion in Solids

## 7.1 Introduction

Diffusion is a frequently used technique for the incorporation of dopant atoms into a semiconductor. The diffusion of impurity atoms into a semiconductor wafer leads to the formation of *p-n* junctions, conduction channels, and source and drain regions (Fig. 7.1). The performance of the devices depends critically on the impurity concentration and the impurity profile. For this reason the diffusion of various impurities in semiconductors has been studied rather extensively.

In semiconductor technology, the semiconductor wafer is placed in a quartz-tube furnace with dopant species contained in a transport gas that passes over the wafer. There are sufficient dopant atoms in the gas ambient so that the dopants are incorporated in the surface of the semiconductor to concentrations near the solid solubility limit. The furnace is generally heated to high temperatures (~900 to 1100°C for Si). At that temperature the dopant atoms diffuse from the surface into the interior of the semiconductor.

In this chapter we present the general relations governing the diffusion of atoms in solids. We start with a simple linear approach that is useful for quick approximate solutions and then progress to a more complete description. The diffusion coefficient increases exponentially with temperature and we describe the origin of its behavior and the diffusion mechanisms. In relation to Chapter 6, we discuss the influence of vacancy concentrations on the diffusion process.

Boron gas ambient in diffusion tube

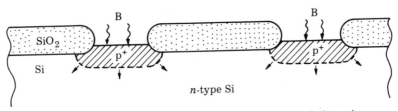

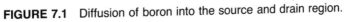

**FIGURE 7.1** Diffusion of boron into the source and drain region.

## 7.2 Fick's First and Second Law of Diffusion

From a macroscopic viewpoint, the description of diffusion processes is relatively straightforward. Atoms will diffuse from regions of high concentration in a solid to regions of low concentration. Following the usual convention, we use the symbol $C$ to indicate the concentration of atoms in units of atoms/cm$^3$, so that in a one-dimensional system $C(x)$ indicates the concentration of atoms as a function of distance, $x$, in the solid.

Let us consider a simple case, shown in Fig. 7.2, where the concentration of atoms decreases uniformly from a value $C(0)$ at the surface to zero at a depth $\lambda$, $C(\lambda) = 0$. If the sample is heated so that the atoms can diffuse, there will be a flow or flux $F$ of atoms into the solid. The flux is the number of atoms crossing a unit area in unit time. The units of flux are generally given in atoms/cm$^2 \cdot$ s.

The flux of atoms can be expressed in our one-dimensional system as

$$F = -D \frac{\partial C}{\partial x} \tag{7.1}$$

where $D$ is the diffusion coefficient ($D$ has units of cm$^2$/s) and $\partial C/\partial x$ is the concentration gradient. [Equation (7.1) is called Fick's first law.] For our case $\partial C/\partial x$ is a constant and is given by

$$\frac{\partial C}{\partial x} = \frac{C(\lambda) - C(0)}{\lambda} = -\frac{C(0)}{\lambda} \qquad C(\lambda) = 0 \tag{7.2}$$

so that the flux $F$ is then

$$F = D \frac{C(0)}{\lambda}. \tag{7.3}$$

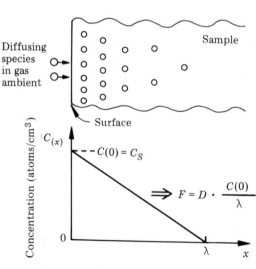

FIGURE 7.2   Diffusion with a linear concentration profile.

Note that there is a negative sign in eq. (7.1) and that $\partial C/\partial x$ is negative for a decrease in the concentration with depth; consequently, the flux into the sample has a positive value. The flux equation indicates that atoms diffuse from a high-concentration region to a low-concentration region. This is, of course, the same phenomenon that governs the diffusion of carriers in semiconductors (Section 2.9).

Let us consider a case of diffusion into a solid where we hold the concentration of atoms at the surface as a constant [$C(0)$ = constant] and have the atoms diffuse deeper into the solid so that the diffusion depth $\lambda$ is a function of time; $\lambda = \lambda(t)$. We will also use the unrealistic but handy linear approximation where the concentration decreases linearly with distance. The concentrations at two times, $t_1$ and $t_2$, are shown in Fig. 7.3.

We would like to find how the diffusion depth $\lambda$ varies with time. One approach is to consider the number of atoms/cm² that diffuse into the sample (shaded area $A$ in the figure):

$$\text{number of atoms in area } A = \int_0^\lambda C(x)d(x) \tag{7.4}$$

$$= \frac{C(0)}{2}\lambda.$$

The flux $F$ is

$$F = \frac{\text{increase in the number of diffused atoms}}{\text{increment in diffusion time}} \tag{7.5}$$

$$= \frac{\partial}{\partial t}\left[\frac{C(0)}{2}\lambda\right] = \frac{C(0)}{2}\frac{\partial\lambda}{\partial t}.$$

Equating the flux $F$ to Fick's first law gives us

$$F = \frac{C(0)}{2}\frac{\partial\lambda}{\partial t} = -D\frac{\partial C}{\partial x} = D\frac{C(0)}{\lambda}. \tag{7.6}$$

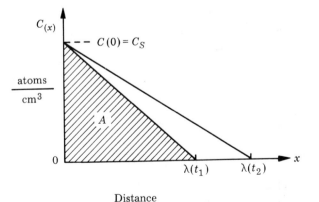

FIGURE 7.3  Diffusion distance as a function of time.

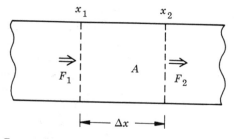

**FIGURE 7.4** Flux $F_1$ entering and flux $F_2$ leaving an element of volume $\Delta x \cdot 1$ cm$^3$.

Upon rearranging we obtain

$$\frac{C(0)}{2} \lambda \, d\lambda = DC(0) \, dt \tag{7.7}$$

and integrating

$$\int_0^\lambda \frac{\lambda}{2} \, d\lambda = \int_0^t D \, dt \tag{7.8}$$

gives

$$\lambda = \sqrt{4Dt}. \tag{7.9}$$

The parameter $\lambda$ is commonly referred to as the characteristic diffusion length and it increases as $(Dt)^{1/2}$. This quantity indicates the distance where the concentration of the diffusing atoms drops down to zero in the case of a linear concentration gradient. For a more general case where the concentration gradient is nonlinear, $\lambda$ signifies the extent of the diffusion.

Let us now consider a volume element of area $A$ and thickness $\Delta x$, with $F_1$ entering and $F_2$ leaving (Fig. 7.4). The increase in diffusing atoms in a volume element $\Delta x$ (with unit area) is, therefore,

$$\frac{\partial C}{\partial t} \Delta x = \frac{\text{increase in diffusing atoms}}{\text{unit time}} = F_2 - F_1$$
$$= -(F_1 - F_2) = -\Delta F. \tag{7.10}$$

For small $\Delta x$, $\Delta x \rightarrow dx$ and $\Delta F \rightarrow dF$. Therefore,

$$\frac{\partial C}{\partial t} = -\frac{\partial F}{\partial x} \tag{7.11}$$

$$\frac{\partial C}{\partial t} = \frac{\partial}{\partial x}\left(D \frac{\partial C}{\partial x}\right). \tag{7.12}$$

This is the so-called continuity equation or Fick's second law, where $C$, the concentration, is generally a function of position $x$ and time $t$. For the case where $D$ is a constant,

$$\frac{\partial C(x, t)}{\partial t} = D \frac{\partial^2 C}{\partial x^2}. \tag{7.13}$$

There are two most frequently used solutions for eq. (7.13), depending on the boundary conditions.

1. *Surface concentration constant, $C_S$ = constant.* For the case where the surface concentration of the diffusing atoms is constant, the boundary conditions are

$$C(0, t) = C_S \tag{7.14}$$

and

$$C(\infty, t) = 0. \tag{7.15}$$

The solution is given by

$$C(x, t) = C_S \, \text{erfc} \left( \frac{x}{\sqrt{4Dt}} \right) \tag{7.16}$$

where erfc is the complementary error function (see Table 7.1). This kind of boundary condition is typical for the process of incorporating dopant atoms into semiconductor substrates known as predeposition. The total amount of the dopant atoms diffusing from an outside source into the substrate, $Q$, is given by

$$Q(t) = \int_0^\infty C_S \, \text{erfc} \left( \frac{x}{\sqrt{4Dt}} \right) dx \tag{7.17}$$

$$= \frac{2 \sqrt{Dt}}{\sqrt{\pi}} C_S.$$

The concentration profile of the diffusing atoms (doping impurities in predeposition) agrees with a linear concentration profile up to a distance about $-0.6\lambda$ (see Fig. 7.5). Since the erfc profile can be reasonably well estimated by the linear

**TABLE 7.1**  Useful Expressions to the Error Function

$$\text{erf}(x) = \frac{2}{\sqrt{\pi}} \int_0^x e^{-a^2} \, da$$

$$\text{erfc}(x) = 1 - \text{erf}(x)$$

$$\text{erf}(0) = 0$$

$$\text{erf}(\infty) = 1$$

$$\text{erf}(x) \approx \frac{2}{\sqrt{\pi}} x \qquad \text{for } x \ll 1$$

$$\text{erfc}(x) \approx \frac{1}{\sqrt{\pi}} \frac{e^{-x^2}}{x} \qquad \text{for } x \gg 1$$

$$\int_0^x \text{erfc}(x') \, dx' = x \, \text{erfc}(x) + \frac{1}{\sqrt{\pi}} (1 - e^{-x^2})$$

*Source:* After Grove (1967).

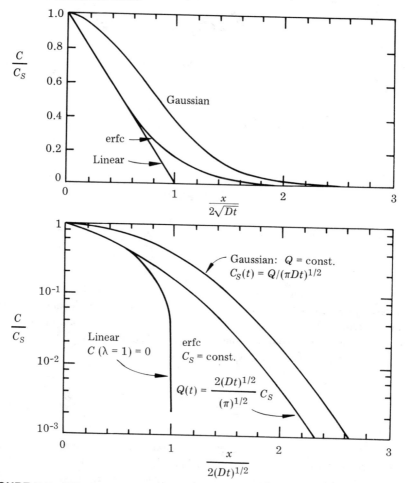

**FIGURE 7.5**   Diffusion functions: normalized concentration versus normalized distance.

profile, the quantity $Q$ can be estimated by the simple geometric considerations from Fig. 7.5:

$$Q \simeq C_S \frac{\lambda}{2} = C_S \sqrt{Dt}. \tag{7.17a}$$

This estimated $Q$ is about 10% less than the more accurate value obtained from eq. (7.17).

  2. *The diffusing amount, Q, is constant.* For the case where the total amount of diffusing atoms, $Q$, is constant, the initial and the boundary conditions are

$$\int_0^\infty C(x, t)\, dx = Q = \text{constant} \tag{7.18}$$

and

$$C(\infty, t) = 0. \tag{7.19}$$

This solution has the form of a Gaussian function,

$$C(x, t) = \frac{Q}{\sqrt{\pi Dt}} \exp\left(-\frac{x^2}{4Dt}\right). \tag{7.20}$$

At $x = 0$,

$$C(0, t) = \frac{Q}{\sqrt{\pi Dt}}. \tag{7.21}$$

This shows that the surface concentration decreases as $(Dt)^{-1/2}$. The boundary condition of a constant $Q$ pertains to the process of the drive-in diffusion in semi-conductor technology. It is clear from the discussion above the $\sqrt{Dt}$ is an important quantity in diffusion, irrespective of the boundary conditions.

## 7.3 Diffusion Coefficient

The diffusion coefficient $D$ is a strong function of temperature. The temperature dependence arises from the fact that some energy (typically, a few electron volts, eV) is required for an atom to jump from one atomic position to another. This energy is often called the activation energy $E_A$. The number of atoms with sufficient energy is proportional to $\exp(-E_A/kT)$, where $k$ is Boltzmann's constant ($k = 8.6 \times 10^{-5}$ eV/K and $T$ is given in kelvin ($kT \approx 0.026$ eV at 300 K). Consequently, the diffusion coefficient can be written

$$D = D_0 \exp\left(-\frac{E_A}{kT}\right) \tag{7.22}$$

where $D_0$ is a preexponential parameter with typical values between $10^{-1}$ and $10^2$ and values of $E_A$ between 1 and 5 eV, depending on the system.

There are many processes that have an exponential dependence on the reciprocal temperature. The Arrhenius plot is commonly used to obtain the activation energy. In this method, the experimental data are plotted against $1000/T$ (K) as shown in the following example (Fig. 7.6).

If eq. (7.22) is followed in the diffusion process, the plot of log $D$ versus $1/T$ should be a straight line. To extract a value of the activation energy, a simple way is to find the values of $1000/T$, where the $D$ values change by a factor of 10.

$$\frac{D_2}{D_1} = 10 = \frac{D_0 e^{-E_A/kT_2}}{D_0 e^{-E_A/kT_1}} = \exp\left[\left(\frac{-E_A}{k}\right)\left(\frac{1}{T_2} - \frac{1}{T_1}\right)\right]$$

$$\ln 10 = 2.3 = -\frac{E_A}{k}\left(\frac{1}{T_2} - \frac{1}{T_1}\right) = \frac{E_A}{k}\left(\frac{1}{T_1} - \frac{1}{T_2}\right)$$

$$E_A = \frac{2.3k}{1/T_1 - 1/T_2} = \frac{2.3 \times 8.6 \times 10^{-5} \text{ eV/K}}{(0.70 \times 10^{-3} - 0.65 \times 10^{-3})/\text{K}}$$

$$= 3.95 \text{ eV}.$$

This relationship can be derived from the following considerations. For any chemical or physical changes to occur, there must be a driving force behind the

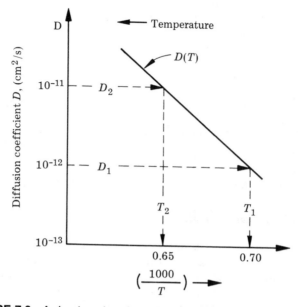

**FIGURE 7.6** Arrhenius plot of diffusion coefficient versus $1000/T$ (K).

changes. In diffusion, the driving force is the gain in free energy, as in any other case. The diffusing atoms tend to move from one region of the system to another along a chemical potential gradient (i.e., from a region of high chemical potential to a region of low chemical potential). In most cases the chemical potential gradient is in the same direction as the concentration gradient of the diffusion species. The concentration gradient is a readily measurable quantity, whereas the chemical potential gradient is not; it is therefore more convenient to regard the concentration gradient as the driving force for diffusion. The diffusing atoms are jumping around in a random fashion, called the random walk process. There are more atoms moving away from a high-concentration region due to a random walk than there are atoms moving away from a low-concentration region. The end result is a net movement of atoms down the concentration gradient, as indicated by the "minus" sign of the Fick's first law [eq. (7.1)].

With a concentration gradient in a system, we know that atoms will move in such a direction to reduce, and finally to remove, the concentration difference in the system. This is the driving force for diffusion. How fast that the atoms move to achieve this is related to the kinetics of the diffusion process.

The exchange frequency $k^0$ between two atoms in the presence of an activation barrier at equilibrium is given by

$$k^0 = v_0 \exp\left(-\frac{E_A}{kT}\right) \tag{7.23}$$

where $v_0$ (usually $10^{13}$ to $10^{14}/s$) is the lattice vibration frequency (see Section 2.10) and $E_A$ is $\Delta E$ in Fig. 7.7. At equilibrium, the frequency of an atom jumping from position A to B is the same as the frequency for an atom jumping from B to

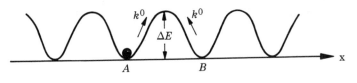

Distance

**FIGURE 7.7** Exchange frequency at equilibrium.

A (see Fig. 7.7). Under the influence of a driving force for diffusion, $\phi_\mu$, the exchange frequencies as established by $\Delta G$, the change in Gibbs free energy (Chapter 6) between the two positions, are no longer the same (see Fig. 7.8). The forward exchange frequency $f^+$ is

$$f^+ = k^0 \exp\left[\frac{(\bar{\lambda}/2)\phi_\mu}{kT}\right] \tag{7.24}$$

and the backward exchange frequency $f^-$ is

$$f^- = k^0 \exp\left[-\frac{(\bar{\lambda}/2)\phi_\mu}{kT}\right]. \tag{7.25}$$

The net forward exchange frequency $f$ is

$$f = f^+ - f^- = 2k^0 \sinh\frac{\bar{\lambda}\phi_\mu}{2kT} \tag{7.26}$$

where $\bar{\lambda}$ is the jump distance. (*Note:* $\bar{\lambda} \simeq a$, lattice parameter.) The driving force essentially provides a bias in the exchange frequency such that the activation barrier

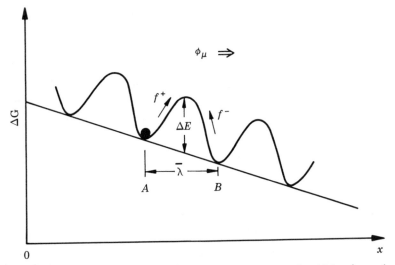

**FIGURE 7.8** Exchange frequency under the influence of a driving force $\phi_\mu$.

in the forward direction is $E_A - (\bar{\lambda}/2)\phi_\mu$, and the activation barrier in the reverse direction is $E_A + (\bar{\lambda}/2)\phi_\mu$. For $\bar{\lambda}\phi_\mu << kT$,

$$f = \frac{k^0\bar{\lambda}}{kT}\,\phi_\mu. \tag{7.27}$$

The driving force $\phi_\mu$ is

$$\phi_\mu = -\frac{\partial\bar{\mu}}{\partial x} \tag{7.28}$$

where $\bar{\mu}$ is the chemical potential, which is the change in free energy with composition, and for an ideal solution $\bar{\mu} = kT \ln C$. Therefore,

$$\phi_\mu \simeq -\frac{\Delta\bar{\mu}}{\Delta x} = \frac{kT}{\Delta x}\ln\frac{C_2}{C_1} \tag{7.29}$$

where $C_1$ and $C_2$ are the concentrations of the diffusion species $(C_2 > C_1)$ and $\Delta x$ is the distance between $C_1$ and $C_2$. The velocity of the diffusing atoms, $v$, is

$$v = \bar{\lambda}f = \frac{k^0\bar{\lambda}^2}{kT}\,\phi_\mu. \tag{7.30}$$

The diffusional flux $F$ is

$$\begin{aligned}
F &= vC \\
&= \frac{k^0\bar{\lambda}^2C}{kT}\,\phi_\mu = \frac{k^0\bar{\lambda}^2C}{kT}\left(-\frac{\partial\bar{\mu}}{\partial x}\right) \\
&= -\frac{k^0\bar{\lambda}^2C}{kT}\frac{\partial\bar{\mu}}{\partial C}\frac{\partial C}{\partial x}.
\end{aligned} \tag{7.31}$$

For an ideal solution,

$$\frac{\partial\bar{\mu}}{\partial C} = kT\frac{\partial \ln C}{\partial C} = \frac{kT}{C} \tag{7.32}$$

Therefore,

$$\begin{aligned}
F &= -\frac{k^0\bar{\lambda}^2C}{kT}\frac{kT}{C}\frac{\partial C}{\partial x} \\
&= -k^0\bar{\lambda}^2\frac{\partial C}{\partial x}.
\end{aligned} \tag{7.33}$$

Since $F$ is also given by

$$F = -D\frac{\partial C}{\partial x}$$

it follows that

$$D = k^0\bar{\lambda}^2. \tag{7.34}$$

Combining eq. (7.23) and (7.33) yields

$$F = -\bar{\lambda}^2 v_0 \exp\left(-\frac{E_A}{kT}\right)\frac{\partial C}{\partial x} \tag{7.35}$$

and

$$D = D_0 \exp\left(-\frac{E_A}{kT}\right).$$

The value of $D_0$ is estimated as follows:

$$D_0 = v^0\bar{\lambda}^2 \simeq v_0 a^2 \simeq (10^{14}/s)(3 \times 10^{-8}\text{ cm})^2 \simeq 10^{-1}\text{ cm}^2/s. \tag{7.36}$$

For nonideal solutions,

$$\bar{\mu} = kT \ln \gamma C$$

where $\gamma$ is the activity coefficient (Shewmon, 1963) and

$$D = k^0\bar{\lambda}^2\left(1 + \frac{\partial \ln \gamma}{\partial \ln C}\right). \tag{7.37}$$

## 7.4 Diffusion of Doping Atoms into Si

The diffusion of impurities into Si wafers is typically done in two steps. In the first step, dopants are introduced into the substrate to a relatively shallow depth of a few thousand angstroms. This process is called predeposition. After the impurities have been introduced into the Si substrate, they are diffused deeper into the substrate to provide a suitable impurity distribution without any additional impurity atoms introduced into the substrate. The second step is called drive-in diffusion. A junction is formed in the substrate after diffusion if the bulk contains the opposite type of impurity atoms. The junction depth $x_j$ is located where the concentration of the diffusing atoms $C(x)$ is equal to the bulk concentration $C_B$ (Fig. 7.9).

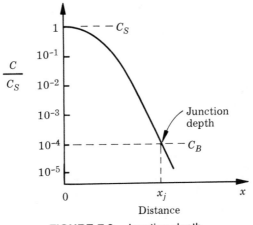

**FIGURE 7.9**   Junction depth.

## 7.4.1 Predeposition

The predeposition step is typically performed by placing the Si wafers in a furnace with flowing inert gas that carries the desired impurities. The temperature is usually between 800 and 1200°C. The impurities are introduced into the ambient from either a solid or a liquid source. The surface concentration $C_S$ of the impurity during predeposition is kept constant, eqs. (7.14) and (7.15) are pertinent boundary conditions, and the solution to the Fick's second law is given by eq. (7.16).

**EXAMPLE 1** _____

Calculate the junction depth $x_j$ and the total amount of dopant introduced into an $n$-type substrate with a bulk concentration $C_B$ of $1 \times 10^{15}$ cm$^{-3}$ after boron predeposition at 975°C for 60 minutes.

**Solution:** The junction depth is defined by the condition

$$C(x_j) = C_B = 1 \times 10^{15} \text{ cm}^{-3}.$$

The solid solubility of boron in Si at 975°C is $3.5 \times 10^{20}$ cm$^{-3}$ (see Fig. 7.10a), and the diffusivity of $B$ in Si is $1.5 \times 10^{-14}$ cm$^2$/s (Fig. 7.10b). From eq. (7.16),

$$C(x, t) = C_S \operatorname{erfc}\left(\frac{x}{\sqrt{4Dt}}\right)$$

so that

$$C(x_j t) = C_B = C_S \operatorname{erfc}\left(\frac{x_j}{\sqrt{4Dt}}\right)$$

$$\frac{x_j}{\sqrt{4Dt}} = \operatorname{erfc}^{-1}\left(\frac{C_B}{C_S}\right) \tag{7.38}$$

$$x_j = \sqrt{4Dt} \operatorname{erfc}^{-1}\left(\frac{C_B}{C_S}\right).$$

For this example,

$$\lambda = \sqrt{4Dt} = (4 \times 1.5 \times 10^{-14} \times 60 \times 60)^{1/2} = 1.47 \times 10^{-5} \text{ cm}$$

and

$$\frac{C_B}{C_S} = \frac{1 \times 10^{15}}{3.5 \times 10^{20}} = 2.9 \times 10^{-6}.$$

The diffusion characteristic length $\lambda$ is a constant ($1.47 \times 10^{-5}$ cm) for a given temperature (975°C) and diffusion time (60 minutes). The junction depth depends on the ratio of the bulk concentration to surface concentration ($C_B/C_S$) and is about three times deeper than $\lambda$ in this example. In using the error

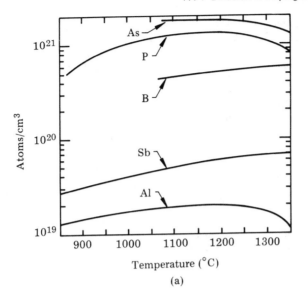

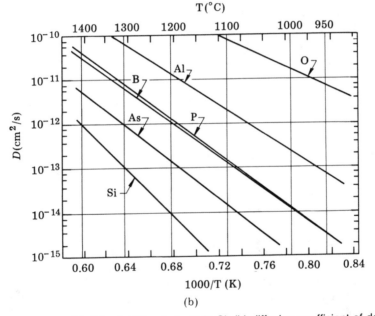

**FIGURE 7.10**    (a) Solid solubility of atoms in Si; (b) diffusion coefficient of dopant atoms in Si. [After Beadle et al. (1985).]

function, it is sometimes helpful to use the values tabulated in Table 7.2. For this example, the expression

$$\text{erfc } (x) \simeq \frac{1}{\sqrt{\pi}} \frac{e^{-x^2}}{x} \quad (x \gg 1)$$

**TABLE 7.2**   Error Function and Complementary Error Function

| $x$ | erf $(x)$ | erfc $(x)$ |
|---|---|---|
| 0.0 | 0.0000 | 1.0000 |
| 0.01 | 0.0113 | 0.9887 |
| 0.1 | 0.1125 | 0.8875 |
| 1.0 | 0.8427 | 0.1573 |
| 1.5 | 0.9661 | 0.0339 |
| 2.0 | 0.9953 | 0.0047 |
| 2.25 | 0.998 537 | 0.001463 |
| 2.50 | 0.999 593 | $4.070 \times 10^{-4}$ |
| 3.00 | 0.999 977 91 | $2.209 \times 10^{-5}$ |
| 3.25 | 0.999 995 70 | $4.300 \times 10^{-6}$ |
| 3.50 | 0.999 999 257 | $7.430 \times 10^{-7}$ |
| 3.75 | 0.999 999 886 | $1.140 \times 10^{-7}$ |

*Source:* After Ghandhi (1983).

is used to estimate the value of $\text{erfc}^{-1} (C_B/C_S)$. Either a trial and error or a graphical method can be used. We found that

$$\text{erfc}^{-1} (2.9 \times 10^{-6}) = 3.3.$$

Then

$$x_j = 1.47 \times 10^{-5} \times 3.3 = 0.49 \ \mu\text{m}.$$

The total amount of dopants introduced into the substrate, $Q(t)$ is given by eq. (7.17). In this example,

$$Q = \frac{\sqrt{4Dt}}{\sqrt{\pi}} C_S = \frac{1.47 \times 10^{-5}}{1.77} \times 1.8 \times 10^{20}$$

$$= 2.9 \times 10^{15} \ \text{atoms/cm}^2.$$

One can also see from this example that the linear diffusion profile is not a good approximation to estimate the junction depth for this example. For the linear approximation to be applicable, the bulk concentration in the substrate; $C_B$, should be larger than 0.5 of the surface concentration (i.e., $C_B/C_S \gtrsim 0.5$). It is clear from Fig. 7.5 that the linear profile is about the same as the erfc profile for $C_B/C_S \gtrsim 0.5$. As the ratio $C_B/C_S$ decreases (in this example, $C_B/C_S$ is $2.9 \times 10^{-6}$), the linear profile approximation should be avoided in estimating the junction depth.

## 7.4.2 Drive-In Diffusion

The predeposition step gives a constant impurity concentration at the surface. This surface concentration, which often equals the solid solubility of the impurity, may be too high for device applications. The impurities introduced by the predeposition step can be redistributed deeper in the substrate to lower the concentrations by a drive-in step. In the drive-in step, the total amount of dopant atoms remains fixed. For these boundary conditions, the Gaussian solution [eq. (7.20)] applies. However, it is important to recognize that for eq. (7.20) to be applicable to the drive-in diffusion, $(\sqrt{Dt})_d$ for the drive-in should be much larger than $(\sqrt{Dt})_p$ for predeposition, such that the predeposition diffusion profile can be represented by a delta function. If this requirement is met, the concentration profile due to the drive-in diffusion is given by

$$C(x,\ t) = \frac{Q}{\sqrt{\pi Dt}} \exp\left(-\frac{x^2}{4Dt}\right) \tag{7.39}$$

where

$$Q = \frac{\sqrt{4(Dt)_{predep}}}{\sqrt{\pi}} \times C_S(predep) = constant.$$

The surface concentration $C_S(t)$ is given by

$$C_S(t) = C(0,\ t) = \left(\frac{Q}{\sqrt{\pi Dt}}\right)_{drive\text{-}in}$$

This equation indicates that the surface concentration during drive-in diffusion decreases so that the total amount of $Q$ is kept constant as the dopant atoms move deeper into the substrate.

## EXAMPLE 2

Calculate the junction depth $x_j$ of the sample in Example 1 after a drive-in diffusion at 1100°C for 4.5 hours.

**Solution:** To apply the Gaussian solution to the drive-in case, we must first make sure that the condition $(\sqrt{Dt})_{drive\text{-}in} \gg (\sqrt{Dt})_{predep}$ is met.

$$(\sqrt{Dt})_{drive\text{-}in} = (2.5 \times 10^{-13}\ \frac{cm^2}{s} \times 4.5 \times 60 \times 60\ s)^{1/2} = 6.4 \times 10^{-5}\ cm,$$

$$(\sqrt{Dt})_{predep} = (1.5 \times 10^{-14}\ \frac{cm^2}{s} \times 60 \times 60\ s)^{1/2} = 7.4 \times 10^{-6}\ cm.$$

From this we can see that $(\sqrt{Dt})_d \gg (\sqrt{Dt})_p$.
From eq. (7.39),

$$C(x,\ t) = \frac{Q}{\sqrt{\pi Dt}}\ e^{-(x^2/4Dt)}$$

$$Q = \left(\frac{2C_S\sqrt{Dt}}{\sqrt{\pi}}\right)_{predep}$$

Therefore,

$$C(x, t) = \frac{2(C_S\sqrt{Dt})_p}{\pi(\sqrt{Dt})_d} e^{-(x^2/4Dt)}$$

$$= C_S'(t)e^{-(x^2/4Dt)}$$

where

$$C_S'(t) = \frac{2(C_S\sqrt{Dt})_p}{\pi(\sqrt{Dt})_d} = \frac{2 \times 3.5 \times 10^{20} \times 7.4 \times 10^{-6}}{3.14 \times 6.4 \times 10^{-5}} = 2.5 \times 10^{19}/cm^3.$$

The surface concentration after the drive-in diffusion has decreased from $3.5 \times 10^{20}/cm^3$ to $2.3 \times 10^{19}/cm^3$.

The junction depth can be obtained by rearranging eq. (7.39):

$$C(x, t)_{x_j} = C_B = (C_S'(t)e^{-(x^2/4Dt)})_{x_j}$$

$$x_j = \left(-4Dt \ln \frac{C_B}{C_S'}\right)^{1/2}$$

$$= \left(-4 \times 3 \times 10^{-13} \times 4.5 \times 60 \times 60 \times \ln \frac{1 \times 10^{15}}{2.5 \times 10^{19}}\right)^{1/2}$$

$$= 4.4 \times 10^{-4} \text{ cm} = 4.4 \text{ } \mu m.$$

$$(7.39a)$$

After the drive-in diffusion, the junction depth has increased from 0.49 $\mu$m to 4.4 $\mu$m.

## 7.5 Junction Depth

We will now discuss the relationship between junction depth and sheet resistance of the diffused layer (Chapter 2). The junction depth $x_j$ of a $p$-$n$ junction is defined by the condition

$$C(x_j, t) = C_B$$

as shown in Fig. 7.9. The impurity distribution in the diffused layer is usually nonuniform in depth; the resistivity of the diffused layer therefore also varies in depth. By knowing the average resistivity of the diffused layer, it is possible to relate the junction depth to the diffused impurity concentration at the surface and the bulk concentration for simple diffusion profiles.

The conductivity of a diffused $n$-type layer is given by

$$\sigma = e\mu_n n \qquad (7.40)$$

where $\sigma$ is the conductivity, $e$ the electronic charge, $\mu_n$ the mobility of the electrons, and $n$ the concentration of the electrons. The average conductivity $\bar{\sigma}$ of an $n$-type diffused layer can be written as

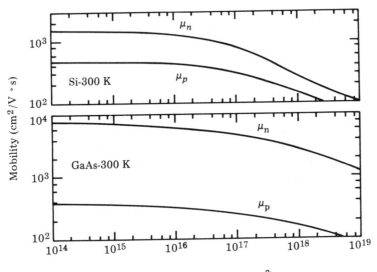

**FIGURE 7.11**   Mobility versus impurity concentration. [After Sze (1981)].

$$\bar{\sigma} = \frac{1}{\bar{\rho}} = \frac{e}{x_j} \int_0^{x_j} \mu_n n(x) \, dx \tag{7.41}$$

where $\bar{\rho}$ is the average resistivity. To evaluate the integral in eq. (7.41), we must know the relationship between the mobility and the impurity concentration and the carrier concentration as a function of depth, $n(x)$. For fully ionized impurities, $n(x)$ may be approximated by $C(x)$, the impurity profile. The mobility of carriers remains relatively unchanged at low impurity concentrations and decreases with increasing impurity concentration (see Fig. 7.11). There are no analytical expressions available for the curves shown in Fig. 7.11. The decrease in mobility with increasing total impurity concentration may be understood from eq. (2.16), $\mu = e\tau/m$. The time between collisions, $\tau$, is reduced when more impurity atoms are incorporated in the lattice. With the assumption that $n(x) \simeq C(x)$, the integral in eq. (7.41) can be evaluated using the data in Fig. 7.11. It should be noted that this assumption does not hold for very highly doped semiconductors.

## EXAMPLE 3

Evaluate the junction depth of a diffused layer where the diffused impurity concentration corresponds to a rectangular profile as shown in Fig. 7.12.

**Solution:** For a rectangular profile, eq. (7.41) can be written as

$$\frac{1}{\bar{\rho}} = \frac{e}{x_j} \int_0^{x_j} \mu_n (C_S - C_B) \, dx = e\mu_n (C_S - C_B).$$

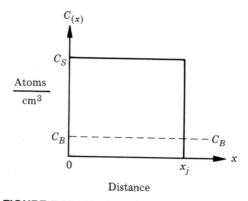

**FIGURE 7.12**   Rectangular impurity profile.

Upon rearranging we obtain

$$\bar{\rho} = R_s x_j = \frac{1}{e\mu_n(C_S - C_B)} \tag{7.42}$$

$$x_j = \frac{1}{R_s e\mu_n(C_S - C_B)} \tag{7.43}$$

where $R_s$ is the sheet resistance (Section 2.5).

The junction depth can then be determined by measuring the sheet resistance $R_s$ of the diffused layer and by knowing the surface concentration $C_S$ and the bulk concentration $C_B$.

---

The rectangular profile is not a realistic profile in IC technology. The erfc and Gaussian functions represent much more common impurity profiles. Equation (7.41) has been evaluated numerically for both the erfc and the Gaussian profile. The results are shown in Fig. 7.13.

The following illustrates how the integration of eq. (7.41) may be performed for an erfc diffusion profile (see Fig. 7.5). For simplicity we assume that the bulk concentration $C_B$ is very small compared to $C_S$; therefore, $x_j$ is near or beyond $\sqrt{4Dt}$. Substituting erfc($x$) for $n(x)$ into eq. (7.41), we have

$$\bar{\sigma} = \frac{1}{\bar{\rho}} = \frac{e}{x_j} \int_0^{x_j} \mu_n C_S \, \text{erfc}\left(\frac{x}{\sqrt{4Dt}}\right) dx.$$

Since $C_B$ is very low, $x_j$ may be taken to be approaching infinity ($x_j \gtrsim \sqrt{4Dt}$) for the purpose of integration. The integral may be approximated by eq. (7.17) or eq. (7.17a), and by assuming that $\mu_n$ is constant, we have

$$\bar{\sigma} = \frac{e\mu_n C_S}{x_j} \sqrt{\frac{4Dt}{\pi}}. \tag{7.44}$$

From $C(x_j, t) = C_B$, we have

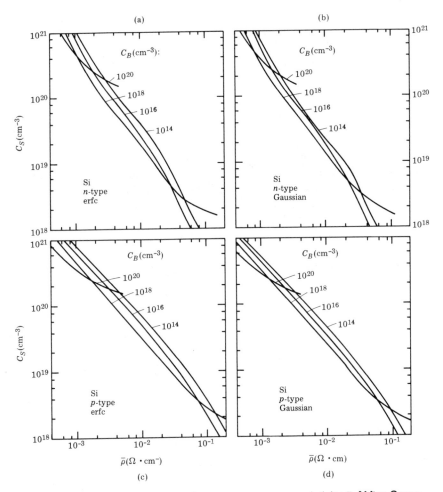

**FIGURE 7.13**  Surface concentration versus average resistivity $\bar{\rho}$. [After Grove (1967).]

$$C_B = C_S \, \text{erfc} \left( \frac{x_j}{\sqrt{4Dt}} \right)$$

$$x_j = \sqrt{4Dt} \, \text{erfc}^{-1} \left( \frac{C_S}{C_B} \right).$$

Substituting into eq. (7.44) yields

$$\bar{\sigma} = \frac{e\mu_n C_S}{\sqrt{4Dt} \, \text{erfc}^{-1} (C_S/C_B)} \sqrt{\frac{4Dt}{\pi}}$$

$$= \frac{e\mu_n C_S}{\sqrt{\pi} \, \text{erfc}^{-1} (C_S/C_B)}.$$

Therefore,

$$\frac{1}{\bar{\rho}} = \frac{1}{R_s x_j} = \bar{\sigma} = \frac{e\mu_n C_S}{\sqrt{\pi}\ \text{erfc}^{-1}\ (C_S/G_B)} \qquad (7.45)$$

or

$$R_s x_j = \frac{\sqrt{\pi}\ \text{erfc}^{-1}\ (C_S/C_B)}{e\mu_n C_S}. \qquad (7.46)$$

One can see that $\bar{\rho}$ depends primarily on the diffusion profile (erfc in this case), as demonstrated previously for a rectangular profile represented by eq. (7.43). When $C_B$ is not small compared to $C_S$, it can be shown that $\bar{\rho}$ (or $R_s x_j$) is again a function of erfc with an expression similar to that in eq. (7.46).

It should be noted that the average resistivity $\bar{\rho}(= x_j R_s)$ is a function of $C_S$, $C_B$, and the shape of the diffusion profile, and independent of the diffusion coefficient and the diffusion time. These simple relationships shown in Fig. 7.13 make the evaluation of the junction depth relatively easy by means of the four-point probe method.

## EXAMPLE 4

The sample in Examples 1 and 2 has a sheet resistance, $R_s$, of 60 $\Omega/\square$ and 75 $\Omega/\square$ after the boron predeposition and drive-in diffusion. Compare the calculated junction depths with those obtained from sheet resistance measurements.

**Solution:** From Example 1, the boron surface concentration $C_S$ is $3.5 \times 10^{-3}$ cm$^{-3}$ after predeposition, and the bulk concentration $C_B$ is $1 \times 10^{15}$ cm$^{-3}$. From Fig. 7.13c the average $\bar{\rho}$ is $2.5 \times 10^{-3}\ \Omega \cdot$ cm.

$$x_j = \frac{\bar{\rho}}{R_s} = \frac{25\ \Omega \cdot \mu\text{m}}{60\ \Omega/\square} = 0.41\ \mu\text{m}.$$

This is close to the calculated $x_j$ [using eq. (7.16) in Example 1] of 0.49 $\mu$m.

The surface concentration after the drive-in diffusion is $2.3 \times 10^{19}$/cm$^3$ (see Example 2). From Fig. 7.13d, the $\bar{\rho}$ value is about $2 \times 10^{-2}\ \Omega \cdot$ cm.

$$x_j = \frac{\bar{\rho}}{R_s} = \frac{200\ \Omega \cdot \mu\text{m}}{75\ \Omega/\square} = 2.7\ \mu\text{m}.$$

This is not too close to the calculated $x_j$ [using eq. (7.39) in Example 2] of 4.4 $\mu$m.

---

This example shows the general validity of using Fig. 7.13 and $R_s$ to estimate the junction depth. For a more accurate determination of $x_j$, better measurement methods are required. These methods include angle lap and stain, spreading resistance measurement, and secondary ion mass spectroscopy (Wolf and Tauber, 1986). It should be noted that we have treated the diffusivities to be constant in previous examples. In general, diffusion can be characterized by constant diffusivities only for impurity concentrations not much larger than the intrinsic carrier concentration, $n_i$. However, as the impurity concentration increases to near the solid solubility limits, the diffusivity becomes concentration dependent, and the diffusion profile can become rather complex (Ghandhi, 1983). A concentration-dependent diffusivity is often encountered in the formation of shallow junctions, where impurity atoms are introduced into a thin region either by diffusion or implantation.

## 7.6 Diffusion Mechanisms

We are interested primarily in diffusion in a crystalline solid; it is natural to think that the lattice type of such a solid should be an important consideration in diffusion mechanisms. Let us consider an impurity atom located between the host atoms (Fig. 7.14a). This impurity atom, called an interstitial, can jump from one interstitial site to the next vacant interstitial site. This jumping path is termed the interstitial mechanism. One can also consider an impurity atom located in a lattice site normally occupied by a host atom jumping to a neighboring vacant site. This path is termed the substitutional or vacancy mechanism, as illustrated in Fig. 7.14b. It is also possible for a substitutional impurity atom to move by pushing a neighboring host atom into an adjacent interstitial site, and take up the vacant substitutional site thus created. This process is called the interstitialcy mechanism. There are other possible mechanisms, but in general, diffusion takes place in semiconductors via one or a combination of the several mechanisms mentioned above.

### 7.6.1 Interstitial Mechanism

The interstitial sites of a diamond or zinc blend lattice are usually vacant. The probability of finding an available site to jump into is very high for an interstitial atom. As the interstitial atom jumps from one site to another, it faces a constriction due to the host atoms (see Fig. 7.14a); the jumping atom needs a little push to squeeze by. This situation is described as having an activation or energy barrier. In this case the barrier is also periodic in the lattice (Fig. 7.6). The jump frequency of an interstitial atom is then given by eq. (7.23):

$$k^0 = v_0 e^{-E_m/kT}$$

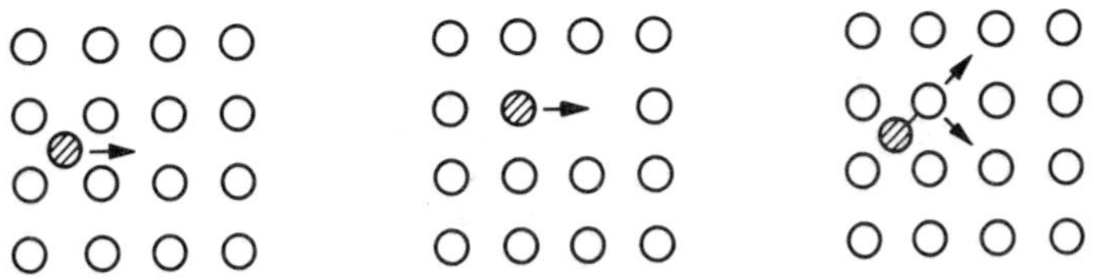

(a) Interstitial        (b) Vacancy        (c) Interstitialcy

**FIGURE 7.14**   Diffusion mechanisms.

where $v_0$ is the lattice vibration frequency, usually on the order of $10^{13}$ to $10^{14}/s$, and $E_m$ is the activation energy for migration. The exponential term $e^{-Em/kT}$ is the Boltzmann factor, which describes the probability that the atom can make a successful jump when enough energy is supplied to overcome the energy barrier, $E_m$. Typical values of $E_m$ range between 0.6 and 1.2 eV for interstitial migration (Ghandhi, 1983). The diffusion coefficient is given by eq. (7.34)

$$D = \overline{\lambda}^2 k^0 = \overline{\lambda}^2 v_0 e^{-Em/kT} \tag{7.47}$$

where $\overline{\lambda}$ is the jump distance, usually taken as the lattice parameter, as described before.

## 7.6.2 Substitutional or Vacancy Mechanism

The jumping of a substitutional atom to a neighboring substitutional site requires a vacancy to be created in the adjacent site (Fig. 7.14b). The probability of such an event is $e^{-Ef/kT}$, where $E_f$ is the formation energy of a vacancy in the lattice. The movement of the atom also involves the breaking of bonds and making new bonds after the jump. The energy barrier for this jumping process is $E_m$ (activation energy for migration). The jumping frequency is given by

$$k^0 = v_0 e^{-Ef/kT} e^{-Em/kT}$$

$$= v_0 e^{-(Ef+Em)/kT}$$

and the diffusion coefficient is

$$D = k^0 \lambda^2 = \lambda^2 v_0 e^{-(Ef+Em)/kT}. \tag{7.48}$$

The value of $(E_m + E_f)$ is between 3 and 4 eV (Fair, 1981).

## 7.6.3 Interstitial–Substitutional Mechanism

Many impurities have both an interstitial solubility ($N_i$) and a substitutional solubility ($N_s$). These interstitial impurity atoms and substitutional impurity atoms can diffuse independently or interdependently. Usually, $N_s$ is larger than $N_i$, but substitutional diffusion is much slower than interstitial diffusion. For independent movements, the combined jumping frequency is

$$v_{\text{effective}} = \frac{v_s N_s}{N_s + N_i} + \frac{v_i N_i}{N_s + N_i} \tag{7.49}$$

where $v_s$ and $v_i$ are the substitutional and interstitial jump frequency. In Si and GaAs, these two components of diffusion appear to move interdependently. They are related by a dissociative reaction of a substitutional impurity atom (S) into an interstitial impurity atom (I) and a vacancy (V):

$$S \rightleftharpoons I + V. \tag{7.50}$$

The movement of impurities becomes primarily controlled by the rate of this dissociation because it produces the faster interstitial diffuser.

In a highly concentrated region, interstitial atoms rapidly diffuse to a lower concentration region, thus driving eq. (7.50) to the right to produce more interstitials; the diffusion rate is enhanced as a result. In a low-concentration region, interstitials tend to stay in a local region due to a smaller concentration gradient; dissociation is therefore reduced, thus retarding the overall diffusion rate. It should also be noted that vacancy sinks such as dislocations are necessary to maintain the equilibrium vacancy concentration locally. If the vacancies cannot be removed, the dissociation rate may also be retarded, driving eq. (7.50) to the left. The interstitial substitutional diffusion mechanism is therefore characterized by a strong dependence of the diffusion rate on impurity and defect concentrations.

## 7.6.4 Interstitialcy and the Kick-Out Mechanism

When the diffusing interstitial atoms have a size comparable to that of the lattice atoms, the interstitialcy mechanism for diffusion may take place. In this case the interstitial impurity atom moves into a host lattice site by pushing a neighboring normal atom into the adjacent interstitial site. This process repeats itself when a self-interstitial atom pushes the substitutionally located impurity atom into an interstitial site (Frank et al., 1984)(see Fig. 7.14c).

The kick-out mechanism is rather similar to the interstitialcy mechanism. In this case, a host self-interstitial atom diffuses around the lattice, and when it comes up to a substitutional impurity atom, the self-interstitial pushes the impurity atom into an adjacent interstitial site. The interstitial impurity then diffuses interstitially until it goes back into a host lattice site by displacing a host atom. It is experimentally difficult to distinguish the kick-out mechanism from the interstitialcy mechanism. The generally accepted view is that the interstitial impurity atoms may tend to diffuse longer distances before returning to the normal lattice sites for the kick-out mechanism, whereas the impurity atoms tend to diffuse interstitially for a relatively short distance before going into the normal lattice sites for the interstitialcy mechanism.

## 7.6.5 Brief Summary on Mechanisms

The diffusion process in Si has been investigated for quite some years; however, the diffusional mechanisms responsible for some impurities are still under debate at present. It is generally accepted that Sb diffusion is dominated by a vacancy-type mechanism but with a nonzero interstitial-type component. Phosphorus and boron diffuse predominantly by an interstitial-type mechanism. Arsenic appears to diffuse by a vacancy mechanism but with a larger interstitial component of diffusion than Sb (Fair, 1981; Fahey, 1985; Tan and Gösele, 1985). The diffusion of Al, Ga, and In is believed to have a strong interstitial component (Tan and Gösele, 1985). Elements such as Li, Na, Ar, He, and H are believed to diffuse by the interstitial mechanism. Most of the transition metals, such as Co, Cu, Au, Fe, Ni, Pt, and Ag, sometimes referred to as fast diffusers, diffuse by the interstitial–substitutional mechanism or the kick-out mechanism (Frank et al., 1984; Tan and Gösele, 1985). The mechanism for Si self-diffusion appears to be vacancy

**TABLE 7.3**   Diffusion Data of Fast Diffusers in Si and Si Self-Diffusion in Intrinsic Si

| Element | $D_0$ (cm²/s) | $E_A$ (eV) | Solubility (cm⁻³) |
|---|---|---|---|
| Li | $2.3 \times 10^{-3}$–$9.4 \times 10^{-4}$ | 0.63–0.78 | $7 \times 10^{19}$ (1200°C) |
| Na | $1.6 \times 10^{-3}$ | 0.72 | $10^{18}$–$9 \times 10^{18}$ (600–1200°C) |
| K | $1.1 \times 10^{-3}$ | 0.76 | $9 \times 10^{17}$–$7 \times 10^{18}$ (600–1300°C) |
| Cu | $4 \times 10^{-2}$ | 1.0 | $5 \times 10^{15}$–$3 \times 10^{18}$ (600–1300°C) |
| Ag | $2 \times 10^{-3}$ | 1.6 | $6.5 \times 10^{15}$–$2 \times 10^{17}$ (1200–1350°C) |
| Au | $1.1 \times 10^{-3}$ | 1.12 | $5 \times 10^{14}$–$5 \times 10^{16}$ (900–1300°C) |
| $(Au)_i$ | $2.4 \times 10^{-4}$ | 0.39 | |
| $(Au)_s$ | $2.8 \times 10^{-3}$ | 2.04 | |
| Pt | $1.5 \times 10^{2}$–$1.7 \times 10^{2}$ | 2.22–2.15 | $4 \times 10^{16}$–$5 \times 10^{17}$ (800–1000°C) |
| Ni | 0.1 | 1.9 | $6 \times 10^{18}$ (1200–1300°C) |
| Cr | 0.01 | 1.0 | $2 \times 10^{13}$–$2.5 \times 10^{15}$ (900–1280°C) |
| Co | $9.2 \times 10^{4}$ | 2.8 | $2.5 \times 10^{16}$ (1300°C) |
| $O_2$ | $7 \times 10^{-2}$ | 2.44 | $1.5 \times 10^{17}$–$2 \times 10^{18}$ (1000–1400°C) |
| $H_2$ | $9.4 \times 10^{-3}$ | 0.48 | $2.4 \times 10^{21} \exp(-1.86/kT)$ at 1 atm |
| Si | $1.5 \times 10^{3}$ | 5.02 | Self-diffusion in intrinsic Si at high temperatures ($T > 1050$°C) |
| Si | 1.5 | 4.25 | Self-diffusion in intrinsic Si at low temperatures ($T < 900$°C) |

*Source:* After Beadle et al. (1985), Fair (1981), and Tsai (1983).

controlled at low temperatures ($<1000$°C); the contribution to self-diffusion from interstitials may become dominant when the temperature increases to $>1000$°C (Wolf and Tauber, 1986). Table 7.3 summarizes some of the diffusion data in Si.

For GaAs, the diffusion mechanisms are even less well characterized. Significant diffusion occurs at temperatures often higher than that of the decomposition temperature of GaAs (about 500 to 600°C in an open system). To prevent vaporization, capping the sample or annealing the sample in an ambient is commonly practiced. This procedure apparently complicates the experimental results and the interpretation of data. Of all the impurities, Zn has been studied the most. Generally speaking, group II elements (Be, Cd, Mg, Hg, and Zn), the p-type dopants, move along the Ga sublattice by the interstitial–substitutional mechanism; since each Ga vacant site, $V_{Ga}$, is surrounded by four As atoms, diffusion of these p-type impurities cannot proceed via Ga vacancies alone. Figure 7.15 shows the diffusion of Zn in GaAs, where the diffusivity depends on concentration. The concentration profiles do not follow the simple erfc function, where the diffusivity is not a function of concentration.

Group VI elements (Se, S, Te), the n-type impurities, also move along the As sublattice by the interstitial–substitutional mechanism. Group IV elements (C, Si, Ge, and Sn) are amphoteric dopants; the carrier type depends on which sublattice they reside in. The detailed mechanisms have not been clarified. At low Si con-

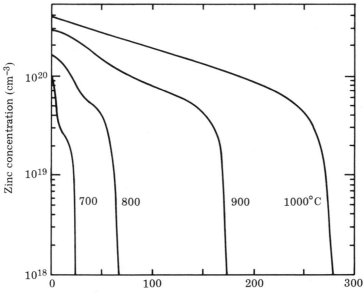

**FIGURE 7.15**   Diffusion profiles of Zn in GaAs at 700, 800, 900, and 1000°C.
[After Casey (1973).]

centrations, there have been experimental observations that the diffusion of Si appears to be proportional to the concentration of Ga vacancies. Under the condition of high Si concentration at the sample surface encapsulated with $SiO_2$, Si atoms are believed to diffuse in pairs (Greiner, 1985). Silicon pairs with one Si atom located in a Ga vacant site, $Si_{Ga}^+$, and the other Si atom located in a neighboring As vacant site, $Si_{As}^-$, form a $Si_{Ga}$–$Si_{As}$ complex. This Si pair is electrically inactive since the electron created by $Si_{Ga}^+$ compensates the hole created by $Si_{As}^-$; however, there is experimental evidence that the Si pairs diffuse much faster than do single Si atoms. Since at high Si concentrations, the probability of pairing is higher than that at low concentrations, the pair diffusion is much more prominent at high concentration. Table 7.4 lists some of the diffusion data in GaAs.

## 7.7 Effect of Fermi-Level Position on Diffusion

Based on the discussion in Section 6.4.3, we note that the total vacancy concentration $[V_T]$ in Si depends on the location of the Fermi level. If diffusion depends on the vacancy concentration, the rate of diffusion should increase accordingly when $[V_T]$ increases, due to the creation of ionized (charged) vacancies. We showed that at a commonly used diffusion temperature of 1400 K (about 1100°C), the charged vacancies $[V^-]$ exceeds the neutral vacancies, $[V^x]$, over the entire band gap (about 0.6 eV at 1400 K) and by a factor of 100 for heavily doped $n$-type Si.

**TABLE 7.4**   Diffusion Data of Impurities in GaAs ($D = D_0 e^{-E_A/kT}$)

| Element | $D_0$ (cm²/s) | $E_A$ (eV) | Solubility (cm⁻³) |
|---|---|---|---|
| As | $7 \times 10^{-1}$ | 3.2 | |
| Ga | $2.9 \times 10^8$ | 6.0 | |
| Ag | $2.5 \times 10^1$ | 2.27 | |
| Au | 2.9 | 2.64 | |
| Be | $7.3 \times 10^{-6}$ | 1.2 | |
| Cd | Concentration dependent | | |
| Cr | $4.3 \times 10^3$ | 3.4 | |
| Cu | $3 \times 10^{-2}$ | 0.53 | $3.7 \times 10^{23}$ $\exp(-1.3/kT)$ |
| Hg | $D(1000°C) = 5 \times 10^{-14}$ | | |
| In | $D(1000°C) = 7 \times 10^{-11}$ | | |
| Li | $5.3 \times 10^{-1}$ | 1.0 | $7.0 \times 10^{21}$ $\exp(-0.57/kT)$ |
| Mg | $2.6 \times 10^{-2}$ | 2.7 | |
| Mn | $6.5 \times 10^{-1}$ | 2.49 | |
| O | $2 \times 10^{-3}$ | 1.1 | |
| S | $1.85 \times 10^{-2}$ | 2.6 | |
| Se | $3.0 \times 10^3$ | 4.16 | |
| Si(low concentration) | $1.5 \times 10^{17}$ | 6.85 | $SiO_2^a$ |
| | 5.03 | 3.48 | $Si_3N_4^a$ } cap |
| Si(high concentration) | 0.11 | 2.5 | $SiO_2$ |
| Sn | $3.8 \times 10^{-2}$ | 2.7 | |
| Te | $D(1000°C) = 10^{-13}$ | | |
| | $D(1100°C) = 2 \times 10^{-12}$ | | |
| Zn | Concentration dependent | | See Fig. 7.15 |

*Source:* After Sharma (1970), Casey (1973), Greiner (1985), and Tan and Gösele (1988).
[a]Deduced from Onuma et al. (1982).

To see how the concentration of vacancies affects the rate of diffusion, we need to know the functional dependence of $D$ (diffusivity) on the various species of ionized vacancies. Let us first examine the case of Si self-diffusion. This type of diffusion experiment is usually performed with a radioactive tracer. Let us further assume that Si self-diffuses via the vacancy mechanism that occurs by the simultaneous migration of a neutral vacancy, $V^x$, and other charged vacancies. From eq. (7.48) we see that the diffusivity is proportional to the concentration of the vacancies (i.e., the diffusing atom needs to find a vacancy to exchange places with). The diffusivity of a tracer atom is then

$$D_{self} = D(V^x)[V^x] + D(V^-)[V^-] + D(V^=)[V^=] + D(V^+)[V^+] \quad (7.51)$$

where $D(V^x)$, $D(V^-)$, $D(V^=)$, and $D(V^+)$ are the diffusivities of the indicated vacancy charge states, and $[V^x]$, $[V^-]$, $[V^=]$, and $[V^+]$ are the vacancy concentrations, now expressed as atomic fractions for convenience (number of vacancies/number of lattice sites). Equation (7.51) is easy to understand: the tracer atoms migrate by exchanging places with vacancies of all charge states simultaneously.

The vacancy can exchange places with any atom, but for the tracer to move, it has to exchange places with an adjacent vacancy. Under intrinsic conditions (i.e., $E_F$ of the sample is located at or near the middle of band gap for a given temperature), the Si self-diffusivity is

$$D_{self}^i = D(V^x)[V^x] + D(V^-)[V^-]_i + D(V^=)[V^=]_i + D(V^+)[V^+]_i \quad (7.52)$$

where the subscript $i$ indicates concentration under intrinsic conditions. The neutral vacancy $[V^x]$ should be unaffected by the Fermi-level position; it follows that $[V^x]_i \equiv [V_x]$, so the subscript $i$ is not needed. Under extrinsic conditions (i.e., $E_F$ of the sample is located far away from the middle of the band gap), the concentration of various vacancy species will change accordingly. We now need to know how the vacancy concentrations change with respect to those under intrinsic condition. For the $V^+$ case, we find that $V^+$ may form with the following reaction of the neutral vacancy $V^x$ with a positively charged hole $h^+$:

$$V^x + h^+ \rightleftharpoons V^+. \quad (7.53)$$

From the mass-action law, the equilibrium ratio is

$$K_1 = \frac{[V^+]}{p[V^x]} \quad (7.54)$$

where $K_1$ is the equilibrium constant and $p$ is the concentration of holes ($h^+$). Under intrinsic conditions, $p = n = n_i$. Therefore,

$$K_1 = \frac{[V^+]_i}{n_i[V^x]_i} = \frac{[V^+]_i}{n_i[V^x]}. \quad (7.55)$$

Combining eqs. (7.54) and (7.55) gives us

$$\frac{[V^+]}{[V^+]_i} = \frac{p}{n_i}. \quad (7.56)$$

This equation may be derived from eq. (6.13) (see the problems). For a vacancy with $-r$ charges, we have the following reaction:

$$V^x + re^- \rightleftharpoons V^{-r}. \quad 7.57)$$

Therefore,

$$K_2 = \frac{[V^{-r}]}{[n]^r[V^x]} = \frac{[V^{-r}]_i}{[n_i]^r[V^x]}. \quad (7.58)$$

It follows, then, that

$$\frac{[V^{-r}]}{[V^{-r}]_i} = \left(\frac{n}{n_i}\right)^r. \quad (7.59)$$

The self-diffusivity under extrinsic conditions is, then,

$$D_{self} = D_i^x + D_i^- \frac{n}{n_i} + D_i^= \left(\frac{n}{n_i}\right)^2 + D_i^+ \frac{p}{n_i} \quad (7.60)$$

where

$$D_i^x = D(V^x)[V^x]_i = D(V^x)[V^x]$$

$$D_i^- = D(V^-)[V^-]_i$$

$$D_i^= = D(V^=)[V^=]_i$$

$$D_i^+ = D(V^+)[V^+]_i.$$

These are diffusivities under intrinsic conditions as defined in eq. (7.52). For the intrinsic case, $n = p = n_i$; eq. (7.60) reduces back to eq. (7.52). For impurity diffusion in Si, a similar analysis applies and we find that

$$D_{imp} = D_i^x + \sum_{r=1} D_i^{-r} \left(\frac{n}{n_i}\right)^r + \sum_{r=1} D_i^{+r} \left(\frac{p}{n_i}\right)^r \qquad (7.61)$$

where $D_i^x$, $D_i^{-r}$, and $D_i^{+r}$ now refer to the impurity diffusivity under intrinsic conditions. Depending on the specific dopant, some of the quantities $D_i^x$, $D_i^-$, $D_i^{+r}$ may be negligibly small. Table 7.5 shows some of the intrinsic diffusivities in Si.

**TABLE 7.5**  Selected Values of Intrinsic Diffusivities in Silicon

| Element | $D_0$ (cm²/s) | $E_A$ (eV) | Remarks |
|---------|---------------|------------|---------|
| Al | 0.5 | 3.0 | |
| | 1530 | 4.1 | $D_i^x + D_i^+$ |
| B | 0.76 | 3.46 | $D_i^+$ |
| Ga | 225 | 4.12 | |
| | 13.1 | 3.7 | $D_i^x + D_i^+$ |
| In, Tl | 16.5 | 3.9 | |
| | 269 | 4.19 | $D_i^x + D_i^+$ |
| As | 12 | 4.05 | $D_i^-$ |
| | 0.066 | 3.44 | $D_i^x$ |
| N | 0.05 | 3.65 | $D_i^x$ |
| P | 3.85 | 3.66 | $D_i^x$ |
| | 4.44 | 4.0 | $D_i^-$ |
| | 44.2 | 4.37 | $D_i^{-2}$ |
| Sb | 12.9 | 3.98 | $2 \times 10^{19} - 5 \times 10^{19}$ |
| | 15 | 4.08 | $D_i^-$ |
| Bi | 396 | 4.12 | $D_i^-$ |
| C | 1.9 | 3.1 | |
| Ge | $6.26 \times 10^5$ | 5.28 | Ge powder |
| Sn | 32 | 4.25 | Tracer measurement |
| Si | 0.015 | 3.89 | $D_i^x$ |
| | 16 | 4.54 | $D_i^-$ |
| | 1180 | 5.09 | $D_i^+$ |

*Source:* After Beadle et al. (1985).

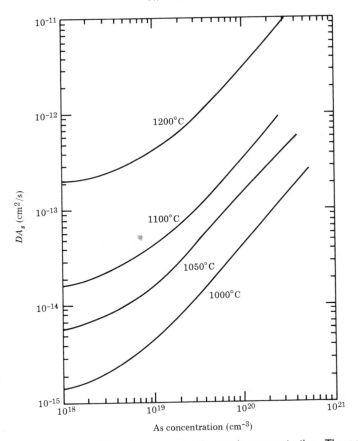

**FIGURE 7.16**  Arsenic diffusivity versus total arsenic concentration. The solid curves are calculated. [After Fair (1981).]

We should now ask the question of whether there is any experimental evidence that the diffusion rates are indeed enhanced due to the Fermi-level position. Figure 7.16 shows one of the many examples of such effects reported in the literature. As shown in the figure, the diffusivity of arsenic in Si is enhanced significantly as the total arsenic concentration increases, and the diffusivity may now be expressed as

$$D_{As} = D_i^x + D_i^- \frac{n}{n_i}$$

$$= 0.066e^{-3.44/kT} + 12e^{-4.05/kT} \frac{n}{n_i}.$$

(7.62)

**EXAMPLE 5** _____

Calculate the diffusion coefficient of arsenic in Si doped with $1 \times 10^{20}$ As/cm$^3$ at 1400 K.

**Solution:** The diffusion coefficient of As into Si is given by eq. (7.62):

$$D_{As} = 0.066 \exp\left(-\frac{3.44}{kT}\right) + 12 \exp\left(-\frac{4.05}{kT}\right) \frac{n}{n_i}.$$

At 1400 K, $n_i \simeq 1.5 \times 10^{19}$ cm$^{-3}$ [from eq. (3.44)]. Therefore,

$$\frac{n}{n_i} = \frac{1 \times 10^{20}}{1.5 \times 10^{19}} \simeq 6.7.$$

Substituting this value into the eq. (7.62), we have

$$D_{As} = 2.3 \times 10^{-14} + 6.7 \times 2.6 \times 10^{-14}$$
$$= 2.0 \times 10^{-13} \text{ cm}^2/\text{s}.$$

This calculated value agrees with the result shown in Fig. 7.16.

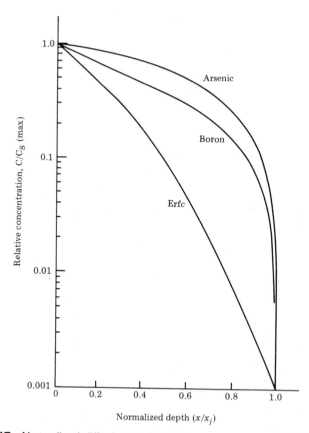

**FIGURE 7.17** Normalized diffusion profiles for boron and arsenic in silicon under extrinsic conditions. [After Ghandhi (1983).]

When the impurity diffusivities are enhanced in the extrinsic regime ($n \gg n_i$), they can no longer be treated as constants. The diffusivities increase as a function of concentration. The continuity equation (7.12) needs to be solved numerically. The concentration profiles are not represented by simple Guassian or error functions as in the case for constant diffusivities shown in Fig. 7.5. The diffusivities shown in Fig. 7.10b are the average values for intrinsic diffusion ($n \simeq n_i$), where the diffusivities are considered to be constant. Figure 7.17 shows the normalized diffusion profiles for boron and arsenic under extrinsic condition where the diffusivities are not constant. The complementary error function is also shown for comparison. For a more comprehensive review of enhanced diffusion due to defect–impurity interaction, readers are referred to the articles by Casey (1973), Fair (1981), and Tan and Gösele (1985).

## 7.8 Electric Field Effect

The diffusion of ionized impurities may be enhanced by the presence of an electric field. An electric field sometimes may be generated internally under high-dopant-concentration conditions. Let us consider the diffusion of an $n$-type dopant into Si with a high surface concentration (about $10^{19}$ to $10^{20}/cm^3$). At the diffusion temperature, the impurities are fully ionized and we assume that the free electrons obtained from the dopants, $n$, is larger than $n_i$ ($n > n_i$, the extrinsic condition). The electrons have much higher mobility than that of the impurity atoms and tend to out run the diffusing atoms, thus creating an electric field $\mathscr{E}$ between the electrons and the positively charged impurity atoms $N_D^+$. The atomic flux $F$ is, then,

$$F = -D \frac{\partial C}{\partial x} - \mu_I C \mathscr{E} \tag{7.63}$$

where $C$ is the concentration of the impurity atoms, $\mu_I$ the mobility of the diffusing atom, and $\mathscr{E}$ the internally generated electric field. Assuming local charge neutrality, we can express the electric field in terms of the Fermi-level position relative to the intrinsic Fermi level $E_i$:

$$\mathscr{E} = \frac{1}{e} \frac{d}{dx} (E_F - E_i) \tag{7.64}$$

$$E_F - E_i = kT \ln \frac{C}{n_i}. \tag{7.65}$$

Therefore,

$$\mathscr{E} = \frac{kT}{e} \frac{\partial}{\partial x} \ln \frac{C}{n_i}. \tag{7.66}$$

Noting that

$$D = \frac{kT\mu_I}{e} \quad \text{(Einstein's relation)} \tag{7.67}$$

then

$$F = -D \frac{\partial C}{\partial x} - D \frac{e}{kT} C \mathscr{E}$$

$$= -hD \frac{\partial C}{\partial x} \tag{7.68}$$

where

$$h = 1 + C \frac{\partial \ln (C/n_i)}{\partial C} = 1 + \frac{C}{n_i} \left[ \left( \frac{C}{2n_i} \right)^2 + 1 \right]^{-1/2}$$

is called the field enhancement factor. The value of $h$ has a maximum value of 2 for $C >> n_i$. When the electric field effect is included, eq. (7.61) should be rewritten as

$$D_{\text{imp}} = h \left[ D_i^x + \sum_{r=1} D_i^{-r} \left( \frac{n}{n_i} \right)^r + \sum_{r=1} D_i^{+r} \left( \frac{p}{n_i} \right)^r \right]. \tag{7.69}$$

Since we note that the electrons out run the impurity atoms, it is interesting to ask if a sample maintains local charge neutrality in the presence of a nonuniform concentration profile. Let's consider first a uniformly doped and fully ionized sample ($n$-type). Since there is no gradient in electron concentration, the ionized donors are compensated electrically by the electron charges everywhere in the sample, and local charge neutrality is therefore maintained.

Now let's consider a sample where the donor concentration is high at the surface and decreases as a function of depth into the sample. The free electrons tend to diffuse into the interior of the sample due to the gradient in electron concentration. Soon an electric field is set up to counterbalance the gradient effect such that the diffusion of electrons stops. One may anticipate that the donor charges near the sample surface are not locally compensated by the displaced free electrons in the interior of the sample. The steeper the gradient is for the electrons to diffuse in; the larger is the electric field generated due to this effect, and the more difficult it is to maintain local charge neutrality. The impurity atom concentration across a $p$-$n$ junction is usually steep, especially in the case of an abrupt junction. The free carriers diffuse away and leave the ionized atoms completely uncompensated, thus forming a space-charge region. This is the basis for the depletion approximation, where any free carriers are considered to be negligibly small in quantity compared to the fixed space-charge density. The depletion approximation usually holds a few "Debye lengths" away from the edge of depletion region in a $p$-$n$ junction. The Debye length $L_D$ may be considered as the distance over which the free carriers rearrange themselves to shield the rest of the sample from a local electric field, and is given by the expression

$$L_D = \left( \frac{\epsilon kT}{e^2 n} \right)^{1/2} \tag{7.70}$$

where the permittivity $\epsilon = \epsilon_r \epsilon_0$, $\epsilon_r$ is the relative dielectric constant, $\epsilon_0$ is the permittivity of free space, $8.85 \times 10^{-14}$ F/cm $= 55.4 e/$V $\cdot$ μm in units conven-

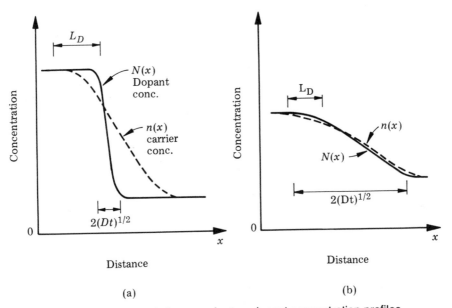

**FIGURE 7.18** Schematical diagrams for two dopant concentration profiles. (a) The diffusion distance, $2\sqrt{Dt}$, is small compared to the extrinsic Debye length; local charge neutrality cannot be assumed. (b) The diffusion distance is large compared to the Debye length; local charge neutrality holds for this condition; $C(x) = n(x)$ in this case. The solid lines depict the dopant profiles and the dashed lines depict the carrier profiles.

ient in semiconductor work, $e$ is the electronic charge, and $n$ is the free carrier concentration; $kT$ has the usual meaning.

Within a few so-called "Debye lengths," free carriers are still present, resulting in partial compensation of the ionized atoms. It is therefore easy to understand that if the impurity diffusion length ($2\sqrt{Dt}$) is short (i.e., very steep impurity gradient) compared to the Debye length, charge neutrality is difficult to maintain, as in the case of the abrupt junction. On the other hand, if the diffusion length is long compared to the Debye length, quasi-charge neutrality can be maintained in the presence of a nonuniform doping profile. As long as the impurity concentration is not too steep, we can approximate $n(x)$ by $C(x)$, as was done with eqs. (7.42) and (7.44) in calculating the average conductivity of a diffused layer (see Fig. 7.18).

# 7.9 Diffusion in Polycrystalline Silicon

Polycrystalline Si layers, referred to as polysilicon or Poly Si, are used as gate material in self-aligned structures; polysilicon is also sometimes used as a conductor in Si VLSI technology (Kamins, 1988). In the self-aligned gate structure, a layer of metal silicide is formed on top of the polysilicon; the composite as a whole acts as the gate in a metal–oxide–semiconductor (MOS) device. This *self-aligned silicide* (salicide) technology has three major advantages: (1) low contact resistivity, due to the intimate silicide contact to the source and drain region, automatically included in the salicide process; (2) low sheet resistance, since a highly conductive silicide is in contact with the polysilicon layer; and (3) no additional lithographic requirements compared to the standard process.

Dopants are incorporated into the polysilicon to reduce the resistivity. Dopants may be introduced by ion implantation, as is the case of salicide process, or by conventional means, such as from a chemical source or a doped-oxide source. Irrespective of the means of impurity introduction, diffusion is generally required. The diffusion of impurities into polycrystalline materials is dominated by grain boundary, or short-circuit, diffusion. Since grain boundaries may be viewed as arrays of dislocations, diffusion along grain boundaries and along dislocations is therefore similar and generally enhanced compared to that of the bulk (Mataré, 1984). In the case of polysilicon, impurity atoms also tend to migrate much faster along the grain boundaries, and tend to segregate at the grain boundaries. This segregation affects the grain growth, which in turn affects the impurity diffusion. Since the microstructure of the polysilicon film varies as a function of deposition temperatures and subsequent annealing, it is difficult to obtain universally useful and consistent diffusion data. On the other hand, the diffusion profile in polysilicon has been observed to follow either the simple erfc function or the Gaussian function. It is therefore possible to extract diffusion coefficients and activation energies by measuring the junction depth and by using eqs. (7.38) and (7.39a).

Polysilicon layers are usually deposited by chemical vapor deposition (CVD) at 600 to 650°C. The main reaction is the pyrolysis of silane ($SiH_4$). Polysilicon deposited below 575°C is usually amorphous or very fine grained in structure. Polysilicon deposited above 625°C has a columnar structure with a grain size between 30 and 300 nm. After high-temperature annealing such as in diffusion, there is significant grain growth and/or recrystallization of the polysilicon. The physical picture of arsenic diffusion into polysilicon is that As diffuses along the grain boundaries as well as into the grains. If the polysilicon grains are initially undoped, diffusion of As is along grain boundaries at temperatures of about 800°C. Above about 900°C, grain growth occurs and As diffuses along grain boundaries as well as into the interior of the grains (Swaminathan et al., 1982). The diffusion into the grains is observed to be very similar to the diffusion into single-crystal Si. Table 7.6 lists some of the diffusion data in polysilicon. Comparing the activation

**TABLE 7.6**  Diffusion Data in Polysilicon

| Element | $D_0$ (cm²/s) | $E_A$ (eV) | $D$ (cm²/s) | $T$ (°C) |
|---------|---------------|------------|-------------|----------|
| As | $8.6 \times 10^4$ | 3.9 | $2.4 \times 10^{-14}$ | 800 |
|  | 0.63 | 3.2 | $3.2 \times 10^{-14}$ | 950 |
| B | $(1.5\text{--}6) \times 10^{-3}$ | 2.4–2.5 | $9 \times 10^{-14}$ | 900 |
|  |  |  | $4 \times 10^{-14}$ | 925 |
| P |  |  | $6.9 \times 10^{-13}$ | 1000 |
|  |  |  | $7 \times 10^{-13}$ | 1000 |

*Source:* After Tsai (1983).

energy for diffusion, we note that the values of $E_A$ for polysilicon are generally a few tenths of an electron volt to about 1 eV lower than those for diffusion in single-crystal Si.

## 7.10 Diffusion in Silicon Dioxide

Silicon dioxide, $SiO_2$, is used as diffusion mask, electrical isolation, and dielectric material in Si technology and as capping layers for implantation–activation annealing in GaAs technology. Silicon dioxide can be grown thermally on Si or deposited by chemical vapor techniques as discussed in Chapter 9. The structure

**TABLE 7.7**   Diffusivities in $SiO_2$

| Element | $D_0$ (cm²/s) | $E_A$ (eV) | $D(900°C)$ (cm²/s) | $C_s$ (cm⁻³) | Source and Ambient |
|---|---|---|---|---|---|
| Boron | $7.23 \times 10^{-6}$ | 2.38 | $4.4 \times 10^{-16}$ | $10^{19}$– $2 \times 10^{20}$ | $B_2O_3$ vapor $O_2 + N_2$ |
| | $1.23 \times 10^{-4}$ | 3.39 | $3.4 \times 10^{-19}$ | $6 \times 10^{18}$ | $B_2O_3$ vapor, Ar |
| | $3.16 \times 10^{-4}$ | 3.53 | $2.2 \times 10^{-19}$ | Below $3 \times 10^{20}$ | Borosilicate |
| Gallium | $1.04 \times 10^5$ | 4.17 | $1.3 \times 10^{-13}$ | | $Ga_2O_3$ vapor, $H_2 + N_2 + H_2O$ |
| Phos-phorus | $5.73 \times 10^{-5}$ | 2.30 | $7.7 \times 10^{-15}$ | $8 \times 10^{20}$ to $10^{21}$ | $P_2O_5$ vapor, $N_2$ |
| | $1.86 \times 10^{-1}$ | 4.03 | $9.3 \times 10^{-19}$ | $8 \times 10^{17}$– $8 \times 10^{19}$ | Phosphosilicate, $N_2$ |
| Arsenic | 67.25 | 4.7 | $4.5 \times 10^{-19}$ | $<5 \times 10^{20}$ | Ion implant, $N_2$ |
| | $3.7 \times 10^{-2}$ | 3.7 | $4.8 \times 10^{-18}$ | $<5 \times 10^{20}$ | Ion implant, $O_2$ |
| Antimony | $1.31 \times 10^{16}$ | 8.75 | $3.6 \times 10^{-22}$ | $5 \times 10^{19}$ | $Sb_2O_5$ vapor, $O_2 + N_2$ |
| Hydrogen ($H_2$) | $5.65 \times 10^{-4}$ | 0.446 | $7 \times 10^{-6}$ | | |
| Helium | $3 \times 10^{-4}$ | 0.24 | $2.8 \times 10^{-5}$ | | |
| Water | $1 \times 10^{-6}$ | 0.79 | $4 \times 10^{-10}$ | | |
| Oxygen | $2.7 \times 10^{-4}$ | 1.16 | $2.8 \times 10^{-9}$ | | |
| Gold | $8.2 \times 10^{-10}$ | 0.8 | $3 \times 10^{-13}$ | | |
| | $1.52 \times 10^{-7}$ | 2.14 | $10^{-16}$ | | |
| Platinum | $1.2 \times 10^{-13}$ | 0.75 | $7.2 \times 10^{-17}$ | | |
| Sodium | 6.9 | 1.3 | $1.8 \times 10^{-5}$ | | |

*Source:* After Tsai (1983).

of thermally grown $SiO_2$ and CVD $SiO_2$ is amorphous. The incorporation of group III and V elements in $SiO_2$ tends to soften the oxide and causes the oxide layer to flow. For example, a few atomic percent (at %) of phosphorus is often incorporated into $SiO_2$ to planarize the oxide layer at about 1000°C, thus improving the step coverage of an oxide layer deposited over three-dimensional structures. This process is called *p*-glass flow (Adams, 1983). Since the incorporation of these elements causes the softening of $SiO_2$, the diffusivity of these elements into $SiO_2$ increases as the oxide softens. For example, at low phosphorus concentration ($<1$ at %) the diffusion of P into $SiO_2$ is slow and independent of the concentration. As the concentration increases to 3 at % or higher, the oxide begins to soften and the diffusivity increases rapidly.

Table 7.7 shows the diffusivities in $SiO_2$ for some elements and molecules. It is interesting to note from the table that the diffusivity of Ga in $SiO_2$ at 900°C is five to six orders of magnitude faster than that of As in $SiO_2$. The use of $SiO_2$ as a capping layer on GaAs for annealing may not be ideal. On the other hand, excess Ga vacancies can be created in GaAs using $SiO_2$ as a cap. This can be used as an advantage in the study of diffusion mechanisms in GaAs, as was done by Greiner (1985).

# REFERENCES

## General References

BEADLE, W. E., J. C. C. TSAI, and R. D. PLUMMER, eds., *Quick Reference Manual for Silicon Integrated Circuit Technology*, Wiley, New York, 1985.

BORG, R. J., and G. J. DIENES, *An Introduction to Solid State Diffusion*, Academic Press, New York, 1988.

CASEY, H. C., JR., "Diffusion in the III-V Compound Semiconductor," in *Atomic Diffusion in Semiconductors*, ed. D. Shaw, Plenum Press, New York, 1973.

FRANK, W., U. GÖSELE, H. MEHRER, and A. SEEGER, "Diffusion in Si and Ge," in *Diffusion in Crystalline Solids*, ed. G. Murch and A. S. Nowick, Academic Press, Orlando, Fla., 1984, p. 63.

GHANDHI, S. K., *VLSI Fabrication Principles, Silicon and Gallium Arsenide*, Wiley, New York, 1983, Chapter 4.

GROVE, A. S., *Physics and Technology of Semiconductor Devices*, Wiley, New York, 1967, Chapter 3.

SHARMA, B. L., *Diffusion in Semiconductors*, Trans. Tech. Publ., Germany, 1970, pp. 87–126.

SHEWMON, P. G., *Diffusion in Solids*, McGraw-Hill, New York, 1963.

SZE, S. M., *Physics of Semiconductor Devices*, 2nd ed., Wiley, New York, 1981, Chapter 1.

TSAI, J. C. C., "Diffusion," in *VLSI Technology*, ed. S. M. Sze, McGraw-Hill, New York, 1983, Chapter 5.

WOLF, S., and R. N. TAUBER, *Silicon Processing for VLSI Era*, Vol. 1, *Processing Technology*, Lattice Press, Sunset Beach, Ca., 1986, Chapter 8.

## Specific References

ADAMS, A. C., "Dielectric and Polysilicon Film Deposition," in *VLSI Technology*, ed. S. M. Sze, McGraw-Hill, New York, 1983, Chapter 3.

FAHEY, P. M., "Point Defects and Dopant Diffusion in Silicon," Ph.D. thesis, Stanford University, 1985.

FAIR, R. B., "Concentration Profiles of Diffused Dopants in Silicon," in *Impurity Doping*, ed. F. F. Y. Wang, North-Holland, Amsterdam, 1981, Chapter 7.

GREINER, M. E., *J. Appl. Phys. 57*, 5181 (1985).

IRVIN, J. C., *Bell Syst. Tech. J. 41*, 387 (1962).

KAMINS, T. I., *Polycrystalline Silicon for Integrated Circuit Applications*, Kluwer, Boston, 1988.

MATARÉ, H. F., *J. Appl. Phys. 56*(10), 2605 (1984).

ONUMA, T., T. HIRAO, and T. SUGAWA, *J. Electrochem. Soc. 129*(4), 837 (1982).

PAI, C. S., E. CABREROS, S. S. LAU, T. E. SEIDEL, and I. SUNI, *Appl. Phys. Lett. 47*(7), 652 (1985).

SWAMINATHAN, B., K. C. SARASWAT, R. W. DUTTON, and T. I. KAMINS, *Appl. Phys. Lett. 40*(9), 795 (1982).

TAN, T. Y., and U. GÖSELE, "Diffusion in Si and Ge," *Appl. Phys. A. 31*, 1 (1985).

TAN, T. Y., and U. GÖSELE, *Appl. Phys. Lett. 52*(15), 1240 (1988).

## PROBLEMS

**7.1.** You want to estimate the diffusion of an impurity into silicon at 1000°C using the linear approximations developed in the text. You are told that the surface concentration $C_0$ of the impurity is independent of temperature with a value of $C_0 = 10^{19}/cm^3$ and the diffusion coefficient $D = 2 \times 10^{-12}$ $cm^2/s$ at 1100°C with an activation energy of 4 eV. Your design calls for a diffusion length of $10^{-4}$ cm.

   **(a)** After diffusion at 1000°C, what is the total amount of impurities/$cm^2$ in the diffused layer?

   **(b)** How long is the diffusion time $t$ at 1000°C?

   **(c)** What would be the flux $F$ of atoms in atoms/$cm^2 \cdot s$ after 4000 s at 1100°C?

**7.2.** Silicon is diffused at 1100°C for 20 minutes with bismuth (Bi). You measure a diffusion length of 1 μm and the total amount of Bi in the sample is $2 \times 10^{15}$ Bi atoms/$cm^2$. Use linear approximation and $D = D_0 \exp(-E_A/kT)$.

   **(a)** What is the diffusion coefficient?

   **(b)** What is the surface concentration?

   **(c)** What is the flux after 10 min?

   You heat another sample to 1200°C for 20 minutes and the surface concentration remains unchanged but the diffusion length increases a factor of 4:

   **(d)** What is the total amount of Bi in the sample?

   **(e)** What is the value of the activation energy $E_A$?

**7.3.** You find that the diffusion coefficient $D$ of boron is a factor of 10 greater than that of As at 1150°C and that the solid solubility (assume equal to the surface concentration) of As is a factor of 10 higher than that of boron: $D_B = 10\,D_{AS}$; $C_0(As) = 10\,C_0(B)$. For the same diffusion time, $t$:
   **(a)** Which species, boron or As, has the largest diffusion length?
   **(b)** Which has the largest number of atoms/cm² diffused into Si?
   **(c)** What would lead to the greatest change in diffusion length: a 20% change in temperature, time, or surface concentration?
   **(d)** From the data in Fig. 7.10, calculate the activation energy in eV for diffusion of As in Si.

**7.4.** You have a silicon wafer and diffuse osmium (Os), which has a diffusion coefficient of $10^{-13}$ cm²/s at 1000°C, an activation energy for diffusion of 3.6 eV, and a surface solubility, $C_0$, of $10^{19}$/cm³. You carry out the diffusion for 20 minutes at 1000°C. The diffusion length $= 2(Dt)^{1/2}$.
   **(a)** What is the diffusion length and total number of Os/cm² in the sample at the end of the diffusion?
   **(b)** Consider the Os to have a single charge. What electric field would be required to give a current equal to the flux at a distance one-half the diffusion length at $t = 20$ minutes?
   **(c)** What temperature would you need to increase the junction depth a factor of 10 at $t = 20$ minutes?

**7.5.** Show that $\bar{p}$ in Fig. 7.13a is only a function of $C_S$, $C_B$, and the diffusion profile for $C_B = 0.2\,C_S$ (erfc distribution).

**7.6.** Show that $[V^+]/[V^+]_i = p/n_i$.

**7.7.** The intrinsic carrier concentration $n_i$ can be calculated using the following formula:

$$n_i = 3.87 \times 10^{16} T^{3/2} \exp\left(-\frac{0.605 + \Delta E}{kT}\right)$$

where $\Delta E = -7.1 \times 10^{-10}\,(n_i/T)^{1/2}$ and $T$ in kelvin. Examine the diffusion calculations in Examples 1 and 2 to check if the use of constant diffusion coefficients is justified.

**7.8.** In an $n$-type semiconductor, the diffusional flux $F$ of the donor atoms in the presence of an electric field is given by

$$F = -eD\,\frac{\partial C}{\partial x} - e\mu C\mathscr{E}$$

(assuming local charge neutrality and full ionization). Show that

$$F = -\frac{d}{dx}(DC)$$

(Hint: $D/D_i = n/n_i$.)

**7.9.** A layer of 1000 Å of undoped polysilicon is first deposited on a $p$-type (100) Si substrate ($N_A = 1 \times 10^{15}/\text{cm}^3$). Phosphorus is then introduced into the polysilicon from a doped-oxide source by depositing a layer of phosphosilicate glass (PSG) on the polysilicon surface. Calculate the junction depth in the substrate after a 1-h 1000°C annealing.

**7.10.** Given a Si sample where the concentration profile of $n$-type dopants follows an exponential function

$$C(x) = C_s e^{-(x/\lambda)}$$

where $\lambda$ is the characteristic diffusion length and $C_s$ is the surface concentration:

(a) Calculate $\lambda$ for a sample where $C_s = 10^{18}$ cm$^{-3}$ and $C(5000 \text{ Å}) = 10^{16}$ cm$^{-3}$.

(b) Calculate the extrinsic Debye length at $10^{18}$ cm$^{-3}$ and $10^{16}$ cm$^{-3}$.

(c) Do you expect local neutrality to hold along this concentration profile [i.e., do you expect that $C(x) = n(x)$]?

**7.11.** Now in another sample the gradient of the concentration is very steep with $\lambda = 50$ Å. Check for local charge neutrality along the concentration profile.

**7.12.** The Poisson's equation states that

$$\frac{d^2\phi}{dX^2} = \frac{\rho}{\kappa_s \epsilon_0} = \frac{-e}{\kappa_s \epsilon_0}(N_D^+ + p - N_A^- - n)$$

where $\phi$ is the potential function, and $n$ and $p$ are the concentrations of electrons and holes. $N_D^+$ and $N_A^-$ are the concentrations of ionized donors and acceptors. When charge neutrality is observed, $\rho = (N_D^+ + p - N_A^- - n) = 0$. There is no electric field gradient or electric field in the sample. When charge neutrality is approximately observed (quasi-charge neutrality, $\rho \neq 0$), there is an internally generated electric field, as we have discussed in the text. When all free carriers are negligibly small in quantity, depletion approximation holds and there is a large electric field such as that across a $p$-$n$ junction. Use the Poisson's equation and check for the charge-neutrality calculations performed for Problems 7.10 and 7.11.

# CHAPTER 8

# Ion Implantation

## 8.1 Introduction

Integrated-circuit technology requires reproducible and controlled dopant concentrations in all active-device components, such as bipolar transistors and field-effect transistors. In a 5 × 5 mm Si chip containing memory arrays and logic elements, there may be a million transistors—all of which must be functional and properly doped with donors and acceptors. Dopants can be introduced by diffusion from gas ambients in quartz-tube diffusion furnaces as discussed in Chapter 7. For dopant control in diffusion, the surfaces must be clean and exposed to a uniform flow of dopants in the gas ambient; that is, the gas source of dopants must provide a uniform flux of dopants impinging on the semiconductor surface, and the transport of dopants across the gas–semiconductor surface should not be impeded by localized patches of contaminants or native oxide layers. From a production-line technology standpoint, the requisite control of surfaces and gas flow is difficult to achieve reproducibly to ensure that wafer after wafer receive the same amount of dopants.

To achieve a uniform and controlled introduction of dopants, ion implantation is used. In ion implantation, a beam of dopant ions of fixed energy—typically between 30 and 100 kiloelectron volts (keV)—is swept (or rastered) across the surface of the semiconductor (Fig. 8.1). The ions have a sufficiently high velocity, about $10^7$ cm/s, so that they penetrate through the surface and come to rest within the semiconductor at depths of 10 to 100 nm below the surface. The penetration depths—the ion ranges—are determined by the energy of the incident ion and are not blocked by surface contamination or native oxide layers. Consequently, the total number $Q_I$ of implanted dopants/cm$^2$ is given by the product of the flux $F_I$ of incident ions/cm$^2 \cdot$ s and the implantation time $t_I$ that the ion beam is incident on the sample:

$$Q_I = F_I t_I. \tag{8.1}$$

A flux $F_I$ of ions with positive charge $q_I$ per ion represents a current that can be measured directly with a current meter connected between the sample and electrical ground (Fig. 8.1). The ion current $q_I F_I$ may vary with time so that in ion-implantation systems, a charge integrator is used to give the total integrated charge

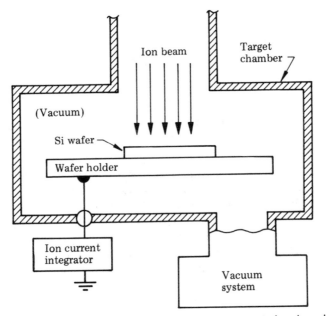

**FIGURE 8.1** Schematic of an ion-implantation system target chamber showing the ion beam incident on a Si wafer and the current integrator.

$\mathcal{Q}_I$ in coulombs of ions implanted in the sample during the time $t_I$ the sample is exposed to the ion beam:

$$\mathcal{Q}_I = \int_0^{t_I} q_I F_I \, dt \tag{8.2}$$

where $q_I = 1.6 \times 10^{-19}$ C for singly ionized ions and $q_I = 3.2 \times 10^{-19}$ C for doubly ionized species. The number $Q_I$ of implanted dopants/cm$^2$ is given by

$$Q_I = \frac{\mathcal{Q}_I}{q_I A} \tag{8.3}$$

where $A$ is the area of the implanted surface. A 100$\mu$A beam of singly ionized ions swept across a 200-cm$^2$ area for 60 s gives

$$Q_I = \frac{100 \times 10^{-6} \times 60}{200 \times 1.6 \times 10^{-19}} = 1.875 \times 10^{14} \frac{\text{dopant ions}}{\text{cm}^2}.$$

By control of the beam current and implantation time, values of $Q_I$ between $5 \times 10^9$ and $5 \times 10^{15}$/cm$^2$ can be obtained for different applications. If the dopants are distributed uniformly over 50 nm (500 Å), the dopant concentrations can be controlled between values of $10^{15}$ and $10^{21}$ dopants/cm$^3$. This illustrates one of the advantages of ion implantation—that dopant concentrations can be obtained over a range of six orders of magnitude by adjusting the ion beam current and implantation time.

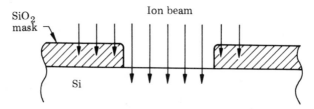

**FIGURE 8.2** Ion implantation into a Si wafer, where portions of the sample are masked by an SiO$_2$ layer whose thickness is greater than the ion range.

Another advantage of ion implantation is that selected areas can be implanted by use of masks which leave well-defined areas of the semiconductor exposed to the beam and other areas masked from the beam (Fig. 8.2). The thickness of the masks must be greater than the penetration depth or range of the ions. However, ion ranges are typically less than 100 nm, so that mask thicknesses of 200 nm serve to prevent the ions from reaching the semiconductor. The masks may be thermally grown oxide layers (SiO$_2$ on Si) or deposited layers of oxides, organic films, or metal layers. The mask layers are patterned by photolithographic techniques, as used in other stages of integrated circuit processes, and the mask material is removed in areas where the semiconductor wafer is to be exposed to the ion beam. After implantation, all the mask material is removed so that the wafer can receive further processing steps.

The penetration of energetic ions into the semiconductor damages the crystal structure. The ions collide with semiconductor atoms and displace them off lattice sites. The disorder can be sufficiently great so that an amorphous layer is formed. High-temperature (600 to 1000°C) processing is required to anneal the disorder and return the implanted semiconductor to a single-crystal state with a minimum number of defects.

## 8.2 Ion-Implantation Systems

The implantation system in concept is quite simple (Fig. 8.3) as it consists of an ion source, an acceleration tube, and a target chamber where the semiconductor wafers are held. Dopant atoms are introduced into the ion-source container from a gas or a vapor supplied by a liquid or solid heated in a small oven connected to the source. The dopant atoms in the source are ionized by energetic electrons emitted from a hot filament. The electrons collide with the cloud of atomic electrons around the positive nucleus of the dopant atom. In the collisions, electrons are knocked out of their atomic orbits and the atom is left in an ionized state with a positive charge of $e$ (singly ionized) or $2e$ (doubly ionized). The positive dopant ions are then extracted through an aperture in the ion source and enter the acceleration tube.

The acceleration tube is an insulating column with a vacuum inside so that the ions from the source can be accelerated without suffering impact collisions with residual gas atoms in the column. The ion source is held at a positive voltage, the

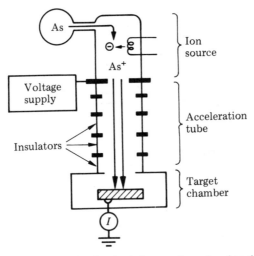

**FIGURE 8.3** Schematic of an ion-implantation system showing the ion source and As ions, the acceleration tube with insulators, and the target chamber.

acceleration voltage potential $V_I$. The ions entering the column are accelerated along the length of column and exit into the target chamber with an energy $q_I V_I$ and velocity $v_I$:

$$\tfrac{1}{2} M_I v_I^2 = q_I V_I \qquad (8.4)$$

where $M_I$ is the mass of the dopant atom. For singly ionized As ions [$M_I$(As) = 74.9 atomic mass units] accelerated through 100 kilovolts (kV), the velocity is

$$v_I = \left( \frac{2 \times 10^5 \text{ eV} \times 1.6 \times 10^{-19}}{74.9 \times 1.66 \times 10^{-27} \text{ kg}} \right)^{1/2} = 5.07 \times 10^5 \text{ m/s}$$

where one atomic mass unit = $1.66 \times 10^{-27}$ kg and 1 eV = $1.6 \times 10^{-19}$ J.

Ion-implantation systems are quite elegant and sophisticated after 20 years of development. Every aspect of the system from ion source to target chamber has been refined to provide controlled implantation conditions. Figure 8.4 shows a schematic layout of a commercial ion-implantation system. Two features are immediately obvious: the extent of the wafer transport area and the presence of an analyzing magnet. A large area is required for wafer transport because in production lines as many as 300 wafers per hour may be implanted. This wafer throughput requires automatic placement of wafers onto wafer holders as well as transport in and out of the target chamber.

An analyzing magnet is required because the ion source produces a variety of ion species but only one species (e.g., singly ionized As ions) is selected for implantation. Ions with charge $q_I$ and velocity $v_I$ will be deflected by a magnetic field directed normal to their path. As in the case of Hall effect measurements discussed in Chapter 2, the force $\mathbf{F}$ on the ions is

$$\mathbf{F} = q(\mathbf{v}_I \times \mathbf{B}) \qquad (8.5)$$

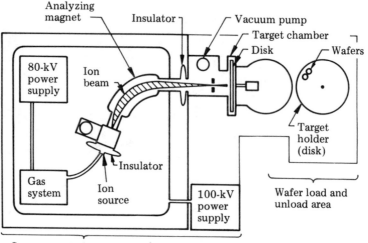

**FIGURE 8.4** Schematic of a commercial ion-implantation system, the Nova-10-160, 10 mA at 160 keV.

where **B** is the magnetic flux density. Charged particles in a homogeneous magnetic field move in a circular path of radius $r$ as indicated in Fig. 8.5 for a magnetic field directed out of the page. The velocity $v_I$ of the particle of mass $M_I$ is unchanged by the deflection in the magnetic field and there is a centripetal force

$$F = \frac{M_I v_I^2}{r}. \tag{8.6}$$

Combining eqs. (8.5) and (8.6), we have

$$r = \frac{M_I v_I}{q_I B}. \tag{8.7}$$

For an ion accelerated through a potential $V_I$, the velocity $v_I = (2\,q_I V_I/M_I)^{1/2}$, so that

$$r = \frac{1}{B}\left(\frac{2M_I V_I}{q_I}\right)^{1/2}. \tag{8.8}$$

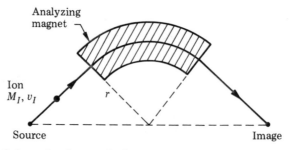

**FIGURE 8.5** Schematic of an analyzing magnet deflecting an ion beam.

For a given acceleration potential and magnetic flux density, the radius of the ion path is directly proportional to the square root of the mass-to-charge ratio. For singly ionized As ions accelerated through 10 kV, in a magnetic field of 0.5 Wb/m² (5000 gauss), the radius is

$$r = \frac{1}{0.5} \left( \frac{2 \times 74.9 \times 1.66 \times 10^{-27} \text{ kg} \times 10^4}{1.6 \times 10^{-19}} \right)^{1/2} = 0.25 \text{ m.}$$

Consequently, magnetic analysis of the ion beam can be carried out with a reasonably compact magnet if the ion energies or ion masses are not too great.

Aside from deflection in the magnetic analyzer for selection of the correct ion species, the ions are focused and steered along their paths by electrostatic lens and deflection plates. With proper design of ion optics, a monoenergetic ion beam of one ion species can be delivered into the target chamber.

## 8.3 Ion-Range Distributions

When an energetic ion penetrates a semiconductor it undergoes a series of collisions with the atoms and electrons in the target. In these collisions the incident particle loses energy at a rate $dE/dx$ of a few to 100 eV per nanometer, depending on the energy and mass of the ion as well as on the substrate material. The energy-loss mechanisms are discussed in the next section, as we are concerned here with the penetration depth or range $R$ of the ions (Fig. 8.6). The range $R$ is determined by the rate of energy loss along the path of the ion,

$$R = \int_{E_0}^{0} \frac{1}{dE/dx} \, dE \tag{8.9}$$

where $E_0$ is the incident energy of the ion as it penetrates the semiconductor. The sign of $dE/dx$ is negative, as it represents the energy *loss* per increment of path, although tabulated values are given as positive quantities.

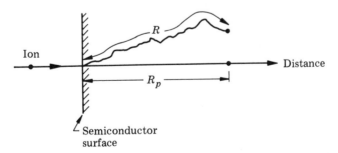

**FIGURE 8.6** An ion incident on a semiconductor penetrates with a total path length $R$ which gives a projected range $R_p$ along the direction parallel to that of the incident ion.

The main parameters governing the range or energy loss rate are the energy $E_0$ and atomic number $Z_1$ of the ion and atomic number $Z_2$ of the substrate if we exclude the effect of the orientation of the crystal lattice. As the incident ion penetrates the semiconductor undergoing collisions with atoms and electrons, the distance traveled between collisions and the amount of energy lost per collision are random processes. Hence all ions of a given type and incident energy do not have the same range. Instead there is a broad distribution in the depths to which individual ions penetrate. The distribution in ranges is referred to as the range distribution or range straggling. Further, in ion implantation it is *not* the total distance $R$ traveled by the ion that is of interest but the projection of $R$ normal to the surface (i.e., the penetration depth or projected range $R_p$) (Fig. 8.6).

In the absence of crystal orientation effects, the range distribution is roughly Gaussian and as a first approximation we describe the projected range distribution $N(x)$ as a one-dimensional Gaussian characterized by a mean value, the projected range $R_p$, and a standard deviation $\Delta R_p$ from the mean:

$$N(x) = \exp\left[ -\frac{1}{2}\left(\frac{x - R_p}{\Delta R_p}\right)^2\right]. \qquad (8.10)$$

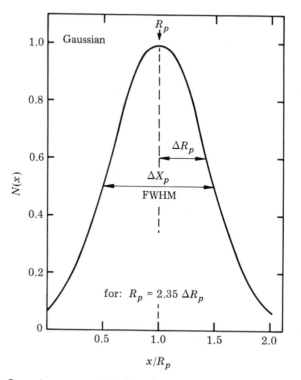

**FIGURE 8.7** Gaussian range distribution for implanted ions with $R_p = 2.35\,\Delta R_p$ and a full width at half-maximum (FWHM) of $\Delta x_p$.

For a Gaussian distribution (Fig. 8.7), the full width $\Delta x_p$ at half-maximum (FWHM) is given by

$$\Delta x_p = 2(2 \ln 2)^{1/2} \Delta R_p = 2.35 \Delta R_p \tag{8.11}$$

and the integral by

$$\int_0^\infty N(x)dx = (2\pi)^{1/2} \Delta R_p. \tag{8.12}$$

In ion implantation, the number $Q_I$ implanted ions/cm² is tightly controlled, and using the relations in eqs. (8.10) and (8.12), the concentration depth distribution $N_I(x)$ in atoms/cm³ is given by

$$N_I(x) = \frac{Q_I}{(2\pi)^{1/2} \Delta R_p} \exp\left[ -\frac{1}{2}\left(\frac{x - R_p}{\Delta R_p}\right)^2 \right]. \tag{8.13}$$

A rough estimate of the average concentration $N_I(\text{ave})$ of implanted ions can be obtained from

$$N_I(\text{ave}) \simeq \frac{Q_I}{\Delta x_p} \approx \frac{Q_I}{R_p} \tag{8.14}$$

where $\Delta x_p \approx R_p$ for medium-mass ions such as As or P implanted into Si. If one implants a dose of $10^{15}$ As ions/cm² into Si at 200 keV so that the projected range $R_p$ is $10^{-5}$ cm (1000 Å), the value of the range straggling $\Delta R_p = 0.4 \times 10^{-5}$ cm and $\Delta x_p \simeq 10^{-5}$, so that the average concentration $10^{15}/10^{-5} = 10^{20}$ As/cm³. Thus one of the features of ion implantation is that high dopant concentrations can be achieved in the near-surface region—an ideal situation in the processing of integrated circuit components.

Projected ranges $R_p$ and range straggle $\Delta R_p$ of dopant ions in Si and GaAs are shown in Fig. 8.8. One sees immediately that the lighter ions such as boron, with their higher velocities, have a greater penetration in silicon than the heavier ions such as As. Further, the ranges of the ions increase nearly linearly with energy. Comparison of $R_p$ and $\Delta R_p$ values show that the expression $R_p \simeq 2.35 \Delta R_p = \Delta x_p$ is a reasonable rule of thumb. The ranges of ions in GaAs are somewhat less than those at the same energy in Si.

Silicon dioxide, $SiO_2$, is often used as a masking material to prevent the penetration of ions into the Si below regions covered by $SiO_2$. The molecular density is $2.2 \times 10^{22}$ $SiO_2$ molecules/cm³ and the mass density is 2.2 g/cm³, a value close to that of Si, 2.33 g/cm³. A very good approximation is to treat the projected ranges of ions in $SiO_2$ as equal to those in Si; $R_p(SiO_2) = R_p(Si)$.

In addition to the straggle in the penetration depth $\Delta R_p$ along a direction normal to the surface, there is a fluctuation, or transverse straggle $\Delta R_t$, in the final position perpendicular to the incident ion direction. Transverse straggle is of importance in defining the penetration at the edge of a mask. The spread of ion distributions implanted into Si with a thick mask ($t > R_p$) with straight edges is shown in Fig.

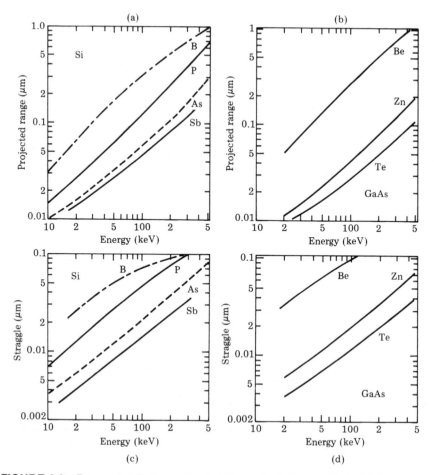

**FIGURE 8.8** Range distributions showing the projected range $R_p$ in (a) Si and (b) GaAs; range straggle $\Delta R_p$ in (c) Si and (d) GaAs. [Adapted from Ghandhi (1983).]

8.9. The lateral distribution is influenced in the region within a few $\Delta R_t$ of the edge of the mask. The values of $\Delta R_t$ are somewhat greater than $\Delta R_p$, but a first approximation is to set $\Delta R_t = \Delta R_p$. Tapered or undercut mask edges will also lead to a lateral distribution.

The description of the ion distribution by use of $R_p$ and $\Delta R_p$ can be improved by the use of higher-order moments. A three-moment approach using two Gaussians with the same $R_p$ but different $\Delta R_p$ values has been used to fit experimental data. A more exact fit is found with a fourth moment approach: $R_p$, $\Delta R_p$, the third moment (skewness), and the fourth moment (kurtosis).

With single-crystal substrates of Si or GaAs, for example, the orientation of the ion beam with respect to the crystallographic axes of the substrate can have a pronounced effect on the range distribution. Figure 8.10 shows the range distribution for 100-keV As implanted with the beam aligned parallel (solid line) to the

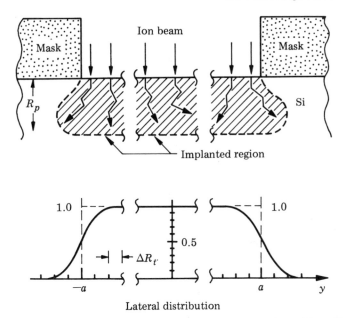

**FIGURE 8.9** Schematic of the lateral distribution of ions implanted into Si with straight-wall oxide masks with a projected range $R_p$ and lateral straggle $\Delta R_t$.

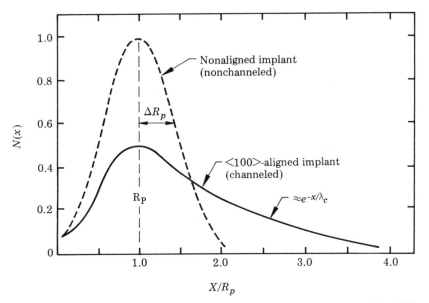

**FIGURE 8.10** Range distributions for channeled ions implanted along the <100> axis of Si. The dashed line shows the Gaussian distribution for incident ions aligned away from any channeling direction.

<100> crystal axis and oriented away (dashed line) from any crystal axes or planes. As is evident in the figure, implantation along crystal axes can lead to a fraction of the total number of ions that penetrate several times $R_p$.

The crystal orientation influence on ion penetration is called channeling or the channeling effect. When an ion trajectory is aligned along atomic rows, the positive atomic potentials of the line of atoms steer the positively charged ion within the open space or channels between the atomic rows. These channeled ions do not make close-impact collisions with the lattice atoms and have a much lower rate of energy loss, $dE/dx$, and hence a greater range than those of nonchanneled ions. The depth distribution of channeled ions is difficult to characterize under routine implantation conditions. The channeling distribution depends on surface preparation, substrate temperature, beam alignment, and disorder introduced during the implantation process itself. The channeled ions that penetrate beyond $R_p$ often have a distribution that falls off exponentially with distance as $\exp(-x/\lambda_c)$, where $\lambda_c \gg R_p$. Since the concentration of implanted atoms near $R_p$ is often large ($\simeq 10^{19}$ to $10^{20}$ atoms/cm$^3$), the junction depth is determined by the penetration of channeled ions.

The channeling effect requires that the incident ions be aligned within a critical angle $\psi_c$ of the crystal axes or planes. The critical angle depends on the ion energy, ion species, and substrate, but is typically less than 5°. Consequently, the substrate holders are often tapered so that the wafers are mounted 7° off normal; this minimizes channeling effects. However, some ions originally incident at angles greater than the critical angle can be scattered into a channeling direction. It is difficult to avoid channeling effects completely unless the implanted region has been made amorphous by a previous implantation.

# 8.4 Energy-Loss Processes

The energy-loss rate $dE/dx$ of an energetic ion moving through a solid is determined by collisions with the substrate atoms and electrons (Fig. 8.11). It is customary to distinguish two different mechanisms of energy loss: (1) nuclear collisions, in which energy is lost in displacements of the substrate atoms, and (2) electronic collisions, in which the energetic ion excites or ejects electrons from atomic orbitals. The energy loss rate $dE/dx$ can be expressed as

$$\frac{dE}{dx} = \left.\frac{dE}{dx}\right|_n + \left.\frac{dE}{dx}\right|_e \tag{8.15}$$

where the subscripts $n$ and $e$ denote nuclear and electronic collisions, respectively.

Nuclear collisions can involve large discrete energy losses and significant angular deflection of the trajectory of the ion (Fig. 8.11). This process is responsible for the production of lattice disorder by the displacement of atoms from their lattice position. Electronic collisions involve much smaller energy loss per collision, negligible deflection of the ion trajectory, and negligible lattice disorder. As shown in

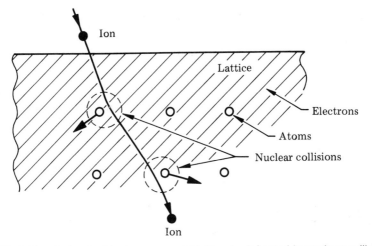

**FIGURE 8.11**    An ion incident on a crystal lattice is deflected in nuclear collisions with the lattice atoms and also loses energy in collisions with electrons.

Fig. 8.12, the relative importance of the two processes depends on the energy $E$ (and atomic number $Z$ of the ion): Nuclear collisions dominate at low energies and electronic collisions at high energies. Values are given in Table 8.1 for the energy $E_1$ in which nuclear collisions are maximum and for the energy $E_2$ at which $dE/dx|_n = dE/dx|_e$.

The values in Table 8.1 indicate that both energy loss processes must be taken into account in determining the range of ions. For light ions in Si such as 100-keV boron, electronic collisions will be dominant over all the trajectory. For 100-keV As, nuclear energy loss can be a factor of 5 to 10 greater than electronic energy loss over much of the trajectory.

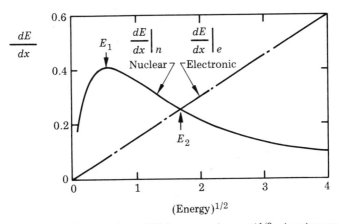

**FIGURE 8.12**    Rate of energy loss $dE/dx$ versus (energy)$^{1/2}$, showing nuclear and electronic loss contributions.

**TABLE 8.1** Characteristic Energies $E_1$ and $E_2$ (keV) Corresponding to the Maximum in $dE/dx|_n$ and $dE/dx|_n = dE/dx|_e$ in Fig. 8.12

| Ion | $E_1$ (Si) | $E_1$ (GaAs) | $E_2$ (Si) | $E_2$ (GaAs) |
|-----|-----------|--------------|-----------|--------------|
| B   | 3   | 7   | 17   | 13   |
| P   | 17  | 29  | 140  | 140  |
| As  | 73  | 103 | 800  | 800  |
| Sb  | 180 | 230 | 2000 | 2000 |
| Bi  | 530 | 600 | 6000 | 6000 |

*Source:* Mayer et al. (1970).

The relative magnitude of the two energy-loss mechanisms in Si is shown directly in Fig. 8.13. Both Figs. 8.12 and 8.13 show that the rate of electronic energy loss is proportional to the ion velocity

$$\frac{dE}{dx}\bigg|_e = k_e(E)^{1/2} \qquad (8.16)$$

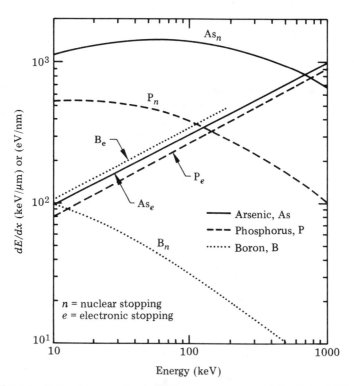

**FIGURE 8.13** Rate of energy loss $dE/dx$ versus energy of As, P, and B ions in Si, showing the contributions of nuclear and electronic energy loss. [Adapted from Seidel (1983).]

where the coefficient $k_e$ does not depend strongly on the ion species. The value of $k_e \simeq 1$ in units of $(eV)^{1/2}$ per nanometer.

In the energy region near the maximum in nuclear energy loss, the value of $dE/dx$ may be treated as energy independent, with a value for stopping in a medium with atomic density $N = 5 \times 10^{22}$ atoms/cm³ given by

$$\left. \frac{dE}{dx} \right|_n = N \frac{\pi^2}{2} e^2 a \frac{Z_1 Z_2 M_1}{M_1 + M_2}$$

or

$$\left. \frac{dE}{dx} \right|_n \simeq \frac{5 Z_1 Z_2 M_1}{M_1 + M_2} \frac{eV}{nm} \qquad (8.17)$$

with the screening radius $a = 1.4 \times 10^{-2}$ nm, $e^2 = 1.44$ eV $-$ nm, where the subscripts 1 and 2 refer to ion and substrate, and where $Z$ and $M$ are the atomic number and mass, respectively. This relation is derived for the rate of energy loss of energetic particles undergoing screened Coulomb collisions with a $1/r^2$ potential (see Feldman and Mayer, 1986). For phosphorus ions ($Z_1 = 15, M_1 = 31$) incident on Si ($Z_2 = 14, M_2 = 28$), eq. (8.17) gives a value for $dE/dx|_n = 550$ eV/nm $= 550$ keV/$\mu$m, a value close to that shown in Fig. 8.13.

In the case where nuclear stopping predominates, the projected range $R_p$ can be estimated by

$$R_p = \int_{E_0}^{0} \frac{1}{dE/dx} dE = \frac{E_0}{dE/dx|_n}. \qquad (8.18)$$

The estimates given in eq. (8.16) to (8.18) are accurate only to 20 to 40% but do give a quick approximation to the range distribution if one uses $\Delta R_p = R_p/2.35$ [eq. (8.11)]. For more accurate values of $R_p$ and $\Delta R_p$, the values given in Fig. 8.8 should be used.

## 8.5 Implantation Damage

As an energetic ion slows down and comes to rest in a semiconductor, it makes a number of collisions with the lattice atoms. In these collisions, sufficient energy $T$ may be transferred from the ion to displace an atom from its lattice site. The displaced atoms can in turn displace other atoms, and so on—thus creating a cascade of atomic collisions. This leads to a distribution of vacancies, interstitial atoms, amorphous regions, and other types of defects as a result of nuclear collisions. As shown in Fig. 8.14, a light ion such as boron in Si loses energy primarily in electronic collisions and has only occasional large-energy transfer collisions along its path. A heavy ion such as Sb loses energy primarily in nuclear collisions and hence produces a dense cascade of collisions and hence disorder along the track. For both heavy and light ions, however, an amorphous layer can be formed, extending from the surface to the ion penetration depth, for sufficiently high num-

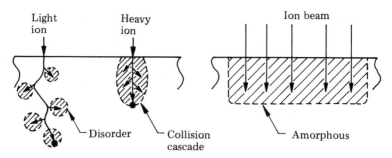

**FIGURE 8.14** Schematic of the disorder produced along the individual paths of light and heavy ions and the formation of an amorphous region.

bers of implanted ions. In this amorphous layer there is no long-range crystallographic order, but the nearest neighbors can maintain their covalent, tetrahedral bonding with some distortion of the bond angles and lengths. In compound semiconductors such as GaAs, the collisions produce an intermixing of the Ga and As atoms so that localized, nonstoichiometric regions, Ga- and As-rich, are formed within the amorphous layer.

The amount of energy $T$ transferred can be determined by considering the kinematics of a collision between an energetic particle, identified by the subscript 1, and a stationary target atom, identified by subscript 2 (Fig. 8.15).

The energy transfers in collisions between two isolated particles can be solved by applying the principles of conservation of energy and momentum. For an incident energetic particle of mass $M_1$, the values of the velocity and energy are $v$ and $E_0$ ($E_0 = \frac{1}{2}M_1v^2$), while the target atom of mass $M_2$ is at rest. After the collision, the values of the velocities $v_1$ and $v_2$ and energies $E_1$ and $E_2$ of the projectile and target atoms are determined by the scattering angle $\theta$ and recoil angle $\phi$. The notation and geometry for the laboratory system of coordinates are given in Fig. 8.15.

Conservation of energy and conservation of momentum parallel and perpendicular to the direction of incidence are expressed by the equations

$$\tfrac{1}{2} M_1 v^2 = \tfrac{1}{2} M_1 v_1^2 + \tfrac{1}{2} M_2 v_2^2 \tag{8.19}$$

$$M_1 v = M_1 v_1 \cos \theta + M_2 v_2 \cos \phi$$
$$0 = M_1 v_1 \sin \theta - M_2 v_2 \sin \phi. \tag{8.20}$$

From these equations, the energy $E_2$ of the recoiling target atom is

$$E_2 = \frac{4M_1 M_2}{(M_1 + M_2)^2} \cos^2\phi \, E_0. \tag{8.21}$$

The energy $T$ transferred in the collision equals $E_2$ and the maximum energy transfer $T_{\max}$ occurs for a head-on collision with $\phi = 0$,

$$T_{\max} = \frac{4M_1 M_2}{(M_1 + M_2)^2} E_0. \tag{8.22}$$

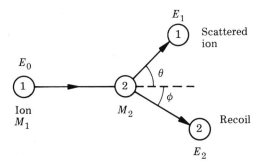

**FIGURE 8.15** Notation for a collision between an incident ion of mass $M_1$ and a target atom of mass $M_2$.

For 100-keV As ions ($M_1 = 75$) incident on Si ($M_2 = 28$), the maximum energy transfer is 79 keV, indicating that substantial amounts of energy can be lost by the incident ion in individual collisions. Even for more glancing collisions, energy transfers can be large, for $\phi = 45°$, $T = 0.395E_0$. These large energy transfers are relatively rare; however, they do serve as an indication that there is sufficient energy available to displace lattice atoms.

The minimum energy $E_d$ required to displace a Si atom from a lattice site is about 15 eV; that is, if a lattice atom receives less energy than $E_d$, it will not be displaced. If the energy transfer is greater than $2E_d$, the displaced atom can displace another lattice atom ($T_{max} = E_0$ for $M_1 = M_2$). As an approximation in the collision cascade we can assume that the average energy required to displace an atom is $2E_d$. After an atom has been displaced, a much lower energy than $E_d$ is required to displace a neighboring atom. However, in this book we use $E_d$.

The number of atoms $N_D$ displaced by an incident ion is then

$$N_D = \frac{F_D}{2E_d} \qquad (8.23)$$

where $F_D$ is the amount of energy deposited in nuclear collisions by the incident ions and secondary recoils. If we assume that all the energy lost by the incident ions in nuclear collisions produces displaced atoms, then $F_D$ is proportional to $dE/dx|_n$. For the case of medium mass or heavier ions, $dE/dx|_n$ is energy independent (for energies around the maximum in the nuclear energy loss curve; i.e., around $E_1$ in Table 8.1), so that

$$F_D = \frac{dE}{dx}\bigg|_n R_p. \qquad (8.24)$$

Where $dE/dx|_n$ dominates over $dE/dx|_e$ and from eqs. (8.18) and (8.23),

$$N_D = \frac{E_0}{2E_d} \qquad (8.25)$$

where $E_0$ is the energy of the incident ion.

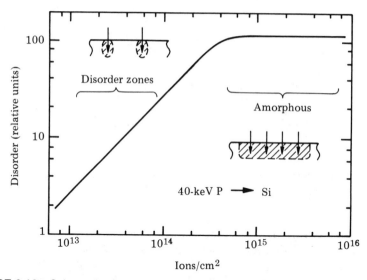

**FIGURE 8.16** Schematic of the amount of disorder versus ion dose for 40-keV phosphorus ions incident on Si. [Adapted from Mayer et al. (1970).]

During ion implantation, each ion produces a region of disorder around the ion track; as the implantation proceeds, the amount of disorder builds up until all the atoms have been displaced and an amorphous layer is produced over a depth $R_p$, as shown in Fig. 8.14. The buildup and saturation of disorder are shown in Fig. 8.16 for 40-keV phosphorus ions incident on Si. In this example, about $4 \times 10^{14}$ phosphorus ions/cm$^2$ are required to form an amorphous layer.

The number $\mathcal{N}_D$ of displaced atoms/cm$^2$ is given by

$$\mathcal{N}_D = Q_I N_D \tag{8.26}$$

where $Q_I$ is the number of implanted ions/cm$^2$. For the case where $dE/dx|_n$ is independent of energy and hence there is a uniform concentration $N_d$ of displaced atoms per cm$^3$, then from eqs. (8.23) to (8.26),

$$N_d = \frac{\mathcal{N}_D}{R_p} = Q_I \left. \frac{dE}{dx} \right|_n \frac{1}{2E_d} . \tag{8.27}$$

In regard to Fig. 8.16, for 40-keV phosphorus in Si, $dE/dx|_n = 500$ eV/nm $= 5 \times 10^9$ eV/cm (Fig. 8.13) and $E_d = 15$ eV, so that for an ion dose $Q_I = 3 \times 10^{14}$ ions/cm$^2$,

$$N_d = \frac{3 \times 10^{14} \text{ ions}}{\text{cm}^2} \cdot \frac{5 \times 10^9 \text{ eV}}{\text{cm}} \cdot \frac{1}{2 \times 15 \text{ eV}} = 5 \times 10^{22} \text{ atoms/cm}^3 .$$

The value of $N_d = N_{Si} = 5 \times 10^{22}$ atoms/cm$^3$, as expected from Fig. 8.16.

Except for low doses or light ions, we can anticipate that an amorphous layer is formed during the implantation process. This assumes that no recovery of lattice order occurs around the ion track. This condition is not met during high-dose-rate implantation, where the flux of energetic ions is sufficient to cause an increase in the substrate temperature. A milliampere (mA) current of 100-keV As ions dissipates 100 W in the outer surface of a Si wafer. With this heat dissipation, the surface of the implanted wafer can be heated to several hundred degrees unless stringent precautions are taken to ensure good thermal contact between the wafer and holder. In such elevated-temperature implantations, annealing of the disorder occurs in the region around the ion track. There can be a flow of vacancies and interstitials into the surrounding material, leading to the formation of extended defects such as dislocation loops.

## 8.6 Annealing of Disorder: Electrical Evaluation

Irrespective of ion species or energy, the number of initially displaced semiconductor atoms is always greater than the number of implanted ions. Even if annealing of disorder occurs during the implantation process, the implanted atom comes to rest in a disordered region. This disordered region may be amorphous, it may contain extended defects, or it may be a region with point defects. In any event, the electrical nature of implanted dopant ions is overwhelmed by the behavior of the disordered regions. Implantation produces a high-resistivity layer containing electron and hole traps and recombination centers.

Of course, the objective of ion implantation, with few exceptions, is to produce a layer with a controlled number of electrically active dopants, with each implanted dopant atom contributing one free carrier to the conduction or valence band. To accomplish this the implanted dopant must occupy a substitutional lattice site and the concentration of electrically degrading defects must be much less than the concentration of dopants. Both requirements—substitutional site location of dopants and reduction of disorder—can generally be met by heating the implanted samples for temperature–time combinations below that in which large-scale diffusion of dopants occurs. This thermal annealing step can achieve nearly complete electrical activity of implanted dopants in Si for anneal temperatures of 900 to 1000°C and times less than 30 minutes. In some cases it is desirable to use higher temperature–time combinations so that the implanted dopants are diffused to depths greater than the original range distribution. Diffusion of implanted species is discussed in the next section; here, we are concerned with electrical activity and recovery of lattice order.

Implantation of donor ions into a p-type semiconductor substrate followed by thermal annealing will form an implanted n-type layer (Fig. 8.17). The depletion layer at the n-p junction serves to isolate the n-type layer from the p-type substrate. Four-point probe measurements of the sheet resistance (Chapter 2) can be used to evaluate the electrical activity of the implanted layer. The measurement current $I$ is confined to the implanted layer.

Lattice disorder can influence carrier mobilities, causing a deviation from the relation between carrier mobility and dopant concentration given in Fig. 3.11. Consequently, Hall effect measurements in conjunction with sheet resistance measurements are made to characterize the implanted layers. To make these

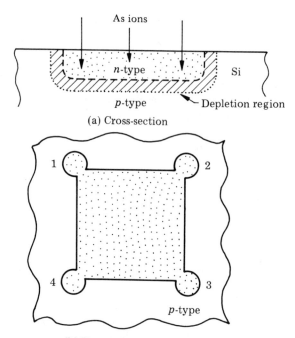

(a) Cross-section

(b) Top view; Hall pattern

**FIGURE 8.17** (a) Cross section and (b) top view of an van der Pauw for Hall effect measurement, formed by implanting As ions into p-type Si.

measurements, square-shaped test areas with contact regions at the corners are formed by lithographic techniques (generally by using oxide masks), as shown in the lower portion of Fig. 8.17. In such a symmetrical structure (often called a van der Pauw pattern) the current is passed through one pair of contacts and the voltage is measured across another pair of contacts. The sheet resistance $R_s$ in $\Omega/\square$ (see Section 2.5) is given by (Beadle et al., 1985)

$$R_s = \frac{\pi}{\ln 2} \frac{V_{34}}{I_{12}} \tag{8.28}$$

where the subscripts indicate the contacts where voltages and currents are measured. The sheet Hall coefficient $R_{sH}$ [see eq. (2.34)] is determined for this configuration by measuring the voltage change $V_{13}$ normal to the current path $I_{24}$ when a magnetic field $B$ is applied perpendicular to the sample. The measured value of $R_{sH}$ is given by

$$R_{sH} = \frac{1}{B} \frac{V_{13}}{I_{24}} \tag{8.29}$$

where measured values are in units of volts, amperes, and webers/m². 

Some insight into these measurements can be gained if we consider a layer uniformly doped to a depth $R_p$ with carriers with a concentration $n_I$ and mobility

$\mu_I$ so that the sheet carrier concentration $\mathcal{N}_s = Q_I$, the number of implanted dopants in a fully annealed sample, is given by

$$\mathcal{N}_s = n_I R_p \tag{8.30}$$

and the sheet conductance $\sigma_s = 1/R_s$ by

$$\sigma_s = \frac{1}{R_s} = e\mathcal{N}_s\mu_I. \tag{8.31}$$

The sheet Hall coefficient (assuming Hall and conductivity mobilities are equal) is then

$$R_{\mathrm{sH}} = \frac{1}{\mathcal{N}_s e} \tag{8.32}$$

so that

$$\mu_I = R_{\mathrm{sH}}\sigma_s. \tag{8.33}$$

In the implanted layer, the carrier concentration $n_I(x)$ and mobility are not uniform in depth so that

$$\mathcal{N}_s = \int n_I \, dx \tag{8.34}$$

$$\sigma_s = e \int n_I(x)\mu_I(x) \, dx \tag{8.35}$$

and the sheet Hall coefficient is also a weighted average of carrier concentrations and mobilities.

Thermal annealing of the implanted layer causes an increase in the sheet conductance as the crystal order is restored. Figure 8.18 shows the fraction $F_I =$

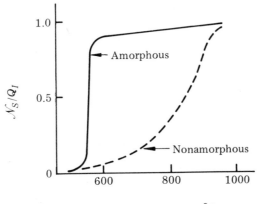

**FIGURE 8.18** Number of carriers per incident ion as a function of anneal temperature in Si for implantations that form an amorphous layer (solid line) and implantations (dashed line) where an amorphous layer is not formed.

$\mathcal{N}_s/Q_I$ of carriers versus anneal temperature for the case of donor-implanted Si, where the implanted layer is amorphous (solid line) or crystalline (dashed line) with residual defects. The latter case represents either a low-dose or elevated-temperature implant. For the amorphous case there is a pronounced increase in the electrical activity in the temperature range 500 to 600°C, where the amorphous layer reorders epitaxially on the substrate. Generally, there is a further increase in the fraction $F_I$ as the anneal temperature is increased to 900 or 1000°C. At these temperatures the residual defects are removed (see dashed curve) and the crystal lattice is restored to its original state before implantation.

Figure 8.18 represents an idealized case. The annealing behavior can be quite complex, reflecting the complex nature of the defects. In general, however, high temperatures are required to achieve complete annealing of the disordered layer. These temperatures are high enough so that one must also consider diffusion of the implanted dopants. In the next sections we discuss the reordering behavior (Section 8.7) of the amorphous layer—a relatively low temperature process—and then (Section 8.8) the diffusion of implanted dopants.

## 8.7 Epitaxial Growth of Implanted Amorphous Silicon

By use of self-ion implantation, Si ions implanted into <100>-oriented Si, one forms an amorphous layer several 100 nm thick on a single-crystal substrate. When the sample is annealed in a furnace at a fixed temperature, one-half the melting temperature (kelvin) of Si—about 550°C, the amorphous layer reorders on the underlying single-crystal substrate. As shown in Fig. 8.19, the thickness of the regrown layer increases linearly in time, thus indicating a constant growth velocity. The velocity is about $10^{-8}$ cm/s at one-half the melting-point temperature and increases rapidly with temperature. The epitaxial reordering process is called "solid phase epitaxy" as it occurs at temperatures well below the melt temperature; in contrast to liquid phase epitaxy where growth occurs from the melt.

Measurements of the growth velocity $v_g$ of the crystal–amorphous interface are shown in Fig. 8.20. The measured velocities extend over nearly 10 orders of magnitude and can be characterized by a single-activation energy $E_A = 2.76$ eV, (Olson and Roth, 1988) so that

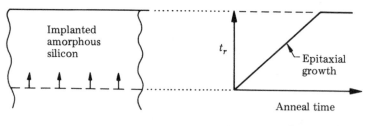

$t_r$ = regrown layer thickness

**FIGURE 8.19**   Solid-phase epitaxial regrowth versus anneal time for an amorphous implanted layer on 100 Si.

$$v_g = v_0 \exp\left(-\frac{E_A}{kT}\right) \qquad (8.36)$$

where the preexponential factor $v_0 = 3.68 \times 10^8$ cm/s. The value of the activation energy is about half that observed in diffusion of dopants in Si. Consequently, regrowth of the implanted amorphous layer can be carried out at times short compared to those required for appreciable dopant diffusion.

During the epitaxial growth process, implanted dopants move onto substitutional lattice sites as the crystal–amorphous interface sweeps by their location. Substitutional concentrations of group III and V dopants can exceed the equilibrium solubility limits of the dopants, as the dopants are effectively frozen in the lattice at the low regrowth temperatures. A second anneal treatment at high temperatures of 900 or 1000°C (carried out to remove residual defects in the regrown layer or to diffuse the dopants) will allow the dopants to move off their substitutional sites. After this high-temperature process, there is a substitutional concentration of dop-

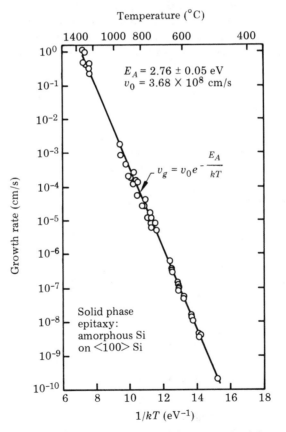

**FIGURE 8.20** Growth rate versus $1/kT$ for solid-phase epitaxial regrowth of implanted amorphous Si on <100> Si. (From Olson and Roth, 1988.)

ants equal to the equilibrium solubility concentration, and the excess dopants above the solubility concentration have formed nonsubstitutional precipitates or clusters.

The presence of high concentrations of implanted dopants influences the epitaxial growth rate. As shown in Fig. 8.21, concentrations of phosphorus at levels greater than 0.1 atomic percent ($5 \times 10^{19}$ cm$^3$) cause an increase in the growth rate. This increase is similar to the increase in the diffusion coefficient of dopants (see Section 7.7), which is attributed to an increase in the vacancy concentration in heavily doped silicon, where the Fermi level is near the conduction or valence-band edges. As one would anticipate, the growth rate is not increased by the implantation of both boron and phosphorus at equal concentrations. In this case, the implanted layer has equal concentrations of donors and acceptors and the Fermi level is near the center of the energy gap.

Implantation of oxygen ions tends to decrease the growth rate. If a sufficiently high concentration of oxygen is implanted, the growth rate will be slowed enough so that the remaining amorphous material recrystalizes in the form of a poly-crystalline layer. The electrical characteristics of this polycrystalline layer are markedly inferior to those of an epitaxial regrown layer.

Polycrystalline layers are formed by implanting other species—notably the inert gas ions (such as Ne, Ar, and Kr), which coalesce to form internal gas bubbles. One can take advantage of these polycrystalline layers by implanting Ar into the back side of the wafers. During high-temperature annealing, impurities such as Cu

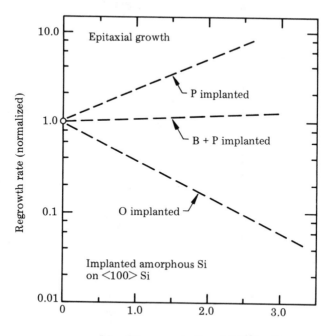

**FIGURE 8.21** Growth rate versus impurity concentration for epitaxial regrowth of implanted amorphous Si on <100> Si.

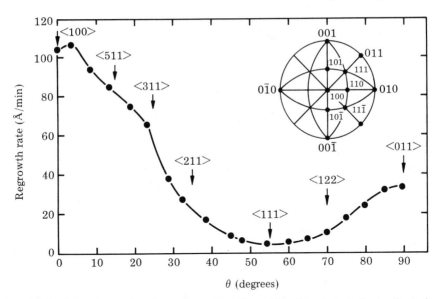

**FIGURE 8.22** Regrowth rate versus orientation of the Si substrate for implanted amorphous Si annealed at 550°C. [From L. Csepregi et al., *J. Appl. Phys. 49*, 3906 (1978).]

that are already present in the wafer can diffuse and be captured (or gettered) at the grain boundaries of the polycrystalline layer. The use of gettering for the removal of unwanted impurities generally produces a marked improvement in carrier lifetimes and *p-n* junction current–voltage characteristics.

The regrowth rate is also strongly dependent on the orientation of the silicon substrate. Figure 8.22 shows regrowth velocities on single-crystal substrates at different orientations. Samples with the <111> direction perpendicular to the surface have the slowest growth rate, while <100> samples have the fastest. This orientation dependence has been explained (Spaepen and Turnbull, 1982) by crystallization proceeding along ledges on the densely packed {111} interfacial planes. The growth velocity is maximum when the crystallographic planes containing the ledges are perpendicular to the surface and decreases when the ledges are inclined to the surface.

# 8.8 Diffusion of Implanted Impurities

The implanted ion distribution can be described by a Gaussian function as given in eq. (8.13):

$$N_I(x) = C(x) = \frac{Q_I}{\sqrt{2\pi}\,\Delta R_p} \exp\left[\frac{-(x - R_p)^2}{2\Delta R_p^2}\right]$$

where $Q_I$ is the implanted dose (ions/cm$^2$), $R_p$ the projected ion range, and $\Delta R_p$ the range straggling. To redistribute the implanted ions, a drive-in diffusion is used, just as in the case of predeposition by diffusion. The concentration profile after the drive-in diffusion can be approximated by another Gaussian function, which includes the diffusion coefficient, $D$:

$$C(x, t) = \frac{Q_I}{\sqrt{\pi} \, (2\Delta R_p^2 + 4Dt)^{1/2}} \exp\left[\frac{-(x - R_p)^2}{2\Delta R_p^2 + 4Dt}\right]. \tag{8.37}$$

This Gaussian function differs from the previous one [eq. (8.13)] only slightly [e.g., the $(2 \Delta R_p^2)$ term is replaced by a $(2 \Delta R_p^2 + 4Dt)$ term]. In this sense the range straggling and the redistribution of the implanted ions due to the drive-in diffusion can be viewed as the spreading of the impurity atoms around a mean location $R_p$, the projected range. It is possible to make a quick estimate of the influence of diffusion during annealing by comparing $\Delta R_p$ with $(Dt)^{1/2}$. Here, however, the diffusion coefficient may be enhanced in the high-dopant-concentration region and $D$ may not be constant over the entire distribution.

When the implanted projected range is close to the surface, the implanted ions may escape from the sample during the drive-in diffusion. The escape of dopant atoms can be reduced or prevented by capping the sample surface with an oxide. With such a boundary condition imposed, the flux at the sample surface is zero, that is,

$$\left.\frac{\partial C}{\partial x}\right|_{x=0} = 0.$$

The solution for this boundary condition can be constructed by adding two Gaussian equations, one identical to eq. (8.37) and the other with $R_p$ placed at an

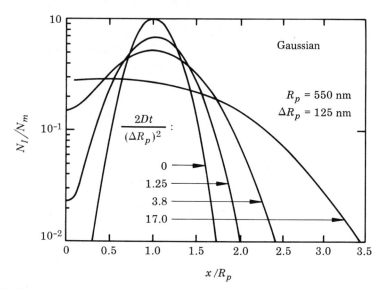

**FIGURE 8.23** Distribution of implanted ions at various stages of anneal for different values of $2Dt/(\Delta R_p)^2$ and the boundary condition that $dC/dx = 0$ at $x = 0$.

image distance of $-R_p$. This operation of adding two Gaussian functions symmetrical with respect to the plane of the sample surface ($x = 0$) is equivalent to reflecting all out-diffusing atoms from the surface. This operation yields

$$N_t(x,\ t) = C(x,\ t) = \frac{Q_I}{\sqrt{\pi}\ (2\ \Delta R_p^2 + 4Dt)^{1/2}}$$

$$\left[\exp - \frac{(x + R_p)^2}{2\ \Delta R_p^2 + 4Dt} + \exp - \frac{(x - R_p)^2}{2\ \Delta R_p^2 + 4Dt}\right]. \tag{8.38}$$

This Gaussian solution [eq. (8.38)] is plotted in Fig. 8.23 for a case where $R_p =$ 550 nm and $\Delta R_p = 125$ nm for various $Dt$ values.

The "reflection" of the out-diffusing impurity atoms causes the concentration near the reflecting surface to build up, and as a result of this the location of the peak concentration tends to move toward the surface. It is clear from Fig. 8.23, however, that the peak of the concentration profile does not move toward the surface until $2Dt$ is several times the value of $(\Delta R_p)^2$.

## EXAMPLE

A Si $p$-channel MOSFET is implanted with boron ions to a dose of $1 \times 10^{12}/\mathrm{cm}^2$ at 30 keV for threshold voltage adjustment. Calculate the peak concentration $N_m/\mathrm{cm}^3$ of the as-implanted profile. After ion implantation, the sample is annealed at 950°C for 30 minutes to activate the implanted boron ions. Calculate the peak concentration $N_m$ and the profile of the implanted species after the activation. Given $R_p = 106$ nm and $\Delta R_p = 39$ nm,

**Solution:**

$$N_m = \frac{Q_I}{\sqrt{2\pi}\Delta R_p{}^2} = \frac{1 \times 10^{12}/\mathrm{cm}^2}{\sqrt{2\pi}\ (3.9 \times 10^{-6})} = 1 \times 10^{17}/\mathrm{cm}^3.$$

The diffusion coefficient of boron in Si at 950°C is about $8 \times 10^{-15}$ cm$^2$/s;

$$2Dt = 2 \times 8 \times 10^{-15} \times 30 \times 60 = 2.9 \times 10^{-11}\ \mathrm{cm}^2$$

$$\Delta R_p^2 = (3.9 \times 10^{-6}\ \mathrm{cm})^2 = 1.5 \times 10^{-11}\ \mathrm{cm}^2.$$

Since $2Dt$ is only about twice the value of $\Delta R_p^2$, the peak location of the implanted species does not change as a result of the activation annealing. The peak concentration is given by

$$N_m = \frac{Q_I}{\sqrt{\pi}\ (2\ \Delta R_p^2 + 4Dt)^{1/2}}$$

$$= \frac{1 \times 10^{12}}{\sqrt{\pi}\ (2 \times 1.5 \times 10^{-11} + 2 \times 2.9 \times 10^{-11})^{1/2}} = 6 \times 10^{16}/\mathrm{cm}^3.$$

The peak concentration has, therefore, decreased from $1 \times 10^{17}/\mathrm{cm}^3$ to $6 \times 10^{16}/\mathrm{cm}^3$ after the activation annealing.

The straggling has also increased from $\Delta R_p$ (39 nm) to $(\Delta R_p^2 + 2Dt)^{1/2}$ (66.3 nm). This increase in straggling means that the impurity profile has spread in both directions about $R_p$, accompanied by a decrease in peak concentration located at $R_p$ to ~0.6 of the initial maximum implanted concentration.

There are cases where the broadening of the implanted ion profile should be kept at a minimum, such as in the control of threshold voltage of MOS devices by implantation and in the formation of very shallow junctions by implantation. In this case the implanted dopant atoms are required to be activated without any significant amount of diffusion during the activation annealing. A process called rapid thermal processing (RTP) can be used to accomplish this requirement. In this process, heat is delivered to the specimen very rapidly, such that the temperature of the specimen can rise uniformly across the wafer to a high processing temperature (800 to 1000°C) in a short time ($\leq 1$ s). The high temperature is then held for a short duration (seconds to minutes) and then the heat is turned off and the specimen is allowed to cool down also rather rapidly. The heat source of such a process is usually an array of lamps or slotted graphite sheets. The advantage of RTP is the rapid rise and fall time at the high processing temperature, so that the activation of the dopant can occur without significant diffusion.

# GENERAL REFERENCES

BEADLE, W. E., J. C. C. TSAI, and R. D. PLUMMER, eds., *Quick Reference Manual for Silicon Integrated Circuit Technology,* Wiley, New York, 1985.

BRICE, D. K., *Ion Implantation Range and Energy Deposition Distributions,* Vol. 1, Plenum Press, New York, 1975.

DEARNALEY, G., J. H. FREEMAN, R. S. NELSON, and J. STEPHEN, *Ion Implantation,* North-Holland, Amsterdam, 1973.

FELDMAN, L. C., and J. W. MAYER, *Fundamentals of Surface and Thin Film Analysis,* North-Holland, Amsterdam, 1986.

GHANDHI, S. K., *VLSI Fabrication Principles,* Wiley, New York, 1983.

LITTMARK, U., and J. F. ZIEGLER, *Range Distributions for Energetic Ions in All Elements,* Pergamon Press, Elmsford, N.Y., 1980.

MAYER, J. W., L. ERIKSSON, and J. A. DAVIES, *Ion Implantation in Semiconductors,* Academic Press, New York, 1970.

OLSON, G. L., and J. A. ROTH, "Kinetics of Solid Phase Crystallization in Amorphous Silicon," *Materials Science Reports,* Vol. 3, pp. 1–78, 1988.

RYSSEL, H., and I. RUGE, *Ion Implantation,* Wiley, New York, 1986.

SEIDEL, T. E., "Ion Implantation," in *VLSI Technology,* ed. S. M. Sze, McGraw-Hill, New York, 1983, Chapter 6.

SPAEPEN, F., and D. TURNBULL, "Crystalline Processes," in *Laser Annealing of Semiconductors,* ed. J. M. Poate and J. W. Mayer, Academic Press, New York, 1982, Chapter 2.

WILLIAMS, J. S., and J. M. POATE, eds., *Ion Implantation and Beam Processing*, Academic Press, New York, 1984.

WILSON, R. G., and G. R. BREWER, *Ion Beams*, R.E. Krieger, Melbourne, Fla., 1979.

ZIEGLER, J. F., ed., *Ion Implantation Science and Technology*, Academic Press, Orlando, Fla., 1984.

## PROBLEMS

**8.1.** For 100-keV implantation of As ions (singly ionized As) into Si at a current of 1 mA ($10^{-3}$ A) over an area of 200 $cm^2$ for 10 min:
  (a) What is the number of implanted ions/$cm^2$?
  (b) What is the value of the projected range $R_p$ and range straggling $\Delta R_p$?
  (c) What is the concentration of implanted As (As/$cm^3$) at $x = R_p$?
  (d) At what depth would the As concentration drop to $10^{-2}$ of its value at $x = R_p$?

**8.2.** For 100-keV implantation of $As_2^+$ ions (singly ionized molecules):
  (a) What is the velocity of the molecules?
  (b) If the magnetic analyzing system has a radius of 0.2 m, what magnetic field is required?
  (c) If the molecule breaks into two As ions when it hits the Si, what is the projected range of the As ions?

**8.3.** For 100-keV As and Sb ions incident on Si (assume that $a = 1.4 \times 10^{-2}$ nm):
  (a) Calculate the nuclear energy loss rate for a $1/r^2$ potential and estimate the range.
  (b) Compare the calculated range values with those given in Fig. 8.8.
  (c) If $\Delta x_p = R_p$, find $\Delta R_p$ and compare with values in Fig. 8.8.

**8.4.** For 100-keV boron in silicon, assume that electronic stopping dominates with $dE/dx = kE^{1/2}$.
  (a) From Fig. 8.13, find a value for $k$.
  (b) Derive an expression for the range.
  (c) Compare your value with that given in Fig. 8.8.

**8.5.** Assume that the diffusion coefficient is $10^{-14}$ $cm^2/s$ for As at 1000°C. You implant $10^{15}$ As ions/$cm^2$ so that $R_p = 10^{-5}$ cm and $\Delta R_p = 4 \times 10^{-6}$ cm.
  (a) What is the maximum As concentration and the depth at which the concentration drops by one-half?
  (b) If the sample is annealed for 30 min at 1000°C, what is the maximum in the As concentration?
  (c) If the silicon contains $10^{18}$ acceptors/$cm^3$ and the junction depth is at $N_D = N_A$, what is the junction depth after implantation and after annealing?

**8.6.** Assume that the As implant leads to a uniform electron concentration of $10^{19}$ electrons/cm$^3$ to a depth of $10^{-5}$ cm with a mobility of 100 cm$^2$/V · s.

   **(a)** What is the sheet resistance?

   **(b)** If a van der Pauw pattern 1 cm square was used with 10 V applied at the contacts, what current would be measured?

   **(c)** If a Hall magnetic field of 1 Wb/m$^2$ is applied, what is the Hall voltage?

**8.7.** If an amorphous layer $5 \times 10^{-5}$ cm thick were formed by self-ion implantation (Si implanted into <100> Si) and annealed at 550°C:

   **(a)** How long would it take to regrow the layer epitaxially?

   **(b)** If $2 \times 10^{20}$ phosphorus atoms/cm$^3$ were present, what time would be required?

   **(c)** If the sample is (110) Si, what time is required?

# Thermal Oxidation of Silicon and Chemical Vapor Deposition of Insulating Films

## 9.1 Introduction

Silicon has become the dominant semiconductor material in integrated-circuit technology because of our ability to grow a stable oxide with good control by relatively easy methods. Silicon dioxide forms the basis of the planar technology. In industrial practice, silicon dioxide layers are frequently formed by thermal oxidation of Si in the temperature range 900 to 1200°C.

The thermal oxidation process is one of the key steps in the fabrication of silicon semiconductor devices and integrated circuits. Thermal oxides are used to mask selectively against dopant diffusion, to passivate device junctions, and for device isolation.

Figure 9.1a shows the cross section of a semirecessed oxide NMOS device (*n*-channel metal–oxide–Si field-effect transistor). The sequence used to grow a layer of recessed oxide is shown in Fig. 9.1b. A thin layer (10 to 20 nm) of $SiO_2$ is grown on the wafer initially, followed by the deposition of silicon nitride ($Si_3N_4$) and patterning of the wafer. Oxidation does not occur in areas covered by $Si_3N_4$, due to the difficulty of oxygen and water vapor to diffuse through $Si_3N_4$. This selective oxidation process results in the recessed oxide structure.

There are two commonly used chemical reactions in oxidation; one is due to oxidation with dry oxygen gas (dry oxidation) and the other is due to oxidation with water vapor (wet oxidation). Wet oxidation generally proceeds with a faster rate than that of dry oxidation. The chemical reaction for dry oxidation is

$$Si(solid) + O_2(gas) \longrightarrow SiO_2(solid). \qquad (9.1)$$

The chemical reaction for wet oxidation is

$$Si(solid) + 2H_2O(vapor) \longrightarrow SiO_2(solid) + 2H_2. \qquad (9.2)$$

The heat of formation of $SiO_2$ at 25°C is generally accepted to be about 70 kcal/g · atom (or 210 kcal/mol). Heats of reaction are reported in units of kcal/g · atom or kcal/mol in the literature, and 23.06 kcal/mol is equal to 1.0 eV/molecule.

Silicon dioxide has more than one crystalline form. The $SiO_2$ layers grown by thermal oxidation, however, have no long-range crystalline order; they are amor-

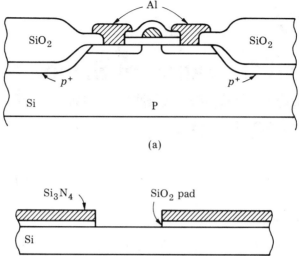

(a)

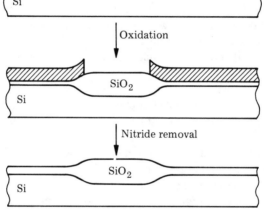

(b)

**FIGURE 9.1** (a) Cross section of a NMOSFET; (b) cross section showing process sequence for recessed oxidation of silicon. [After Jaeger (1988).]

phous in structure and are often referred to as glass or fused silica. Amorphous $SiO_2$ tends to return to the crystalline form at temperatures below 1710°C. The rate of transformation is, however, negligibly small at about 1000°C or lower.

The density of thermal $SiO_2$ (2.15 to 2.27 g/cm³) is less than that of crystalline $SiO_2$ (2.65 g/cm³). Because of its low density and therefore, more open structure, a number of impurities may diffuse through thermal $SiO_2$ interstitially. Some of the diffusion data in $SiO_2$ (such as water, oxygen, and sodium) are summarized in Table 7.7.

In growing $SiO_2$ from Si, due to the changes in density and molecular weight, a film of $SiO_2$ with a thickness of $x_0$ consumes a layer of crystalline Si about $0.45x_0$ in thickness; the oxide–Si interface is therefore displaced from the original Si surface and located deeper into the Si wafer. It has also been observed that for oxidation both with oxygen and with water vapor, the oxidizing species move through the oxide layer to react with Si. The principal effect of a displaced and deeper interface and of an inward migration of the oxidizing species through the oxide layer is to provide a clean oxide–Si interface, free of contaminants due to air exposure prior to oxidation (interfacial impurities tend to end up on the oxide surface).

## 9.2 Reaction Kinetics

The oxidation model proposed by Deal and Grove (1965) is shown schematically in Fig. 9.2. According to this model, for the oxidizing species to reach the silicon surface, there are three steps in series:

1. The oxidizing species must be transported from the bulk of the gas to the oxide–gas interface. The flux is given by

$$F_1 = h_G(C_G - C_S) \qquad (9.3)$$

where $h_G$ is the mass-transfer coefficient and $C_G$ and $C_S$ are the concentration of the oxidant in the bulk of the gas and in the gas at the oxide surface, respectively. We now want to relate the concentrations of the oxidant in the gas to the concentrations of the oxidant in the solid (oxide) through Henry's law. The idea behind Henry's law is that the equilibrium concentration of a species in a solid is proportional to the partial pressure of that species in the surrounding gas. Applying Henry's law to the given situation, the concentration of the oxidant at the oxide surface (but within the oxide) is

$$C_o = K_H P_s \qquad (9.4)$$

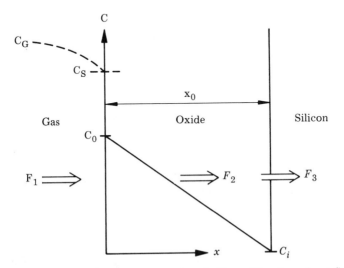

FIGURE 9.2  Model for the thermal oxidation of silicon. Direction of gas flow is normal to plane of paper. [After Grove (1967).]

where $C_o$ is the actual concentration of the oxidant at the oxide surface (within the oxide), $K_H$ the Henry's law constant, and $P_s$ the partial pressure of the oxidant right next to the oxide surface. The concentration of the oxidant at the oxide surface (within the oxide), $C^*$, that *would be* in equilibrium with the partial pressure in the bulk of the gas $P_G$ is given by

$$C^* = K_H P_G. \tag{9.5}$$

The driving force for $F_1$ is therefore proportional to the deviation from equilibrium (i.e., $C^* - C_o$). Using the ideal gas law,

$$PV = N_m kT. \tag{1.3}$$

The gas-phase oxidant concentrations can be expressed as

$$C = \frac{N_m}{V} = \frac{P}{kT} \tag{9.5a}$$

where $P$ is the partial pressure of the oxidant in the gas, $N_m$ the number of molecules of the oxidant in the gas, and $C$ the gas-phase concentration expressed in units of molecules/volume. The concentration of oxidant in the bulk of the gas $C_G$ is

$$C_G = \frac{P_G}{kT} \tag{9.5b}$$

and the concentration of the oxidant at the oxide surface (but in the gas)

$$C_S = \frac{P_S}{kT}. \tag{9.5c}$$

Substituting into eq. (9.3) yields

$$F_1 = h_G(C_G - C_S) = \frac{h_G(P_G - P_S)}{kT}.$$

Using eqs. (9.4) and (9.5), $F_1$ can be expressed as

$$F_1 = \frac{h_G(P_G - P_S)}{kT} = h(C^* - C_o) \tag{9.6}$$

where

$$h = \frac{h_G}{kT K_H}.$$

2. The oxidizing species must diffuse across the oxide layer already present. The diffusional flux $F_2$ is given by

$$F_2 = D\frac{C_o - C_i}{x_o} \tag{9.7}$$

where $D$ is the diffusivity, $C_i$ the concentration of the oxidant at the oxide–Si interface, and $x_o$ is the thickness of the oxide at a given time, $t$.

3. The oxidizing species must then react with the Si at the oxide–Si interface. The flux to incorporate the oxidant into Si, $F_3$ is given by

$$F_3 = k_s C_i \tag{9.8}$$

where $k_s$ is the chemical reaction rate constant for oxidation at the oxide–Si interface.

It is important to recognize at this point that these three steps are in series; eventually, the slowest step among the three will control the entire process of oxidation. When a steady-state condition is established, all fluxes are equal (i.e., $F = F_1 = F_2 = F_3$). Solving for $C_i$ and $C_o$, we have

$$C_i = \frac{C^*}{1 + k_s/h + k_s x_o/D} \tag{9.9}$$

and

$$C_o = \frac{(1 + k_s x_o/D)C^*}{1 + k_s/h + k_s x_o/D}. \tag{9.10}$$

In thermal oxidation of Si, step 1 ($F_1$) is rarely the rate-limiting step. We can assume that $h$ is much larger than $k_s$. With this assumption, eqs. (9.9) and (9.10) can be further simplified to

$$C_i \simeq \frac{C^*}{1 + k_s x_o(t)/D} \tag{9.11}$$

and

$$C_o \simeq C^*. \tag{9.12}$$

Let us consider the situation where step 2 ($F_2$) is the rate-limiting step. For this to occur, the diffusivity of the oxidizing species, $D$, is very small compared to $k_s x_o(t)$, thus causing $C_i \to 0$. This is the case where the oxidation process is said to be diffusion controlled. In the opposite case where $D$ is very large compared to $k_s x_o(t)$, $C_i \to C^* \simeq C_o$. The oxidation process is said to be interfacial reaction controlled. Combining eqs. (9.7), (9.11), and (9.12) and eliminating $C_i$, the flux $F$ is given by

$$F = \frac{DC_o k_s}{D + k_s x_o}. \tag{9.13}$$

Thus the rate at which the oxide layer grows is

$$\frac{dx_o}{dt} = \frac{F}{N} = \frac{DC_o k_s}{N} \frac{1}{D + k_s x_o} \tag{9.14}$$

where $N$ is the number of oxidant molecules incorporated into a unit volume of oxide. For dry oxidation, $N = 2.3 \times 10^{22}$ cm$^{-3}$; for wet oxidation $N = 4.6 \times 10^{22}$ cm$^{-3}$ since two $H_2O$ molecules are incorporated into one molecule of $SiO_2$.

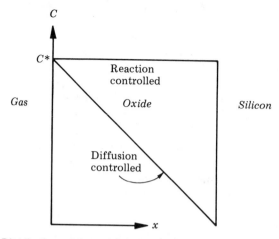

**FIGURE 9.3** Distribution of the oxidizing species in the oxide layer for the two limiting cases of oxidation. [After Grove (1967).]

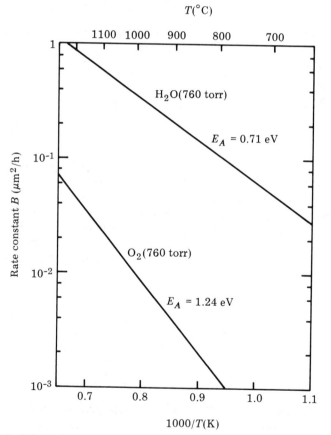

**FIGURE 9.4** Effect of temperature on the parabolic rate constant. [After Grove (1967).]

Inspection of equation (9.14) reveals that when $D \gg k_s x_o$, $F = C_o k_s$ (reaction controlled); when $D \ll k_s x_o$, $F = DC_o/x_o$ (diffusion controlled). The two limiting cases are illustrated schematically in Fig. 9.3. Using the boundary condition of $x_o(t = 0) = x_i$, the general solution for eq. (9.14) is

$$x_o^2 + Ax_o = B(t + \tau) \tag{9.15}$$

where $A = 2Dk_s^{-1}$ (cm), $B = 2DC_o N^{-1}$ (cm$^2$/s), and $\tau = x_i^2/B + x_i/(B/A)$.
Solving for the oxide thickness as a function of oxidation time, we obtain

$$x_o = \frac{A}{2} \left\{ \left( 1 + \frac{t + \tau}{A^2/4B} \right)^{1/2} - 1 \right\}. \tag{9.16}$$

For short oxidation times, $t + \tau \ll A^2/4B$, we obtain the linear oxidation law, $x_o = (B/A)(t + \tau)$. For long oxidation times, $t \gg A^2/4B$, we obtain the parabolic oxidation law, $x_o^2 = Bt$. It should be noted that for short-time oxidation, the oxide layer is thin, leading to a large concentration gradient and large diffusional flux. Under such a circumstance, the interfacial oxidation reaction should be the rate-limiting step [$D/x_o(t) > k_s$ or $k_s \rightarrow 0$] for oxidation. For long-time oxidation, the oxide thickness is large, resulting in a small concentration gradient and small diffusional flux [$D/x_o(t) < k_s$ or $D \rightarrow 0$]; the rate-limiting step is switched from a reaction-controlled mechanism to a diffusion-controlled mechanism.

The parabolic rate constant $B$ and the linear rate constant $(B/A)$ are plotted against $1000/T$ in Figs. 9.4 and 9.5, respectively. It can be seen that the activation for the rate constant $B$ is 1.24 eV for dry oxidation and 0.71 eV for wet oxidation. These values are in good agreement with the activation energies for diffusion of $O_2$ and $H_2O$ in fused silica (see Table 7.7), suggesting that the parabolic rate is associated with the diffusion of the oxidants. The temperature dependence of the linear rate constant $(B/A)$ is about 2 eV (Fig. 9.5), close to the bond-breaking energy of Si, indicating that $B/A$ is associated with the interfacial reaction rate constant $k_s$. The general relationship for silicon oxidation [eq. (9.16)] and its two limiting forms are shown in Fig. 9.6. The rate constants are listed in Tables 9.1 and 9.2.

In solving eq. (9.14), an initial oxide thickness $x_i$ is assumed. It should be noted that for wet oxidation the value of $x_i = 0$ (Nicollian and Brews, 1982). For dry oxidation, the initial condition $x_i = 23$ nm must be used in eq. (9.15). This is

**TABLE 9.1**  Rate Constants for Oxidation of Silicon in Wet Oxygen
[$P_{H_2O} = 640$ torr; $x_o^2 + Ax_o = B(t + \tau)$]

| Oxidation Temperature (°C) | A (μm) | B (μm²/h) | B/A (μm/h) | τ (h) |
|---|---|---|---|---|
| 1200 | 0.05 | 0.720 | 14.40 | 0 |
| 1100 | 0.11 | 0.510 | 4.64 | 0 |
| 1000 | 0.226 | 0.287 | 1.27 | 0 |
| 920 | 0.50 | 0.203 | 0.406 | 0 |

*Source:* After Deal and Grove (1965).

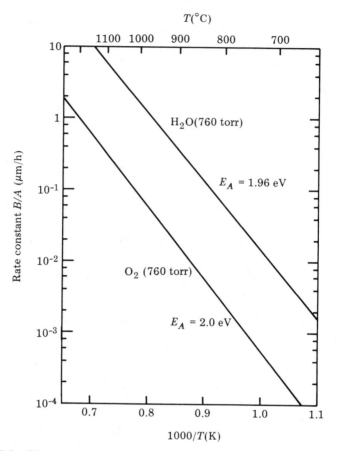

**FIGURE 9.5** Effect of temperature on the linear rate constant. [After Grove (1967).]

**TABLE 9.2** Rate Constants for Oxidation of Silicon in Dry Oxygen
$[P_{O_2} = 760 \text{ torr}; x_o^2 + Ax_o = B(t + \tau)]$

| Oxidation Temperature (°C) | $A$ ($\mu$m) | $B$ ($\mu$m$^2$/h) | $B/A$ ($\mu$m/h) | $\tau$ (h) |
|---|---|---|---|---|
| 1200 | 0.040 | 0.045 | 1.12 | 0.027 |
| 1100 | 0.090 | 0.027 | 0.30 | 0.076 |
| 1000 | 0.165 | 0.0117 | 0.071 | 0.34 |
| 920 | 0.235 | 0.0049 | 0.0208 | 1.40 |
| 800 | 1.340 | 0.0011 | 0.0030 | 9.0 |

*Source:* After Deal and Grove (1965).

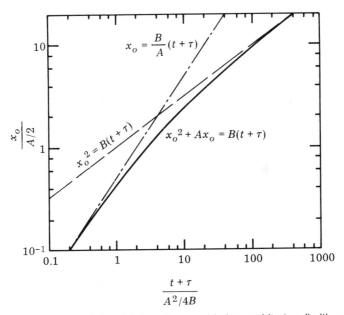

**FIGURE 9.6**  General relationship for silicon oxidation and its two limiting forms. [After Grove (1967).]

because of the fact that an extrapolation of the oxide thickness $x_o$ versus oxidation time $t$ at $t = 0$ yields an intercept of 23 nm. The initial nonlinear growth of 23 nm is believed to be a space-charge effect. Figure 9.7 shows the oxide thickness as a function of time at different temperatures in wet (95°C $H_2O$) oxygen and dry oxygen. These curves obey eq. (9.16); and it is clear that oxidation proceeds at a much faster rate in $H_2O$ than in $O_2$ in both the linear and parabolic regions. Some of the properties of thermally grown silicon dioxide are listed in Table 9.3.

**TABLE 9.3**  Properties of Thermal Silicon Dioxide

| | | | |
|---|---|---|---|
| DC resistivity ($\Omega \cdot$ cm), 25°C | $10^{14}$–$10^{16}$ | Melting point (°C) | ~1700 |
| Density (g/cm³) | 2.27 | Molecular weight | 60.08 |
| Dielectric constant | 3.8–3.9 | Molecules/cm³ | $2.3 \times 10^{22}$ |
| Dielectric strength (V/cm) | $5$–$10 \times 10^6$ | Refractive index | 1.46 |
| Energy gap (eV) | ~8 | Specific heat (J/g · °C) | 1.0 |
| Etch rate in buffered HF (nm/min)[a] | 100 | Stress in film on Si (N/m²) | $2$–$4 \times 10^8$ Compression |
| Infrared absorption peak (μm) | 9.3 | | |
| Linear expansion coefficient (°C⁻¹) | $5.0 \times 10^{-7}$ | Thermal conductivity (W/cm · °C) | 0.014 |

*Source:* After Wolf and Tauber (1986).
[a]Buffered HF:28 ml HF, 170 ml $H_2O$, 113 g $NH_4F$.

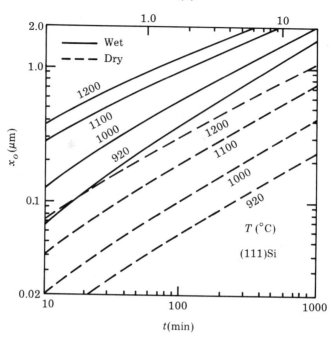

**FIGURE 9.7** Oxide thickness as a function of time with temperature as parameter. Solid lines represent wet oxygen; dashed lines represent dry oxygen. Silicon surface orientation is (111). [After Nicollian and Brews (1982).]

## 9.3 Factors Affecting Oxidation Kinetics

Besides the dry versus wet oxidation and the oxidation temperature, there are three factors commonly observed to affect the oxidation rate of Si: (1) orientation of the substrate, (2) pressure effects, and (3) impurity effects. The orientation dependence of the oxidation rate arises primarily from the total number of available Si atoms for oxidation as a function of the substrate orientation. Since the available number of Si atoms is related to the oxidation reaction at the interface between the oxide and the Si wafer, only the linear oxidation rate is expected to change as a function of orientation. In the diffusion-limited regime, the oxidation rate is independent of the substrate orientation. Figure 9.8 shows the oxidation kinetics on (100) and (111) Si at 1000°C. The oxidation rate for (111) Si is faster than that for (100) Si initially in the linear region. As the oxidation kinetics change from the linear rate to the parabolic rate, the difference between the two orientations diminishes.

In both the linear regime and the parabolic regime, the oxide thickness is related to the rate constant $B$, which is given by $2DC_oN^{-1}$ or more correctly, $2DC^*N^{-1}$. The equilibrium oxidant concentration $C^*$ is directly proportional to the partial pressure of the oxidant in the gas stream [eq. (9.5)]. The oxidation rate is therefore expected to increase with increasing pressure of the oxidant, as shown in Fig. 9.9.

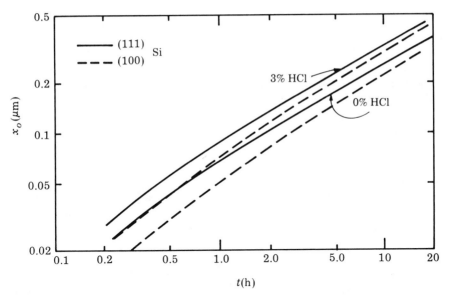

**FIGURE 9.8** Oxide thickness $x_o$ versus oxidation time $t$ for the oxidation of (100)- and (111)-oriented $n$-type silicon ($N_D = 10^{15}$ cm$^{-3}$) in 0 and 3% HCl–O$_2$ mixtures at 1000°C. [After Nicollian and Brews (1982).]

Boron and phosphorus can cause the linear and parabolic oxidation rate to increase when the Si substrates are doped with these impurities to high concentrations. The effect of boron and phosphorus will be important when oxidizing over diffused $p$- and $n$-type layers. In this case the oxide will be thicker than elsewhere on the wafer (Deal and Sklar, 1965). The explanation of the increase in oxidation rate with increasing boron concentration is that boron is preferentially incorporated

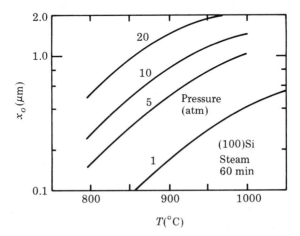

**FIGURE 9.9** Wet oxide growth at increased pressures for (100)-oriented Si substrate with a resistivity of 3 to 10Ω · cm. The oxidation time is 60 minutes. [After Su (1981). Reprinted, permission of *Solid State Technology*.]

into the silica during growth, thereby weakening the bond structure of the silica. Silicon and oxygen bonds are easily broken; oxygen and $H_2O$ molecules enter the silica more easily and diffuse through more easily. The cause of increased oxidation with increasing phosphorus concentration is apparently different from that for boron. There is very little incorporation of phosphorus into the silica. The influence of phosphorus is primarily on the reaction limited regime of Si oxidation, and is possibly due to a Si bond-breaking mechanism at the interface to facilitate the oxidation reaction.

Chlorine is another impurity that has large effects on the characteristics of the oxidation of Si. The presence of chlorine in the oxidation ambient increases the oxidation rate. Chlorine is added to a dry oxygen carrier gas in small amounts, up to 6% by volume, either as chlorine gas, anhydrous HCl, which oxidizes to form chlorine and water vapor, or as an organic molecule, such as trichloroethylene, which oxidizes to release chlorine. Anhydrous HCl is the most common way to introduce chlorine into the oxidation ambient at present. Figure 9.8 shows the effect of chlorine on the oxidation rates of (111) and (100) Si. There are several reasons for the enhanced oxidation rate with the addition of chlorine or HCl (Hirabayashi and Iwamura, 1973): (1) the enhanced diffusion of $O_2$ and $H_2O$ molecules in the oxide containing HCl (the diffusivities of $O_2$ and $H_2O$ are about two to three times higher than those in dry oxide); (2) enhanced reactions at the Si–$SiO_2$ interface due to the catalytic action of HCl; and (3) $H_2O$ molecules created by the reaction

$$2HCl + \tfrac{1}{2} O_2 \rightleftharpoons H_2O + Cl_2.$$

The oxidation rate is increased due to the enhanced diffusivity of $H_2O$ molecules through the growing oxide.

There are other impurities that may affect the oxidation kinetics of Si. For example, the implantation of about one monolayer of Ge (about $1 \times 10^{15}$ cm$^{-2}$) has been found to considerably increase the initial steam oxidation rate of Si (about 1.5 times faster). This is due primarily to the segregation of Ge at the $SiO_2$–substrate interface. As the thickness of oxide increases to the diffusion-limited regime, the oxidation rate is no longer enhanced (Holland et al., 1987).

## 9.4 Oxide Passivation by Chlorine

The addition of chlorine also appears to suppress the formation of stacking faults during oxidation. When wafers are oxidized in dry $O_2$, mixed with a proper amount of HCl, suppression of stacking fault generation and shrinkage of existing stacking faults take place, thus reducing the possibility of stacking faults intersecting *p-n* junctions.

The presence of HCl in oxidizing atmospheres has the effect of passivating the oxide layer. One of the main causes of electrical instabilities observed in MOS structures is the migration of positively charged ions within the oxide. In particular, sodium and other alkali ions can move easily in $SiO_2$ even near room temperature under bias and temperature stressing. In bipolar devices, the presence of heavy metallic impurities in silicon is a primary cause of collector–emitter shorts. It has been observed that the addition of a small amount of HCl to the water vapor or oxygen in the high-temperature oxidation of silicon greatly reduces the influence of mobile ions and metallic impurities on device performance. The pressure of chlorine in the furnace atmosphere promotes the formation of volatile metal chloride via the reaction

$$M(metal) + Cl \longrightarrow MCl.$$

The volatile compounds are then exhausted at the exit end of the oxidation furnace. Thus both mobile ions and heavy metallic impurities are removed. When chloride is incorporated into the oxide near the Si–SiO$_2$ interface, the positively charged ions interact with chlorine to become "trapped" and neutralized as mobile ions. The oxide layer is therefore passivated for more stable operation as a dielectric layer.

## 9.5 Redistribution of Impurities During Thermal Oxidation

It has been found that impurities in the bulk of a Si wafer will redistribute near a growing thermal oxide. Some of the impurity atoms will be incorporated into the oxide layer. The concentration of impurity atoms within the Si located near the Si–SiO$_2$ interface may increase or decrease depending on the distribution coefficient. Because of the redistribution, the oxidation rate and the interface properties may change accordingly. The redistribution of dopant impurities is characterized by the segregation coefficient $m$, defined as

$$m = \frac{\text{equilibrium concentration of impurity in Si}}{\text{equilibrium concentration of impurity in SiO}_2}. \tag{9.17}$$

When $m > 1$, the oxide rejects the impurity; when $m < 1$, the oxide takes up the impurity. In addition to the segregation coefficient, the redistribution behavior also depends on the diffusion rate of the impurity in the oxide. Figure 9.10 shows four different cases of impurity redistribution in Si due to thermal oxidation. These cases can be classified into two groups. In one group the oxide takes up the impurity ($m < 1$), and in the other the oxide rejects the impurity ($m > 1$). In each group the redistribution depends on how rapidly the impurity can diffuse through the oxide. For the case of $m < 1$, the silicon surface is depleted of the impurity near the interface (Fig. 9.10a). Rapid diffusion of the impurity through the SiO$_2$ will increase the amount of depletion (Fig. 9.10b). For the case $m > 1$, the diffusion rate of the impurity through the SiO$_2$ plays an even more significant role. When diffusion through the oxide is rapid, so many impurities may escape from the solid to the gaseous ambient that the overall effect will be a depletion of the impurity (Fig. 9.10c and d).

Redistribution may still occur even if $m = 1$. This is because the Si–SiO$_2$ interface is advancing into the Si substrate, converting the Si into SiO$_2$ about twice as thick as the amount of Si consumed. Thus more impurity atoms are required to

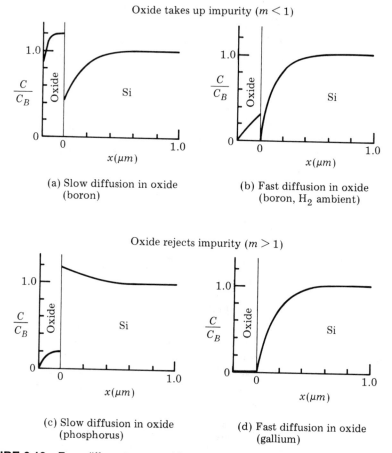

**FIGURE 9.10** Four different cases of impurity redistribution in silicon due to thermal oxidation. [After Grove et al. (1964).]

bring up the impurity concentration in the larger volume of $SiO_2$, resulting in the depletion of the impurity near the Si surface. It should be noted the boron in Si usually has an experimental value of $m < 1$, whereas $n$-type dopants (P, Sb, and As) usually have experimental values of $m > 1$ under normal growth rates of the oxide layer.

## 9.6 Properties of the Si–SiO₂ Interface

Associated with the thermally oxidized silicon system are several types of electrical charges which can adversely affect the device performance, reliability, and yield. These charges are summarized in Fig. 9.11. They include (1) the interface trapped charge ($Q_{it}$), (2) the oxide fixed charge ($Q_f$), (3) the oxide trapped charge ($Q_{ot}$), and (4) the mobile ionic charge ($Q_m$) (Deal, 1980).

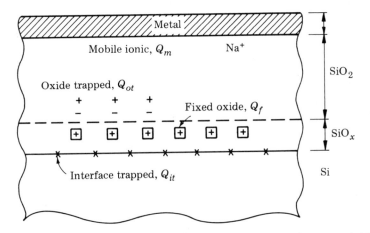

**FIGURE 9.11** Charges associated with the Si–SiO$_2$ system. [After Deal (1980).]

The interface trapped charge, $Q_{it}$, is located at the Si–SiO$_2$ interface with either positive or negative charges. Historically, these states are called fast states or surface states. The origin of these states is due primarily to the disruption of the periodicity of the lattice at the Si–SiO$_2$ interface. On clean Si surfaces in ultrahigh vacuum, $Q_{it}$ is very high (about $10^{15}$ states/cm$^2$)—of the order of the density of surface atoms. For well-prepared SiO$_2$ on Si, most of the interface trapped charge can be neutralized by a low-temperature (450°C) hydrogen annealing. The value of $Q_{it}$ can be as low as $10^{10}$ cm$^{-2}$; typical values range between $10^{10}$ and $10^{13}$ cm$^{-2}$.

The oxide fixed charge, $Q_f$, is positive, fixed, and cannot be easily charged or discharged. These positive charges are located within 3 nm of the Si–SiO$_2$ interface. The value of $Q_f$ is not greatly affected by the oxide thickness or by the type or concentration in the silicon; however, it is a strong function of the substrate orientation and the annealing process. It has been suggested that incomplete oxidation of the Si near the Si–SiO$_2$ interface is the origin of the fixed oxide charge. The ratio of $Q_f$ under a given oxidation condition for silicon with (111), (110), and (100) orientation is approximately 3 : 2 : 1. It is also interesting to note that regardless of the previous history of a sample, the final heat treatment determines the value of $Q_f$. Figure 9.12 shows the oxide fixed charge density as a function of annealing, indicating that low $Q_f$ can be obtained by either a high-temperature dry oxidation (1200°C) followed by cooling in an inert ambient, or a low-temperature oxidation (<1000°C) followed by an anneal at the same temperature in either dry nitrogen or argon and then cooling down in the inert gas ambient. It has also been observed that the addition of HCl in O$_2$ reduces fixed charge density. Typical values of $Q_f$ ranges between $10^{10}$ and $10^{12}$ cm$^{-2}$.

The oxide trapped charge, $Q_{ot}$, is located within the oxide layer; both electrons and holes can be trapped in the oxide. These charges are caused by ionizing radiation (such as x-rays), avalanche injection, or by high currents induced through the oxide, thus generating electron–hole pairs in the oxide. In the absence of an

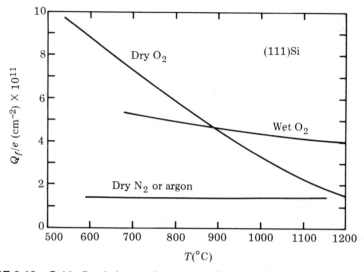

**FIGURE 9.12** Oxide fixed charge density as a function of annealing temperature for annealing times of 1 hour or less, showing oxidation–annealing paths. Based on (111) silicon samples. [After Deal et al. (1967).]

electric field, the electrons and holes may recombine. In the presence of an electric field, charge buildup within the oxide may occur. Typical values of $Q_{ot}$ ranges between $10^{10}$ and $10^{13}$ cm$^{-2}$.

The mobile ionic charge, $Q_m$, is due to impurities such as Na$^+$, Li$^+$, and K$^+$. These ions can diffuse easily in the oxide (see Table 7.7). These mobile ions may be neutralized or "gettered" by the addition of chlorine in the oxidizing ambient. The value of $Q_m$ may vary between $10^{10}$ and $10^{13}$ cm$^{-2}$.

The presence of these charges causes the threshold voltage $V_T$ of MOSFET devices to shift to unacceptable values (see Chapter 5). From a processing point of view, the control of these charges is absolutely necessary for reliable operation of MOS devices.

## 9.7 Deposition of Insulating Layers

In addition to thermal oxidation of silicon to form silicon dioxide, insulating layers such as SiO$_2$ and Si$_3$N$_4$ can be deposited by a number of chemical vapor deposition (CVD) processes. Chemical vapor deposition systems are either hot wall systems, where the reactors are resistance heated, or cold wall systems, where the reaction zones are heated by radio-frequency (RF) susceptors. These systems can be operated at atmospheric as well as at low pressures (commonly called low-pressure CVD). The reduction of pressure in the reaction chamber increases the mean free path of the reactant species; this leads to more uniform layers with fewer pinholes from slice to slice. The growth rate (about 10 to 15 nm/min) is slower than that for atmospheric pressure reactors (about 50 to 100 nm/min). In some systems, a

**TABLE 9.4**  Characteristics and Applications of CVD Reactions

| Process[a] | Advantages | Disadvantages | Applications |
|---|---|---|---|
| APCVD (low temperature) | Simple reactor, fast deposition, low temperature | Poor step coverage, particle contamination | Low-temperature oxides, both doped and undoped |
| LPCVD | Excellent purity and uniformity, conformal step coverage, large wafer capacity | High temperature, low deposition rate | High-temperature oxides, both doped and undoped, silicon nitride, poly Si, W, $WSi_2$ |
| PECVD | Low temperature, fast deposition, good step coverage | Chemical (e.g., $H_2$) and particulate contamination | Low-temperature insulators over metals, passivation (nitride) |

*Source:* After Wolf and Tauber (1986).
[a]AP, atmospheric pressure; LP, low pressure; PE, plasma enhanced.

plasma is used to enhance the chemical reactivity, thus requiring much lower deposition temperatures (room-temperature deposition is possible). The characteristics of some of the CVD systems are shown in Table 9.4.

Deposited insulating layers are usually used to provide protection to semiconductor surfaces, to serve as implantation or diffusion masks, or to serve as interlayer dielectrics between two levels of metallization. These layers need to be pinhole free, adherent to the deposited surface, and relatively stress free to survive subsequent thermal cycles.

## 9.7.1 Silicon Dioxide Films

Silicon dioxide films can be formed by pyrolytic oxidation of various alkoxysilanes. The most commonly used compound is tetraethylorthosilane (TEOS), which is a liquid at room temperature (boiling point = 167°C). The vapor of TEOS is transported to the reaction zone by a bubbler and proceeds to react at about 700°C (commonly in a cold wall reactor):

$$Si(C_2H_5O)_4 + 12O_2 \rightleftharpoons SiO_2 + 8CO_2 + 10H_2O. \qquad (9.18)$$

This reaction produces water as a by-product, and the water may be incorporated into the $SiO_2$ layer, resulting in a poor-quality insulating film.

Better-quality $SiO_2$ films can be produced by reacting silane with oxygen at lower temperatures (typically at 450°C) in a cold wall CVD system at atmospheric pressure:

$$SiH_4 + O_2 \rightleftharpoons SiO_2 + 2H_2. \qquad (9.19)$$

**TABLE 9.5** Properties of Deposited Silicon Dioxide

| Deposition | Plasma | $SiH_4 + O_2$ | TEOS | $SiCl_2H_2 + N_2O$ |
|---|---|---|---|---|
| Temperature (°C) | 200 | 450 | 700 | 900 |
| Composition | $SiO_{1.9}(H)$ | $SiO_2(H)$ | $SiO_2$ | $SiO_2(Cl)$ |
| Step coverage | Nonconformal | Nonconformal | Conformal | Conformal |
| Thermal stability | Loses H | Densifies | Stable | Loses Cl |
| Density (g/cm³) | 2.3 | 2.1 | 2.2 | 2.2 |
| Refractive index | 1.47 | 1.44 | 1.46 | 1.46 |
| Stress ($10^8$ N/m²)[a] | 3C-3T | 3T | 1C | 3C |
| Dielectric strength ($10^6$ V/cm) | 3–6 | 8 | 10 | 10 |
| Etch rate (nm/min) (100:1 $H_2O$:HF) | 40 | 6 | 3 | 3 |

*Source:* After Wolf and Tauber (1986).

[a]C, compressive stress; T, tensile stress.

One other advantage of this low-temperature reaction is to allow films to be deposited over aluminum metallization. Consequently, these films can be used for passivation coatings over the final device and for insulation between Al levels. Table 9.5 lists some of the properties of deposited silicon dioxide by various processes. One of the important properties of deposited $SiO_2$ layer is the stress in the film. Low stresses in the as-deposited film and during thermal cycling improve the stability of the films without cracking up or breaking the substrate. The built-in stress in the silica film may be adjusted by the incorporation of $P_2O_5$ in the layer.

## 9.7.2 Phosphosilicate Glass and Stress Adjustment

Phosphosilicate glass (PSG) is silicon dioxide doped with phosphorus; it can be produced by the simultaneous pyrolysis of silane and phosphine in oxygen:

$$SiH_4 + O_2 \longrightarrow SiO_2 + 2H_2$$
$$4PH_3 + 5O_2 \longrightarrow 2P_2O_5 + 6H_2. \tag{9.20}$$

Water may also be a reaction by-product in the pyrolysis of phosphine. There appears to be no limit on the amount of incorporation of $P_2O_5$ into $SiO_2$. However, the silica film becomes hygroscopic with increasing $P_2O_5$ incorporation. For this reason, phosphosilicate glass used for passivation is limited to 2 to 8 wt % phosphorus.

The incorporation of $P_2O_5$ into silica films [grown by reactions (9.19) and (9.20)] causes a reduction in the built-in stress in the layer. The typical built-in stress in undoped silica is about $3 \times 10^8$ N/m² tension (see Table 9.5). This level of stress may be reduced to zero by incorporating 20% $P_2O_5$ by weight in the film. This reduction in stress improves the film stability.

The thermal expansion coefficient of silica is found to increase with the incorporation of $P_2O_5$. Consequently, PSG films can be designed to provide a desirable thermal match to the underlying semiconductor for better film stability during thermal cycling. For example, the thermal expansion coefficient of undoped silica increases from $6 \times 10^{-7}/°C$ to $5.9 \times 10^{-6}/°C$ (thermal expansion coefficient of GaAs) with about 20 to 24 wt % $P_2O_5$ in the film. This thermal matching greatly improves the stability of PSG as a capping layer for GaAs. Generally, PSG is a better capping layer to block the migration of elements such as Zn and Sn than silica; it is also effective in reducing sodium ion transport through oxide layers for MOS technology. The primary reason for this blocking capability is due to the dense structure of PSG compared to that of silica. PSG is also occasionally used as a dopant source for diffusion.

### 9.7.3 Silicon Nitride

Silicon nitride has a stoichiometric composition of $Si_3N_4$. Large deviations from stoichiometry are often encountered in deposited films. Silicon nitride films are extensively used in both Si and GaAs technology because of their ability to block diffusion of water and sodium. One example is shown in Fig. 9.1, where a $Si_3N_4$ layer is used to grow a recessed oxide on Si. Silicon nitride is chemically deposited by the following reactions:

$$3SiH_4 + 4NH_3 \overset{700-900°C}{\underset{\text{atmospheric pressure}}{\rightleftharpoons}} Si_3N_4 + 12H_2 \qquad (9.21)$$

$$3SiCl_2H_2 + 4NH_3 \overset{700-800°C}{\underset{\text{reduced pressure}}{\rightleftharpoons}} Si_3N_4 + 6HCl + 6H_2. \qquad (9.22)$$

Table 9.6 lists some of the properties of silicon nitride deposited by low-pressure CVD (LPCVD) and plasma-enhanced CVD (PECVD). The built-in stresses in LPCVD-produced SiN films are very high in tension; the highly stressed films may be used to generate dislocations on the back side of wafers for impurity gettering purposes (see Section 6.8). The built-in stresses in PECVD-produced SiN films are lower and typically compressive in nature, although the stresses may change sign from compression to tension, depending on the growth condition. SiN films up to 1 μm thick can be grown with compressive stress, with good adhesion to the semiconductor.

## 9.8 Etching of Insulating Films

Etching processes are necessary in integrated-circuit processing technology. Traditionally, etching is done with wet chemicals. More recently, dry etching methods are being used extensively for VLSI processing. Dry etching is an etching method performed in a low-pressure gaseous discharge; therefore, the chemical reactions responsible for the etching action occur in the gas phase with the assistance of the plasma.

**TABLE 9.6**   Properties of Silicon Nitride

| Property | LPCVD 900°C | PECVD(LP) 300°C |
|---|---|---|
| Composition | $Si_3N_4$ | $Si_xN_yH_z$ |
| Si/N ratio | 0.75 | 0.8–1.0 |
| Density $(g/cm^3)$ | 2.8–3.1 | 2.5–2.8 |
| Refractive index | 2.0–2.1 | 2.0–2.1 |
| Dielectric constant | 6–7 | 6–9 |
| Dielectric strength (V/cm) | $1 \times 10^7$ | $6 \times 10^6$ |
| Bulk resistivity $(\Omega \cdot cm)$ | $10^{15}$–$10^{17}$ | $10^{15}$ |
| Energy gap (eV) | 5 | 4–5 |
| Stress at 23°C on Si $(10^8 \ N/m^2)^a$ | 10T | 2C-5T |
| Thermal expansion | $4 \times 10^{-6}/°C$ | $> 4 < 7 \times 10^{-6}/°C$ |
| Step coverage | Fair | Conformal |
| $H_2O$ permeability | Zero | Low–none |
| Thermal stability | Excellent | Variable >400°C |
| Solution etch rate $(nm/min)^b$ | | |
| BHF, 20–25°C | 1–1.5 | 20–30 |
| 49% HF, 23°C | 8 | 150–300 |
| 85% $H_3PO_4$, 155°C | 1.5 | 10–20 |
| 85% $H_3PO_4$, 180°C | 12 | 60–100 |
| Plasma etch rate (nm/min) | | |
| 70% $CF_4$/30% $O_2$, 150 W, 100°C | 20 | 50 |
| $Na^+$ penetration (nm) | <10 | <10 |
| $Na^+$ retained in top 10 nm (%) | >99 | >99 |

*Source:* After Wolf and Tauber (1986).

[a]T, tensile stress. C, compressive stress.

[b]BHF, buffered HF.

## 9.8.1 Wet Chemical Etching

Silica films are readily etched by hydrofluoric acid with the following reaction:

$$SiO_2 + 6HF \rightleftharpoons H_2SiF_6 + 2H_2O. \tag{9.23}$$

In practice, the etching of $SiO_2$ layers is done with a diluted HF solution buffered with $NH_4F$ (BHF; see Table 9.3). The addition of $NH_4F$ in the etching solution serves to replenish F ions for a more consistent etching rate from run to run as well as to reduce attack of the photoresist by the HF. The etching rates of $SiO_2$ formed by different methods with various HF solutions are reported in Tables 9.3 and 9.5.

The addition of $P_2O_5$ into the $SiO_2$ film increases the etching rate of the doped oxide (PSG). Typically, the etching rate of thermally grown $SiO_2$ in dry oxygen is about 100 nm/min in BHF at room temperature. The etching rate of densified PSG films (heat treated at 1000 to 1200°C after deposition) with 8 mol % $P_2O_5$ is about 250 to 300 nm/min in the same solution.

Both HF and BHF can be used to etch SiN films. The etching rates are sensitive to the deposition condition (see Table 9.6). When SiN films are used in conjunction with $SiO_2$ films, SiN films may be etched selectively with phosphoric acid. Typically, the etching rate of LPCVD-grown SiN in boiling $H_3PO_4$ acid is about 10 to 12 nm/min, the etching rate of thermally grown $SiO_2$ is about 1 to 2 nm/min, and the etching rate of Si is under 0.3 nm/min. It should be noted that the etching rate of SiN is sensitive to the oxygen content in the films; increasing oxygen in SiN causes reduced etching rates in $H_3PO_4$ but increased rates in BHF (Wolf and Tauber, 1986). The etching rates of SiN films are reported in Table 9.6.

## 9.8.2 Dry Etching

In dry etching, the ions, atoms, and radicals are created in the plasma by ionization and fragmentation. These species provide etching action either by physical sputtering (erosion by ion bombardment) or by catalyzing chemical reactions, or both. The species in the plasma are produced by several types of impact reactions. For example, Ar and oxygen ions are produced by simple ionization with electron impact:

$$Ar + e^- \rightleftharpoons Ar^+ + 2e^-. \tag{9.24}$$

Radicals such as $CF_3^+$ are produced by a dissociative ionization:

$$CF_4 + e^- \rightleftharpoons CF_3^+ + F + 2e^- \tag{9.25}$$

or

$$CF_4 + e^- \rightleftharpoons CF_3^+ + F^- + e^-. \tag{9.26}$$

Electron impact may also produce molecular fragmentation without ionization:

$$CF_3Cl + e^- \rightleftharpoons CF_3 + Cl + e^-. \tag{9.27}$$

Dry etching or plasma-assisted etching can take different forms. The ion etching techniques, which include ion milling and sputter etching, produce etching by physical sputtering. The reactive etching techniques, which include plasma etching and reactive ion etching, rely on chemical reactions to produce volatile compounds and sometimes also on combined ion bombardment effects. The directionality of the ions provides the possibility of etching features with very high aspect ratios.

Gas composition is a dominant factor in determining etching rate and selectivity in dry etching. Generally speaking, halogen-containing gases are used to remove Si, $SiO_2$, $Si_3N_4$, Al, and Al alloys (i.e., Al–Si and Al–Cu); an oxygen plasma is used to remove photoresist and organic layers.

The dry etching reaction of $SiO_2$ may have the following steps:

$$CF_4 + e^- \rightleftharpoons CF_x + F + e^- \tag{9.28}$$

$$CF_x + SiO_2 \rightleftharpoons SiF_4 + (CO, CO_2, COF_2). \tag{9.29}$$

**TABLE 9.7** Etching Rate and Selectivity in Dry Etching

| Etched Material, $M$ | Gas | Etch Rate (nm/min) | Selectivity[a] M/Resist | M/Si | M/SiO$_2$ |
|---|---|---|---|---|---|
| Al, Al–Si, Al–Cu | $BCl_3 + Cl_2$ | 50 | 5–8 | 3–5 | 20–25 |
| Polysilicon | $Cl_2$ | 50–80 | 5 | — | 25–30 |
| SiO$_2$ | $CF_4 + H_2$ | 50 | 5 | 20 | — |
| PSG | $CF_4 + H_2$ | 80 | 8 | 32 | — |

*Source:* After Mogab (1983).

[a]$M$, material intended to be etched.

The etching of Si may follow:

$$Si + CF_x \rightleftharpoons C \text{ absorbed on Si} \tag{9.30}$$

$$Si + 4F \rightleftharpoons SiF_4. \tag{9.31}$$

Increasing the amount of oxygen in $CF_4$ favors the etching of Si relative to that of $SiO_2$ since oxygen tends to remove C atoms absorbed on the Si surface. Increasing the amount of hydrogen tends to favor the etching of $SiO_2$ relative to that of Si due to the removal of F in the plasma by forming HF.

Table 9.7 lists some of the typical etching rates and selectivity for dry etching processes for Si VLSI technology. Table 9.6 also lists the dry etching rates for SiN films. For VLSI technology, dry etching is favored because of the highly selective and directional etching ability to meet the requirements for dimensional accuracy.

## 9.9 Stress in Thin Films

The elastic stress present in thin films is an inherent part of the deposition process, and can be either tensile or compressive. The sign and the magnitude of the film stress depend on the processing parameters (i.e., substrate temperature, type of the substrate, rate of deposition, and method of deposition). Stresses of about $10^8$ to $10^9$ N/m$^2$ are often observed in thin films. These stresses may cause film fracture, delamination (peeling), and occasionally, substrate fracture.

The sign of the stress can easily be determined by observing the sign of the curvature (concave or convex) of a film–substrate composite structure (Fig. 9.13). For a composite structure with a convex curvature (curved like the exterior of a circle, Fig. 9.13a), the film is under a compressive stress (i.e., the film is compressed from its relaxed length to fit onto the substrate). For a composite structure with a concave curvature (curved like the interior of a circle, Fig. 9.13b), the film is under a tensile stress (i.e., the film is stretched from its relaxed length to fit onto the substrate).

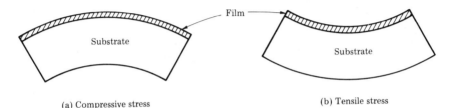

(a) Compressive stress          (b) Tensile stress

**FIGURE 9.13** Stress in thin films. Convex curvature (a) indicates compressive stress in the film, and concave curvature (b) indicates tensile stress in the film.

To improve the mechanical stability of thin-film structures, it is desirable to reduce the stresses in the film to a minimum value. This is one of the reasons to incorporate $P_2O_5$ into silica film—to reduce stresses and improve the structure stability (Section 9.7.2).

Stress in a film consists of two major components, a "thermal" component arising from the difference in the thermal-expansion coefficients of the film and the substrate, and an "intrinsic" component resulting from the growth and structure of the film (Chopra, 1969). The thermal component, called thermal stress, $\sigma_{th}$, is given by

$$\sigma_{th} = Y_F(\alpha_F - \alpha_S) \Delta T \tag{9.32}$$

where $Y_F$ is the Young's modulus of the film, $\alpha_F$ and $\alpha_S$ are the thermal expansion coefficients of the film and the substrate, respectively, and $\Delta T$ is the temperature of the substrate during film growth or deposition minus the temperature at measurement.

The intrinsic stress of the film is much less well understood. The sign and magnitude of the intrinsic stress are strongly affected by the structure, and therefore the deposition parameters, of the film. This is the reason for a variety of stress levels observed in similar films deposited under various conditions, as shown in Tables 9.5 and 9.6.

One of the most convenient ways to determine the stress in thin films is to measure the curvature of the film–substrate composite structure. The following expression derived from bending plate theory gives the stress in a film (Chopra, 1969):

$$\sigma = \frac{dY_S t_S^2}{3L^2 t_F(1 - \mu)} \left(1 + \frac{Y_F \, t_F}{Y_S \, t_S}\right) \tag{9.33}$$

where $d$ is the deflection of the substrate, $Y_F$ and $Y_S$ are the Young's moduli of the film and the substrate, respectively, $t_S$ and $t_F$ are the thicknesses of the substrate and the film, respectively, $L$ is the length of the substrate and $\mu$ is Poisson's ratio. For a very thin film, eq. (9.33) reduces to

$$\sigma = \frac{Y_S t_S^2}{6r(1 - \mu)} \tag{9.34}$$

where $r$ is the curvature of the bent substrate.

If the stress in a thin film is higher than the adhesion between the film and the substrate, peeling or delamination of the film would occur. If the stress in the film is higher than the fracture strength of the film but less than the film–substrate adhesion, the film itself may fracture. The stress in thin-film structures is a critical parameter for the mechanical stability of the deposited layers.

## REFERENCES

### General References

ADAMS, A. C., "Dielectric and Polysilicon Film Deposition," in *VLSI Technology*, ed. S. M. Sze, McGraw-Hill, New York, 1983, Chapter 3.

GHANDI, S. K., *VLSI Fabrication Principles, Silicon and Gallium Arsenide*, Wiley, New York, 1983, Chapters 8 and 9.

GROVE, A. S., *Physics and Technology of Semiconductor Devices*, Wiley, New York, 1967, Chapter 2.

JAEGER, R. C., *Introduction to Microelectronic Fabrication*, Modular Series on Solid State Devices, ed. G. W. Neudeck and Robert F. Pierret, Addison–Wesley, Reading, Mass., 1988, Chapter 3.

MOGAB, C. J., "Dry Etching," in *VLSI Technology*, ed. S. M. Sze, McGraw-Hill, New York, 1983, Chapter 8.

NICOLLIAN, E. H., and J. R. BREWS, *MOS Physics and Technology*, Wiley, New York, 1982, Chapter 13.

WOLF, S., and R. N. TAUBER, *Silicon Processing for the VLSI I Era*, Vol. 1, *Processing Technology*, Lattice Press, Sunset Beach, Calif., 1986, Chapters 6 and 7.

### Specific References

CHOPRA, K. L., *Thin Film Phenomena*, McGraw-Hill, New York, 1969, Chapter 5.

DEAL, B. E., *IEEE Trans. Electron Devices ED-27*, 606 (1980).

DEAL, B. E., and A. S. GROVE, *J. Appl. Phys. 36*, 3770 (1965).

DEAL, B. E., and M. SKLAR, *J. Electrochem. Soc. 112*(4), 430 (1965).

DEAL, B. E., M. SKLAR, A. S. GROVE, and E. H. SHOW, *J. Electrochem. Soc. 114*, 266 (1967).

GROVE, A. S., O. LESTIKS, and C. T. SAH, *J. Appl. Phys, 35*,2695 (1964).

HIRABAYASHI, K., and J. IWAMURA, *J. Electrochem. Soc. 120*(11), 1595 (1973).

HOLLAND, O. W., C. W. WHITE, and D. FATHY, *Appl. Phys. Lett. 51*(7), 520 (1987).

SU, S. C., *Solid State Technol. 24*, 72 (1981).

## PROBLEMS

**9.1.** (a) An $SiO_2$ layer, $4 \times 10^{-6}$ cm thick is formed by heating a sample of Si in oxygen for 100 s. What is the value of the diffusion coefficient in

$cm^2/s$, and at 100 s what is the value of the flux of oxygen in $atoms/cm^2 \cdot s$?

(b) Does the flux $F$ increase, decrease, or remain unchanged as
  (1) the temperature increases ($t$ = const.)?
  (2) the time increases (temp. = const.)?

(c) If the value of $D$ increases in a factor of 10 when the temperature is increased from 800°C to 900°C, what is the value of $E_A$ in eV?

(d) After you form an oxide layer, the $SiO_2$ molecules in the center of an oxide layer are marked by radio-tracer atoms. If you double the thickness of the oxide layer, will the marked layer with radio-tracer atoms be located in the middle of the oxide layer, in the top half of the oxide, or in the bottom half of the oxide layer?

**9.2.** (a) Using the data in the text, how long would it take to grow a layer of $SiO_2$ that is 1000 Å thick at 900°C in steam using the *linear* growth law, and how much Si is consumed?

(b) The diffusion coefficient of As is approximately $10^{-15}$ $cm^2/s$ at 900°C. Would the As in Si diffuse away from or accumulate at the $SiO_2$–Si interface? Compare the value of the diffusion length of As to the thickness of consumed Si.

(c) If the 1000-Å oxide layer were grown in dry $O_2$ at 1100°C, how much time would be required (use parabolic rate constant)?

(d) The diffusion coefficient of As is about $4 \times 10^{-14}$ $cm^2/s$ at 1100°C. Compare the diffusion length of As for the time in part (c) with the amount of Si consumed.

**9.3.** You implant Si with As ions at 100 keV and after annealing at 700°C you find that residual disorder exists to a depth of $2R_p$. To consume the disordered Si, you grow an oxide layer in dry $O_2$ at 1100°C. Will the implanted As be consumed, accumulate at the $SiO_2$–Si interface, or diffuse ahead of the interface?

**9.4.** A (100) silicon wafer has 500 nm of oxide on its surface. How long will it take to grow an additonal 1 $\mu$m of oxide in wet oxygen at 1100°C?

**9.5.** A thin film of Pt (500 nm) is deposited on a Si plate (100 $\mu$m thick). Calculate the deflection of the plate if the stress in the Pt film is $2 \times 10^9$ $N/m^2$ in tension.

**9.6.** A Si wafer (111) in orientation is oxidized at 1100°C in wet oxygen, followed by an oxidation in dry oxygen also at 1100°C. The wafers are then cooled in an inert ambient. What is the expected fixed oxide charge at the interface?

**9.7.** The diffusivity of $H_2O$ in $SiO_2$ is smaller than that of oxygen (see Table 7.7). Why, then, is the oxidation rate of Si faster in a wet atmosphere?

# CHAPTER 10

# Metallization and Phase Diagrams

## 10.1 Introduction

The proper choice of metal for contact to Si (or to compound semiconductors such as GaAs) is a major area of research and development in fabrication of integrated circuits at present. As device dimensions approach micrometer or submicrometer sizes in very large scale integration (VLSI), only limited (a few tens of nanometers) penetration of the metal into the Si can be tolerated. In this chapter we discuss the basic concepts underlying the choice of metals for contacts and interconnects—the topic called "metallization." We begin with a discussion of phase diagrams.

A phase diagram is a graphical representation of phases that are present in equilibrium as a function of variables such as composition, temperature, and pressure. The simplest kind of equilibrium is one that shows the phases of a one-component system as a function of temperature and pressure (from now on we will simply call them phase diagrams, remembering that all phase diagrams are obtained under equilibrium conditions). One of the best-known one-component phase diagrams is the pressure ($P$)–temperature ($T$) diagram of $H_2O$.

For electronic applications, the binary and ternary phase diagrams are important to us. A binary phase diagram has three major variables: pressure, temperature, and composition. A generalized binary phase diagram is, therefore, three-dimensional. Since most of the applications for phase diagrams are under atmospheric pressure, a two-dimensional binary phase diagram is often used, with the temperature and the composition as variables and the pressure fixed at 1 atm pressure.

## 10.2 Lever Rule and Phase Rule

Let us start with a binary system where the two components are similar in crystal structure and atomic diameter (the atomic diameters are taken to be given by the closest distances of approach of the atoms in the crystals of the elements); both elements tend to dissolve in one another in all proportions in the liquid as well as in the solid state. The phase diagram for such a system is relatively simple, such as the Ge–Si phase diagram shown in Fig. 10.1. The four conditions for extended substitutional solid solutions are commonly known as the Hume–Rothery rules and require that the two components be similar in electronegativity, valence as well as crystal structure, and atomic diameter.

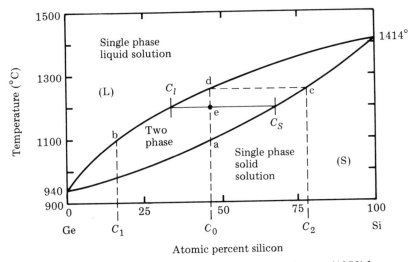

**FIGURE 10.1**  Ge–Si phase diagram. [After Hansen (1958).]

The details of phase changes can be demonstrated through the use of Fig. 10.1. Let the composition of a piece of the Ge–Si alloy be $C_0$ (about 47 at % Si); below 1100°C it is a homogeneous solid solution. As the temperature is raised just above 1100°C, some of the alloy melts. The first liquid has the composition $C_1$ (16 at % Si); it is more rich in Ge than the original alloy. As the temperature increases further, more liquid forms. At each temperature the composition of the liquid moves along the line $bd$, the "liquidus" line. The composition of the remaining solid follows the line $ac$, the "solidus" line. As the temperature is raised to point $d$ (about 1260°C), the last bit of solid has the composition $C_2$ (78 at % Si). Above 1260°C, the entire liquid solution has the initial composition $C_0$ (47 at % Si).

The region of the phase diagram between the solidus and liquidus lines is a two-phase region where a liquid phase and a solid phase coexist at equilibrium. The compositions $C_l$ and $C_s$ of the liquid and solid of the phases are read directly from the abscissa; they are the values along the liquidus and solidus lines where the constant-temperature line intersects. The relative amount of these two phases can be determined by use of the "lever rule." Applying the lever rule to an alloy with an initial composition $C_0$, the atom fraction of the solid phase at 1200°C is given by

$$\text{atom fraction of solid} = \frac{C_l - C_0}{C_s - C_l} = 0.37$$

$$\text{atom fraction of liquid} = 1.00 - 0.37 = 0.63.$$

Therefore, at 1200°C a solid alloy with an initial composition of $Si_{0.47}Ge_{0.53}$ ($C_0$) will separate into a solid phase and a liquid phase at equilibrium. The solid phase has a composition of $Si_{0.67}Ge_{0.33}$ ($C_s$) with an atom fraction of 37% of the total initial atoms. The liquid phase has a composition of $Si_{0.34}Ge_{0.66}$ with an atom fraction of 63% of the total initial atoms. The line joining $C_l$ and $C_s$, called a tie line, is used to indicate the compositions of the equilibrium phases in a two-phase region.

The relationship between the number of components, $C$, of a system and the numbers of degrees of freedom, $F$, is given by the Gibbs phase rule:

$$P + F = C + 2 \tag{10.1}$$

where $P$ is the number of phases, $F$ the number of degrees of freedom, and $C$ the number of components of the system. Use of the phase rule is demonstrated in Fig. 10.1.

In the single-phase solid solution region of the phase diagram (any point below the solidus line),

$$F = C - P + 2 = 2 - 1 + 2 = 3.$$

There are three degrees of freedom. With pressure fixed at 1 atm, the other two degrees of freedom are temperature and composition. Within limits of this region of phase diagram, both temperature and composition can be assigned arbitrarily without the occurrence of a second phase.

In the two-phase region (the area between the liquidus and the solidus),

$$F = C - P + 2 = 2 - 2 + 2 = 2.$$

When both the pressure and the temperature are held fixed, compositions of the liquid phase and the solid phase are also fixed; there is no degree of freedom in composition. The relative amounts of the two phases can vary by changes in the total composition. When both the pressure and the compositions of the two phases are held fixed, there is no degree of freedom in temperature.

## 10.3  Eutectic Systems

There are three eutectic systems, Au–Si, Al–Si, and Pb–Sn, that are used extensively in integrated-circuit fabrication. The semiconductor Si is very different from Au or Al, and Au (or Al) and Si have little or no mutual solubility in the solid state (remember the Hume–Rothery rules). This type of binary systems often leads to simple eutectic (from the Greek word meaning "easily melted") phase diagrams. Figure 10.2 shows the Au–Si phase diagram.

The melting points of Au and Si are above 1000°C [$T_m(\text{Si}) = 1414°C$ and $T_m(\text{Au}) = 1063°C$], yet the melting point of a mixture of Au and Si (82 at % Au, 18 at % Si) is 363°C (the eutectic point). This deep depression in the melting point has been used for over 30 years—almost from the very start of the development of silicon transistors—as a means of attaching the Si sample to the metal support (Fig. 10.3).

The metal support is usually molybdenum (Mo) because the thermal expansion coefficient change in length or volume with temperature of Mo is close to that of Si. There is a thin Au metal layer on the Mo. The Si sample—usually referred to as a chip—is placed on the Au-covered Mo and the system is heated just above

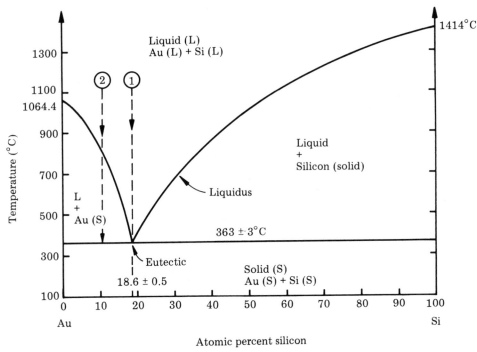

FIGURE 10.2    Au–Si phase diagram. [After Hansen (1958).]

the eutectic temperature of 363°C. The area where the Au and Si are in contact melts and forms a solid bond between Si and Mo when the sample cools.

One can get some appreciation of the phase diagram by following the cooling of liquid mixtures. If one starts with a liquid of the eutectic composition (line 1 in Fig. 10.2) and cools the liquid, it will solidify at 363°C into a solid with crystallites of Au and of Si forming the characteristic eutectic structure of lamellae. If one cools a more Au-rich solution (line 2) (10 at % Si), the solid component Au will crystallize out when the temperature reaches the liquidus curve. As the temperature continues to fall, more and more Au crystallizes and the liquid composition follows the liquidus and becomes enriched in Si. As the temperature drops to 363°C, the liquid has the eutectic composition; then Si can crystallize together with Au into a eutectic structure.

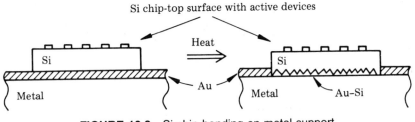

FIGURE 10.3    Si chip bonding on metal support.

We can apply the Gibbs phase rule to eutectic phase diagrams such as the Au–Si phase diagram. At the eutectic point,

$$F = C - P + 2 = 2 - 3 + 2 = 1.$$

Since this one degree of freedom is used up by fixing the pressure, we learn that the eutectic composition and temperature are fixed by the system at a given pressure. At a point inside the two-phase region of liquid and solid Au,

$$F = C - P + 2 = 2 - 2 + 2 = 2.$$

With one degree of freedom being used up by fixing the pressure, the remaining degree of freedom can be either temperature or composition.

The Pb–Sn eutectic system that has a eutectic composition of 26 at % Pb at 183°C is used by IBM as solder-ball connections between the Si chip and the substrate. As shown in Fig. 1.17, the Si chip faces down (flipped over and hence called a "flip chip") and the solder makes a connection between the transistor elements and the terminals on the ceramic layer.

The Al–Si system is used extensively as metallization for Si technology and is discussed next.

## 10.3.1 Aluminum Metallization Scheme

To make electrical connections to the Si devices, a layer (usually 1 to 2 μm thick) of metal is deposited on a "patterned" Si wafer. The metal makes contact with the Si only in the openings in the oxide layer. Regions of the metal are also removed to define a pattern in the metal layer following procedures (photoresist, mask, etching) similar to those outlined in Section 1.5. The metal layer is then heated to ensure an intimate contact between the Si and metal layer. Most semiconductor manufacturers use aluminum (Al) as the contact metal because it has a high electrical conductivity and it forms a protective oxide layer on the top surface. The concept is sketched in Fig. 10.4 using Al metallization as an example.

While aluminum metallization is the most important metallization system of Si IC technology, there is one major problem associated with the use of aluminum (Al). Silicon is soluble in Al, and during heat treatment, Si diffuses into Al, forming pits in the Si. The pitting phenomenon is shown schematically in Fig. 10.5. The

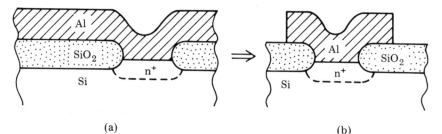

(a)                                              (b)

**FIGURE 10.4** Patterning of Al contact to Si. (a) Deposit Al film. (b) Pattern and etch Al.

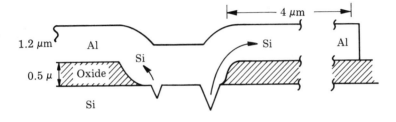

$T_{anneal} \cong 450°C$

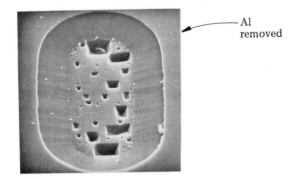

**FIGURE 10.5**    Pit formation of Al contact to Si.

solubility of Si in Al is between 0.5 and 1.0 at percent at temperatures of 450 to 500°C (below the eutectic temperature; see Fig. 10.6) commonly used in Si IC processing.

There have been numerous investigations of the diffusion of Si into Al. One method, shown in Fig. 10.7a, measures the concentration distribution and diffusion length of Si in polycrystalline Al films. At a temperature of 446°C, Si had diffused 140 μm along the film in 40 min. By measuring the concentration of Si as a function of distance for a fixed temperature it was possible to fit the distribution with an error function and then to obtain a value of the diffusion coefficient (Fig. 10.7b). The values of the diffusion coefficient were measured for different diffusion temperatures and were plotted versus reciprocal temperature (in Fig. 10.7c). The slope of the line gives the activation energy ($E_A = 0.79$ eV). The diffusion coefficient of Si in Al can be written as

$$D = D_0 e^{-E_A/kT} = 25 \times 10^{-4} e^{-0.79/kT}.$$

The high diffusivity of Si into Al is responsible for the Al "spikes." Chemical etching to remove the Al layer also dissolves the regions in Si where the Al has penetrated (the spike region). The spikes are then revealed as pits in the Si, Fig. 10.8a. It is often possible to reduce the spiking problem by increasing the processing temperature rapidly—rapid thermal processing (RTP)—such that intimate contact is formed with only a limited amount of Si dissolving into the Al layer. Figure 10.8b shows an improved uniformity of the contact-area morphology for short annealing times.

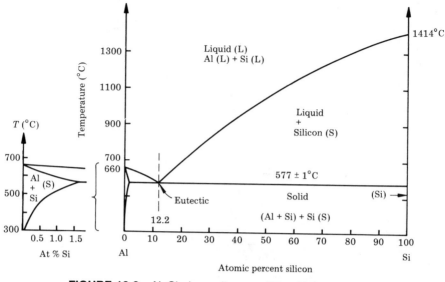

**FIGURE 10.6** Al–Si phase diagram. [After Hansen (1958).]

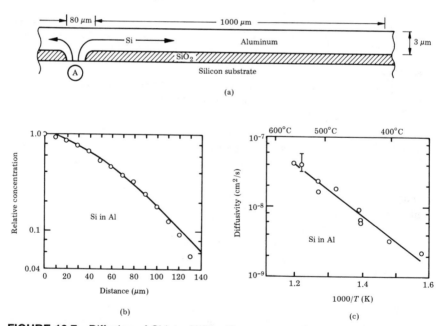

**FIGURE 10.7** Diffusion of Si into Al film. The concentration of Si in Al at point A is determined by the solubility of Si at the diffusion temperature (446°C in this case). From the measurement of the concentration profile (b), diffusion coefficients, (c) are obtained. [After McCaldin and Sankur (1971).]

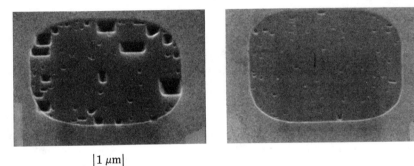

|1 μm|

(a)                                            (b)

**FIGURE 10.8**  Scanning electron micrographs (tilt angle = 45°) taken from 5 × 5 μm² contact holes after annealing with RTP and the removal of the aluminum layer; (a) 425°C for 10 min; (b) after RTP at 350°C for 10 s. (The Al layers were removed.) [After Pai et al. (1985).]

In semiconductor devices made up to the mid-1970s, the formation of pits was not too troublesome because the amount of Si dissolved in the Al could be controlled by designing the Al pattern correctly (referred to as Al design rules in pattern layouts). Recently, as the dimensions of the transistors has grown smaller in large-scale integration (LSI), the Al spikes in the Si have been deep enough to destroy operation of the transistor. To overcome this difficulty, it is common practice to deposit an Al film containing 3 to 5 at % of Si. This procedure also has problems in that the excess Si in the Al can precipitate during heat treatment and form crystallites of Si on the Si surface, as shown in Fig. 10.9. These crystallites appear light colored in the photograph, contain Al atoms in solution and are $p$-type in conductivity. In Fig. 10.9 the Al layer was etched away. These $p$-type Si islands in contact with $n$-type Si substrate may lead to undesirable electrical behavior.

|1 μm|

**FIGURE 10.9**  Silicon crystallites regrown on Si after dissolution of Si into Al and subsequent recrystallization during slow cooling (Al film removed). [After Pai et al. (1985).]

## 10.3.2 Electromigration

Aluminum metallization is used not only for making contacts but also serves as interconnects. As the dimensions of the devices shrink, the interconnects are also reduced in thickness and in width, thus increasing the current densities that are carried through these conducting lines. As the current density increases to about $1 \times 10^6$ A/cm$^2$, the interconnects begin to fail due to a phenomenon called electromigration (d'Heurle and Ho, 1978). Pure Al interconnects would develop cracks along the conductor after carrying a current of about $1 \times 10^6$ A/cm$^2$ for less than 100 hours at about 220°C (Learn, 1974). The increase in resistance and finally the development of cracks are due to the mass transport from one end of the interconnect to the other. The driving force for the mass transport in a metallic conducting line is caused primarily by the frictional force between metallic ions and the flowing charge carriers, called the "electron wind" force. Atoms are transported mainly through the grain boundaries, creating voids on one end and hillocks on the other end. There are in general two ways to reduce electromigration: (1) reduce grain boundaries by increasing grain size of the conducting lines, or (2) reduce grain boundary diffusion by stuffing the grain boundaries with an alloying element (see Section 11.6). A convenient and effective way is to alloy the Al interconnect with about 4 wt % Cu and 1.7 wt % Si. This alloying process increases the mean lifetime of Al interconnects to about 5600 hours from less than 100 hours. The resistance to electromigration may be improved further by incorporating a refractory metal (such as Ti) to laminate the Al–Cu–Si interconnect structure. The interaction between Al and the refractory metal leads to an intermetallic compound formation and offers more resistance to hillock formation along the conducting lines. The role of electromigration in integrated circuit failure is discussed in Section 15.9.

# 10.4  Silicide Metallization

From the discussion in Section 10.3 it is obvious that the interface between the Si substrate and the Al metallization is unstable. For modern electronic devices, much more stable and uniform metallized contacts are required. The interfacial morphology can be much improved by using silicide contacts because the Si substrate reacts uniformly (no pits) with the deposited metal layer to form silicides. Silicides are metal–Si compounds; in Si-processing applications these compounds are formed by thermal annealing. It is essential that these reactions take place in the solid state, so that the reaction front can remain relatively planar and uniform. The silicides make good electrical contacts to Si with known Schottky barrier heights. Most of the silicides are good electrical conductors, with resistivities of 20 to 50 $\mu\Omega$ cm. (see Table 10.1).

## 10.4.1 Pt Silicide

We use the Pt–Si system to illustrate the formation of Pt silicides. Figure 10.10 shows the Pt–Si phase diagram. The Pt–Si phase diagram is considerably more complicated than the Al–Si phase diagram. There are many equilibrium compounds

**TABLE 10.1**  Typical Resistivity of Common Silicides Used in Integrated-Circuit Technology

| Silicide | Typical Resistivity ($\mu\Omega \cdot cm$) | Silicide | Typical Resistivity ($\mu\Omega \cdot cm$) |
|----------|--------------------------------------------|----------|--------------------------------------------|
| $Co_2Si$ | 70 | $NiSi_2$ | 35 |
| $CoSi$ | 150 | $Pd_2Si$ | 25 |
| $CoSi_2$ | 15 | $PtSi$ | 35 |
| $CrSi_2$ | 500 | $TaSi_2$ | 50 |
| $MoSi_2$ | 100 | $TiSi_2$ | 15 |
| $NbSi_2$ | 50 | $VSi_2$ | 15 |
| $Ni_2Si$ | 25 | $WSi_2$ | 100 |
| $NiSi$ | 20 | $ZrSi_2$ | 35 |

*Source:* After Nicolet and Lau (1983).

in this phase diagram. If the composition of the liquid is the same as the solid when the equilibrium compound melts, this compound is a congruent phase. The phases $Pt_2Si$ and $PtSi$ are congruent. The compound $Pt_3Si$ is not a congruent phase because phase separation occurs when $Pt_3Si$ melts. Upon melting at about 870°C,

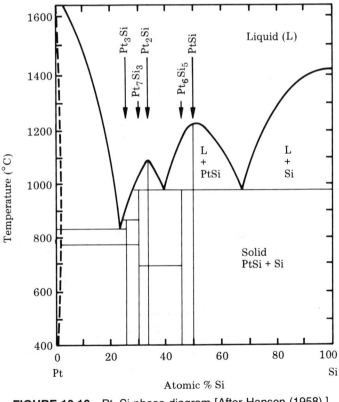

**FIGURE 10.10**  Pt–Si phase diagram [After Hansen (1958).]

solid $Pt_3Si$ transforms into a mixture of liquid with a composition of about $Pt_{78}Si_{22}$ and a solid with a composition of $Pt_7Si_3$. This type of transformation,

$$Pt_3Si(solid) \longrightarrow liquid + Pt_7Si_3(solid) \qquad (10.2)$$

is called a peritectic reaction. If we divide the Pt–Si phase diagram in three parts using Pt, $Pt_2Si$ (a congruent phase), PtSi (a congruent phase), and Si as the terminal phases, then the portions of the Pt–Si phase diagram between Pt and $Pt_2Si$, $Pt_2Si$ and PtSi, and PtSi and Si viewed separately are very similar to the simple Al–Si eutectic phase diagram.

When a thin film of Pt (100 nm, for example) is deposited on Si and the sample is heated to 300°C (below any melting point), a compound is formed as shown in Fig. 10.11. There are several interesting issues to be addressed here:

1. There are at least five equilibrium phases in the Pt–Si system, yet the $Pt_2Si$ phase is always the first phase to form for thin-film samples. The second and final phase to form is the PtSi phase, after the Pt layer has been completely reacted to form $Pt_2Si$ (see Fig. 10.11). Only one phase forms at a time.
2. The temperatures at which these reactions take place are low compared to the lowest eutectic temperature in the phase diagram.
3. The interface between the PtSi and Si is much more uniform than that between Al and Si after heat treatment.
4. The PtSi–Si interface is stable against further annealing—no further solid-state reactions take place. PtSi is the end phase in equilibrium with the Si substrate. This is expected from the average composition of the initial sample (about 100 nm of Pt on about 100 μm of Si), which is located very close to the Si side of the phase diagram.

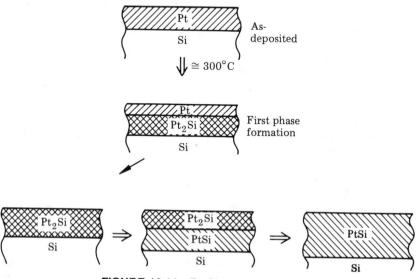

**FIGURE 10.11** Pt–Si thin-film reaction.

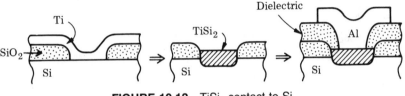

**FIGURE 10.12**  TiSi$_2$ contact to Si.

The first two experimental observations are characteristics of thin-film reactions. Thin-film reactions are very fast and start at low temperatures. Usually, only one phase forms at a time. The first phase to form generally cannot be understood from thermodynamical considerations. There are, however, empirical rules to predict the first phase formation for silicide formation on thin-film samples. The phase sequence is discussed in more detail in Chapter 11.

Figure 10.12 shows silicide contact formation in an oxide window. This silicide contact can be used as a Schottky barrier or as an ohmic contact, depending largely on the silicide and the doping concentration of the Si substrate in direct contact with the silicide. Barrier heights of silicides are summarized in Table 10.2.

## 10.4.2 Salicide Process

The salicide process is an advanced processing technique for VLSI to reduce parasitic effects such as source and drain series resistance and to reduce contact resistance between the contacting silicide and the junction. These beneficial effects are realized, without additional lithographic steps, by making contacts to the $n^+$ and $p^+$ region with a *self-*aligned *silicide* layer (salicide process). This process may be used to its fullest advantage in fabricating complementary MOS devices (CMOS) on VLSI chips. Separate ion implantations and diffusions are used to form wells for both the *p*-channel MOS transistor (PMOS) and the *n*-channel MOS transistor (NMOS). Figure 10.13 shows the basic salicide processing steps, starting from the standard MOS process. Before arriving at this stage, steps of separate ion implantations and drive-in diffusion to form the wells, field, and gate oxide for-

**TABLE 10.2**  Schottky Barrier Heights of Silicides on *n*-Si

| Silicide | Barrier Height (eV) | Silicide | Barrier Height (eV) |
|----------|---------------------|----------|---------------------|
| CoSi, CoSi$_2$ | 0.65 | PtSi | 0.88 |
| CrSi$_2$ | 0.57 | RhSi | 0.69 |
| ErSi$_2$, GdSi$_2$ | 0.38 | TaSi$_2$ | 0.59 |
| MoSi$_2$ | 0.55 | TiSi$_2$ | 0.61 |
| NiSi, NiSi$_2$ | 0.66 | WSi$_2$ | 0.65 |
| Pd$_2$Si | 0.75 | ZrSi$_2$ | 0.55 |
| Pt$_2$Si | 0.79 | | |

*Source:* After Nicolet and Lau (1983).

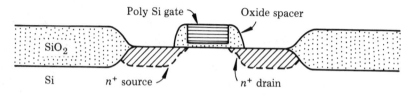

(a) *n*-channel MOS with poly Si gate.

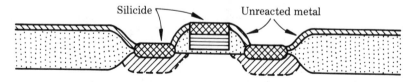

(b) Deposit metal, anneal to form silicide.

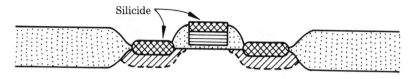

(c) Selectively remove unreacted metal.

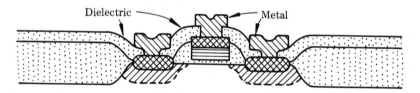

(d) Deposit and pattern dielectric layer and metal.

**FIGURE 10.13** Basic salicide processing steps.

mation, poly Si gate formation, $n^+$ and $p^+$ ion implantation for source and drain, and the formation of oxide spacer for the poly Si gate have already been done. The following steps, shown in Fig. 10.13, are used to complete the salicide process:

**(a)** *n*-Channel MOS with the poly Si gate has been formed.
**(b)** Ti deposition over the wafer, followed by annealing to form $TiSi_2$ in the source–drain region and on the poly Si gate. Since $TiSi_2$ is formed in the oxide windows defining the source–drain region; the silicide is said to be self-aligned.
**(c)** Etch away unreacted Ti.
**(d)** Apply dielectrics (such as P-glass, PSG) and final metallization by Al.

In the final stages of processing, the integrated-circuit structures are held at temperatures of 450 to 500°C. This processing temperature may cause interactions between the Al overlayer and the silicide contacting layer. It is for this reason that a diffusion barrier is sometimes placed in between to prevent failure. Aluminide formation and diffusion barriers are discussed in Chapter 11.

## 10.5 Ternary-Phase Diagrams

The interactions between Al and a silicide are dictated by the phase equilibria between three components (i.e., Al, Si, and the silicide-forming metal). Information on phase equilibria is contained in a ternary-phase diagram.

To represent the phase behavior of three component systems on a two-dimensional diagram, it is necessary to consider both the pressure and the temperature as fixed. The amounts of the three components can be shown on a triangular plot, as indicated in Fig. 10.14. The corners of the triangle labeled A, B, C correspond to the pure components A, B, and C. The side opposite A implies the absence of A. Thus the horizontal lines across the triangle show increasing percentages of A, from zero at the base to 100% at the apex. In a similar manner, the percentages of B and C are given by the distances from the other two sides to the remaining two apexes. At point p, the composition is 20% A, 40% B, and 40% C.

Figure 10.15a is a schematic ternary-phase diagram showing a tie line between phase B and phase AC (a compound formed by A and C). Any ternary alloy with a composition lying on this tie line will have two phases (B and AC) in equilibrium.

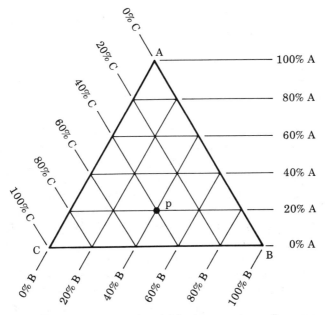

**FIGURE 10.14**   Triangular plot for ternary-phase diagrams.

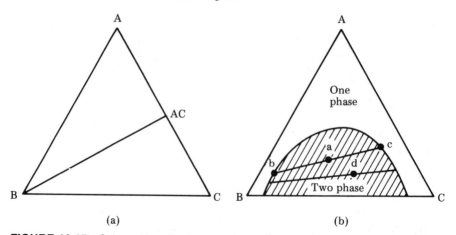

**FIGURE 10.15** Schematics of a ternary-phase diagram: (a) single tie line; (b) a two-phase region.

Tie lines in a ternary-phase diagram are not necessarily horizontal lines as they are in a binary phase diagram. Binary phase diagrams are plots of temperature versus composition; horizontal tie lines indicate the composition of equilibrium phases at any temperature of interest (see Fig. 10.1). On the other hand, the temperature of a ternary-phase diagram is fixed; all three apexes and the sides of a triangular ternary-phase diagram indicate compositions.

Figure 10.15b shows a schematic ternary-phase diagram with a two phase region enclosed in the shaded region. A composition at point $a$ in the two-phase region has two phases at equilibrium. The tie line connecting $b$ and $c$ through $a$ indicates that the phases in equilibrium have composition $b$ and $c$. Applying the phase rule to the two-phase region gives

$$F = C - P + 2 = 3 - 2 + 2 = 3.$$

Since $P$ and $T$ are fixed, one degree of freedom in composition is left. Thus the composition of both phases cannot be arbitrarily fixed. If one is fixed, the tie line from that composition fixes the composition of the second phase. We have shown one tie line for composition $a$. Other tie lines, such as the one through $d$, also exist.

A ternary-phase diagram can be calculated if free-energy data are available for all phases. From the Gibbs phase rule, a maximum of three phases can be in equilibrium at fixed temperature and pressure. Regions of three-phase equilibrium form triangles in the isotherm map of the phase diagram. These regions are found by determining the stable two-phase tie lines. The tie lines are established by thermodynamical calculations or in some cases by experimental investigations. It should also be pointed out that while the ternary-phase diagrams are isotherm maps, they do not necessarily vary significantly over a large temperature range.

The lever rule can also be applied to ternary-phase diagrams to obtain the mole or atom fractions of the phases in equilibrium. Figure 10.16 shows three phases, R, S, and L, forming a three-phase equilibrium triangle. The compositions are

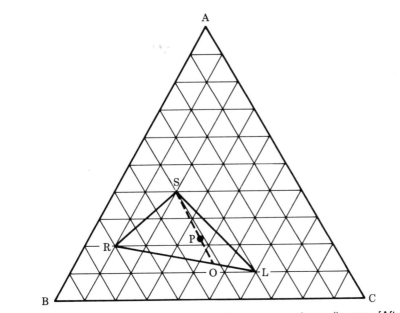

**FIGURE 10.16**  Application of the lever rule for a ternary-phase diagram. [After Rhines (1956).]

$$R = 0.2A + 0.7B + 0.1C$$
$$S = 0.4A + 0.4B + 0.2C$$
$$L = 0.1A + 0.3B + 0.6C.$$

Let us assume that a mixture is made up with two parts of phase R, three parts of phase S, and five parts of phase L. The total composition of this three-phase mixture is

$$0.2 \times 0.2 + 0.3 \times 0.4 + 0.5 \times 0.1 = 0.21A$$
$$0.2 \times 0.7 + 0.3 \times 0.4 + 0.5 \times 0.3 = 0.41B$$
$$0.2 \times 0.1 + 0.3 \times 0.2 + 0.5 \times 0.6 = 0.38C.$$

The composition of the mixture is given by the point P inside the equilibrium triangle. To calculate the fraction of R, S, and L using the lever rule, a straight line is drawn from any corner through P to intersect the opposite side. In Fig. 10.16, a straight line (dashed) SPO is drawn. The quantity of the S phase in the mixture P is

$$\% \text{ S} = \frac{PO}{SO} \times 100\%.$$

The quantity of O in the mixture P is

$$\% \text{ O} = \frac{SP}{SO} \times 100.$$

The composition O represents the mixture of phases R and L, so that

$$\% \ R = \frac{OL}{RL} \frac{SP}{SO} \times 100$$

and

$$\% \ L = \frac{RO}{RL} \frac{SP}{SO} \times 100.$$

By measuring the lines SPO and ROL, we obtain for AC $\simeq$ 83 mm,

$$SP \simeq 14 \text{ mm}$$
$$PO \simeq 7 \text{ mm}$$
$$RO \simeq 26 \text{ mm}$$
$$OL \simeq 12 \text{ mm}.$$

Substituting into

$$\% \ S \simeq \frac{7}{21} \times 100 \simeq 30\%$$

$$\% \ R \simeq \frac{12}{38} \times \frac{14}{21} \times 100 \simeq 20\%$$

$$\% \ L \simeq \frac{26}{38} \times \frac{14}{21} \times 100 \simeq 50\%.$$

these results obtained by the lever rule are the same as the originally assumed quantities of the phases.

## 10.6 Metal–SiO$_2$ Interactions

The calculation of tie lines for ternary phase diagrams is best exemplified by examining the interaction between a metal and SiO$_2$, since there are three elements involved. In Si technology, refractory metal films are often brought in contact with various dielectrics, such as SiO$_2$. The thermal stability of the film–SiO$_2$ structure is an important issue in microelectronics. For a ternary system involving a metal, Si and O, the metal is stable in contact with SiO$_2$ if a tie line exists between the metal and SiO$_2$ in the ternary phase diagram.

Figure 10.17 shows an example of the determination of tie lines of the Ti–Si–O ternary phase diagram at 700°C (Beyers, 1984). Let us first examine the possibility of a tie line between TiSi$_2$ and SiO$_2$. The free energy of reaction between two possible crossing tie lines must be considered. Table 10.3 shows the free-energy change for all possible reactions between TiO$_x$–Si couples and the TiSi$_2$–SiO$_2$ couple (i.e., reactions between the crossing solid tie line and the dashed tie lines). Since negative free-energy changes were found for all cases, the reaction will proceed to the right, indicating that SiO$_2$ is stable in contact with TiSi$_2$, hence a

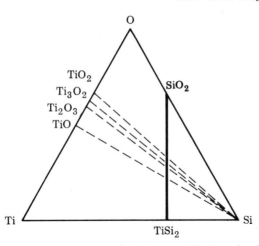

**FIGURE 10.17** Stable (solid) and unstable (dashed) tie lines for the Ti–Si–O system determined from the chemical reactions listed in Table 10.3. [After Beyers (1984).]

stable tie line connects $TiSi_2$ and $SiO_2$ (the solid tie line). Within the triangle with corners of $SiO_2$, $TiSi_2$, and Si, all three phases are in equilibrium. No stable tie line can cross the solid tie line between $SiO_2$ and $TiSi_2$, since at the crossing point four phases are at equilibrium, in violation of the Gibbs phase rule. The dashed lines shown in Fig. 10.17 are therefore not stable tie lines.

Figure 10.17 shows a calculated tie line (solid line). Ternary phase diagrams can be constructed from empirical data. For example, the Ti–O–Si phase diagram was constructed by examining the reaction of a thin Ti film deposited on a thick $SiO_2$ substrate. After completion of the reaction of Ti with $SiO_2$, it was found that $SiO_2$ is in equilibrium with $Ti_5Si_3$ and TiO. These three phases are connected by tie lines as shown in Fig. 10.18a. The remaining tie lines can be inferred using the Gibbs phase rule. This ternary-phase diagram shows no tie line between Ti and $SiO_2$, indicating that Ti in contact with $SiO_2$ is unstable and Ti will decompose $SiO_2$ to form perhaps TiO and $Ti_5Si_3$, kinetics permitting. It is reported in the literature that refractory metals such as V, Ti, Nb (Pretorius et al., 1978), and Zr (Wang and Mayer, 1988) react with $SiO_2$, whereas Mo and W do not. These results suggest that there are no tie lines between V, Nb, Zr, and $SiO_2$ in those relevant

**TABLE 10.3** Free-Energy Calculations Showing That a Stable Tie Line Exists Between $TiSi_2$ and $SiO_2$ at 700°C

| Tie-Line Reactions | $\Delta G$ (kcal/mol) |
|---|---|
| $2TiO + 5Si = 2TiSi_2 + 1SiO_2$ | $-25$ |
| $2Ti_2O_3 + 11Si = 4TiSi_2 + 3SiO_2$ | $-51$ |
| $2Ti_3O_5 + 17Si = 6TiSi_2 + 5SiO_2$ | $-94$ |
| $1TiO_2 + 3Si = 1TiSi_2 + 1SiO_2$ | $-23$ |

*Source:* After Beyers (1984).

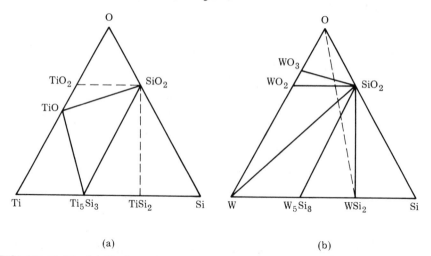

**FIGURE 10.18** (a) Tie lines determined (solid) and inferred (dashed) from the observed Ti–SiO$_2$ reaction product $T = 950°C$; (b) W–O–Si ternary-phase diagram. [After Beyers (1984).]

ternary-phase diagrams, but that a tie line exists between Mo and SiO$_2$ and between W and SiO$_2$ in their respective ternary-phase diagrams. The W–Si–O ternary phase diagram is shown in Fig. 10.18b, where a tie line connects W and SiO$_2$. This phase diagram indicates that W and WSi$_2$ are stable in contact with SiO$_2$. For the Ti–Si–O system, TiS$_2$ and Ti$_5$Si$_3$ are stable in contact with SiO$_2$, but not Ti. For this reason, W is a more suitable metal to fill the via holes in the SiO$_2$ interlevel dielectric layer between two levels of metallization.

## 10.7 Oxidation of Silicides

The utilization of silicides as interconnects has several advantages: (1) metallic electrical conductivity, (2) possibility of reduced electromigration rates, and (3) possibility of producing a protective coating of SiO$_2$ under heat treatment in an oxidizing ambient. Typically, silicides are not used alone as interconnects, but with an underlayer of poly Si. Upon thermal treatment in an oxidizing ambient, certain silicides form a SiO$_2$ overcoat, while the silicide–Si interface below is morphologically preserved (Fig. 10.19). The possibility that a silicide can be oxidized can be predicted by examining the pertinent ternary-phase diagrams. For example, the Ti–Si–O phase diagram shown in Fig. 10.18a indicates a tie line between SiO$_2$ and TiSi$_2$. This means that SiO$_2$ is thermodynamically stable on TiSi$_2$. Hence if Si can be transported through TiSi$_2$ sufficiently rapidly, only SiO$_2$ should form on TiSi$_2$ in an oxidizing ambient without breaking down the silicide into metal oxides.

The growth of SiO$_2$ on silicide also follows the linear-parabolic growth law, similar to the growth of SiO$_2$ on Si (Jiang et al., 1986; d'Heurle et al., 1987),

$$x_0^2 + Ax_0 = B(t + \tau) \tag{9.15}$$

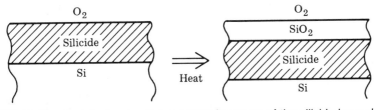

**FIGURE 10.19**   Sketch showing the apparent inertness of the silicide layer during oxidation. [After d'Heurle et al. (1987).]

where $x_0$ is the oxide thickness, $t$ the oxidation time, $B$ the parabolic rate constant, $B/A$ the linear rate constant, and $\tau$ a fitting parameter that takes into account the initial oxide growth (Section 9.2). The parabolic rate constant, $B$, is similar under similar oxidizing conditions (wet or dry; see Tables 9.1 and 9.2), regardless of the substrate (i.e., single-crystalline Si, poly Si, or a silicide on Si). This implies that the diffusion process of the oxident through the oxide is the same for all combinations of Si substrate and silicides. The physical properties of the oxide grown in the parabolic regime should be similar. The linear rate constant, $B/A$, is much higher than that of Si for most silicides with low resistivity (highly metallic). The generally higher, but different linear constants for silicide oxidation is related to the differences in the oxidation process of various silicides at the silicide–$SiO_2$ interface, to the differences in the transport of atomic species across the silicide toward the Si substrate, or to both. In this respect, the inertness of a silicide layer during oxidation shown in Fig. 10.19 is only apparent; in reality, complicated reactions occur at the Si–silicide interface and the silicide–$SiO_2$ interface. These reactions account for the different and higher (than that for Si) linear constants for silicide oxidation. Figure 10.20 shows the oxidation kinetics for various silicides; results for the oxidation of Si are also shown for comparison. One can see that silicides generally oxidize faster. On the other hand, the oxidation rate of $Ir_3Si_5$ is very similar to that of (100) Si, due perhaps to the fact that $Ir_3Si_5$ is a semiconductor with an energy gap almost the same as that of Si (d'Heurle et al., 1987).

## 10.8  Metal–GaAs Interactions

The formation of reliable metallized contacts to GaAs is one of the most challenging problems in GaAs integrated-circuit fabrication. Metallization to GaAs serves as either an ohmic contact or a Schottky contact, as in the case of Si technology. However, because of the special properties of GaAs, difficulties in contact formation are encountered. These special properties are: (1) GaAs decomposes into Ga and $As_2$ (gas) at about 850 K (580°C) (Pugh and Williams, 1986); (2) the Schottky barrier height on $n$-GaAs is large (~0.8 eV) irrespective of the metallization; (3) it is difficult to dope $n$-GaAs to a carrier concentration higher than $5 \times 10^{18}/cm^3$; and (4) annealing up to 850°C for 30 minutes is needed to activate implanted dopants. One can immediately see that dopant activation requires special annealing procedures, such as capping or use of As overpressure to reduce or avoid

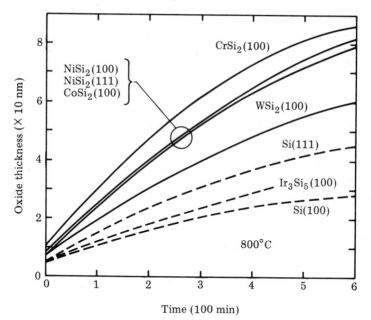

**FIGURE 10.20** Kinetics of oxide growth for several silicides at 800°C. Nominally dry $O_2$. [After d'Heurle et al. (1987).]

GaAs decomposition. It is also obvious that ohmic contacts are difficult to achieve, especially on $n$-GaAs, since the barrier heights are always high, with carrier concentrations limited to $\leq 5 \times 10^{18}/cm^3$.

For contacts to be reliable, they need to be (1) stable against further processing such as heat treatment, and (2) reproducible. Reproducibility constitutes consistently low contact resistance if the contact is ohmic, or constant barrier height if the contact is rectifying (Schottky barrier). The basic knowledge of the metal–GaAs ternary-phase diagrams can be essential to an understanding of stable contacts and the identification of suitable materials for stable contacts to GaAs. Several laboratories have begun the investigation of ternary phase diagrams to address metallization problems in GaAs and other III–V compounds (Pugh and Williams, 1986; Beyers et al., 1987; Juan et al., 1985; Sands et al., 1986). In the following, we outline the findings based on the work of Beyers et al. (1987). One can identify four simple types of M–Ga–As phase diagrams, with the following assumptions and simplifications:

1. Only one $MGa_x$ and $MAs_y$ binary compound exist.
2. Assume that no $MGa_xAs_y$ ternary phases exist.
3. Neglect the homogeneity range of each phase.
4. Neglect the formation of liquid phases.
5. Neglect arsenic sublimation.
6. Assume that equilibrium between the solid phases is reached during processing.

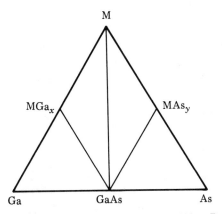

**FIGURE 10.21**    Type I phase diagram of M–Ga–As. [After Beyers et al. (1987).]

*Type I: GaAs dominant.* The tie line between M and GaAs indicates that GaAs is stable in contact with the elemental metal (see Fig. 10.21). GaAs is also stable in contact with $MGa_x$ and with $MAs_y$. Metals such as Au, Ag, and W together with Si and Ge have type I diagrams, as shown in Fig. 10.22. These phase diagrams have more than one $MGa_x$ or $MAs_y$ phase, but the key feature is the existence of a tie line between the pure element and GaAs, as examplified in type I diagrams.

Gold has a substantial solubility for Ga (about 10%) and some for As (about 1%) at 300°C. The metal in equilibrium with GaAs is actually a ternary solid solution (i.e., Au saturated with Ga and As, the black area near the Au apex). Strictly speaking, GaAs is not stable in contact with pure Au, especially in an open system where As (gas) can escape. Gold films saturated with Ga and As will be stable in contact with GaAs in a closed system. One way to provide a closed system is to cap the metallic layer on GaAs with a layer of $Si_3N_4$. According to the Au–Ga–As phase diagram, GaAs is stable in contact with the $\beta'$, $\gamma'$, AuGa, and the $AuGa_2$ phases, in addition to the Au-rich ternary alloy. The tie lines also

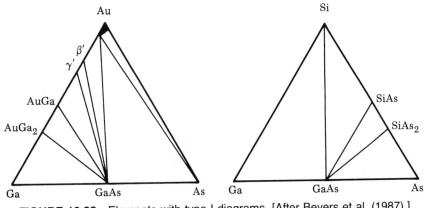

**FIGURE 10.22**    Elements with type I diagrams. [After Beyers et al. (1987).]

indicate that elemental As can be stable in contact with Au saturated with As and Ga.

The Si–Ga–As phase diagram is very similar to that of the Au–Ga–As phase diagram; GaAs is stable in contact with Si, SiAs, and $SiAs_2$ in a closed system. Elements such as Ge, Sn, Ag, and W also have a type I ternary phase diagram.

*Type II: $MGa_x$ dominant.* The existence of the $MGa_x$–$MAs_y$ tie line precludes the existence of the M–GaAs tie line (see Fig. 10.23a). The elemental metal is therefore not stable in contact with GaAs.

*Type III: $MAs_y$ dominant.* The type of diagram is the complement of the type II diagrams (see Fig. 10.23b). When a metal in contact with GaAs is annealed in a closed system, the phases in equilibrium are $MAs_y$, GaAs, and elemental Ga.

Few metals have type II or type III diagrams. The O–Ga–As diagram is essentially a type II diagram (see Fig. 10.24a). It has been reported that tantalum reacts with GaAs to form TaAs and liquid Ga at 650°C; the Ta–Ga–As system thus appears to be a type III diagram.

*Type IV: No phase dominant.* This type of phase diagram has no tie line between binary compounds and the remaining element, as shown in Fig. 10.24b. The phase diagram of Pt–Ga–As belongs to this category, as shown in Fig. 10.24c.

It is quite possible that ternary phases exist for some systems. This type (type V) of phase diagram is more complex than those four simple ones mentioned above because of the presence of ternary phases. Figure 10.24d shows a schematic type V diagram. The existence of $M_zGaAs$ implies that the elemental metal cannot be stable in contact with GaAs because there is no direct tie line between the metal and GaAs. $M_zGaAs$ is stable in contact with GaAs only if no $MGa_x$–$MAs_y$ tie line intervenes. The phase diagrams of Pd–Ga–As (Fig. 10.25), Ni–Ga–As, and Co–Ga–As are examples of type V.

The last category (type VI) of ternary-phase diagrams involves M (group III)–Ga–As (Fig. 10.26a) and M (group V)–Ga–As systems (Fig. 10.26b). Some

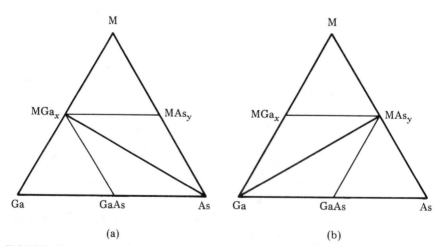

(a)                                    (b)

**FIGURE 10.23** Ternary phase diagrams: (a) type II phase diagram, $MGa_x$ dominant; (b) type III diagram, $MAs_y$ dominant. [After Beyers et al. (1987).]

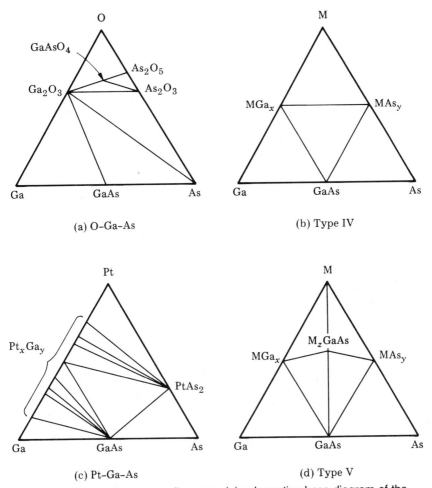

**FIGURE 10.24**  Ternary phase diagrams: (a) schematic phase diagram of the O–Ga–As system, essentially a type II diagram; (b) type IV phase diagram, no phase dominant; (c) type IV phase diagram for the Pt–Ga–As system; (d) schematic phase diagram with a type V ternary phase. [After Beyers et al. (1987).]

of the group III and V elements form complete series of solid solution with GaAs (dark areas at corners in Fig. 10.26). The In–Ga–As phase diagram is an example of type VI.

## 10.9 Stable Metallization for GaAs

According to the Au–Ga–As phase diagram shown in Fig. 10.22a, Au deposited on a GaAs substrate should be stable against annealing since there is a tie line between Au and GaAs. (the Au–Ga alloy is treated as pure Au for simplicity in this case). However, Au–GaAs interactions have been observed and reported in

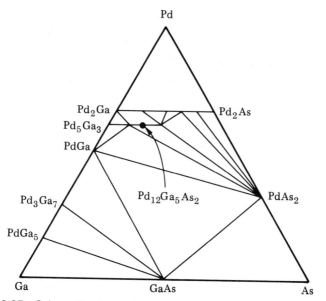

**FIGURE 10.25** Schematic phase diagram of Pd–Ga–As. [After El-Boragy and Schubert (1981).]

the literature repeatedly. Why is this so? The reason is that the phase diagrams shown in these cases are obtained in a closed system; that is, there is no mass exchange between the system and its surroundings. This is true if the system is encapsulated and the external pressure is greater than the vapor pressure of the volatile species (Williams et al., 1986). The environment in which most experi-

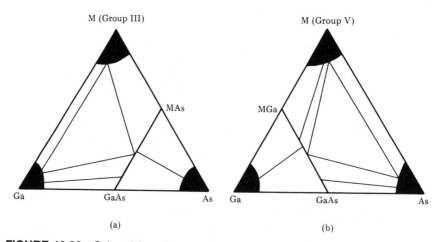

**FIGURE 10.26** Schematics of type VI phase diagrams: (a) MAs dominant; (b) MGa dominant. The black corners represent single-phase regions. [After Beyers et al. (1987).]

ments of Au–GaAs interactions were done is an open system (i.e., in a dynamic vacuum or in a flowing reducing gas ambient). The volatile species ($As_2$ in this example) are either pumped away or escape into the atmosphere. This means that the actual vapor pressure of the group V species is very small, causing a decomposition of GaAs driven by the entropy change of the sublimation of the group V species (Williams et al., 1986). At a particular temperature, the change in Gibbs free energy for a reaction is

$$\Delta G_R = \Delta H_R - T\Delta S_R \tag{10.3}$$

where $\Delta H_R$ is the change in enthalpy and $\Delta S_R$ is the change in entropy. Even if $\Delta H_R$ is positive for a reaction involving the sublimation of species, there will be a temperature beyond which $\Delta G_R$ becomes negative since $\Delta S_R$ is positive. For reactions in which a gas-phase species is evolved as a product at a partial pressure other than 1 atm pressure,

$$\Delta G = \Delta G_R^\circ + nRT \ln P \tag{10.4}$$

where $\Delta G_R^\circ$ is the free-energy change at 1 atm, $P$ the partial pressure of the gaseous product, $R$ the gas constant, and $n$ the number of moles of the product. It follows then that

$$\Delta G_R = \Delta H_R^\circ - T(\Delta S_R^\circ - nR \ln P). \tag{10.5}$$

The temperature at which $\Delta G_R$ become zero is given by

$$T = \frac{\Delta H_R^\circ}{\Delta S_R^\circ - nR \ln P}. \tag{10.6}$$

This critical temperature is a function of partial pressure of the gaseous product. The driving force for the decomposition is essentially due to the term $nR \ln (P)$, the entropy change due to sublimation. The equilibrium vapor pressure for that reaction at temperature $T$ is given by $P = \exp (\Delta S_R^\circ/nR - \Delta H_R^\circ/nRT)$ [from eq. (10.6)].

Figure 10.27 shows the plot of $\Delta G_R$ as a function of temperatures for three reactions involving GaAs with a background pressure of $10^{-8}$ torr of $As_2$ or $As_4$. These three reactions are

(**1**) $GaAs \longrightarrow Ga + \frac{1}{2}As_2$
(**2**) $Au + GaAs \longrightarrow AuGa + \frac{1}{4}As_4$
(**3**) $AuGa + GaAs \longrightarrow AuGa_2 + \frac{1}{4}As_4$.

From Fig. 10.27 it is clear that GaAs will decompose [reaction (1)] at about 870 K (about 600°C) if it is annealed in a dynamic vacuum with a pressure of $10^{-8}$ torr. If Au is in contact with GaAs, reaction (2) will take place at about 520 K (about 250°C) at a pressure of $10^{-8}$ torr:

$$Au + GaAs \longrightarrow AuGa + \frac{1}{4}As_4.$$

However, this reaction *will not* take place if sample is placed in a "closed" system as predicted by the Au–Ga–As phase diagram (Fig. 10.22a).

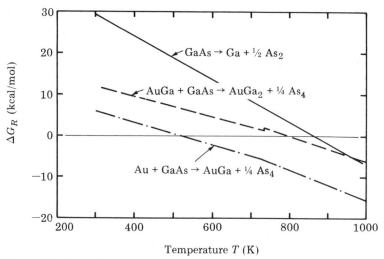

**FIGURE 10.27** Plot of $\Delta G_R$ as a function of temperature for reactions involving GaAs with a background pressure of $10^{-8}$ torr of $As_2$ or $As_4$. [After Pugh and Williams (1986).]

It can be seen from Fig. 10.27 that AuGa in contact with GaAs is thermodynamically more stable than pure Au in contact with GaAu. The critical temperature for reaction (3) to occur is about 800 K (about 530°C), almost 300°C higher than that for reaction (2). The most stable compound in contact with GaAs is $AuGa_2$, since no other Au-containing compound could be formed between $AuGa_2$ and GaAs, as shown in Fig. 10.22a. This means that $AuGa_2$ is stable in contact with GaAs even *in vacuo* until the system is heated above the noncongruent sublimation temperature, at which point the GaAs will decompose thermally.

Based on this reasoning, in a closed system any metallization having a tie line to GaAs in the ternary-phase diagram should be stable until melting or decomposition or further changes in the phase diagram at higher temperatures. For example, Si in contact with GaAs is thermodynamical stable in a closed system, just as $AuGa_2$ is, as shown by the tie lines in Fig. 10.22a.

# REFERENCES

## General References

D'HEURLE, F. M., and P. S. HO, "Electromigration in Thin Films," in *Thin Films—Interdiffusion and Reactions,* ed. J. M. Poate, K. N. Tu, and J. W. Mayer, Wiley, New York, 1978.

HANSEN, M., *Constitution of Binary Alloys,* McGraw-Hill, New York, 1958.

MURAKA, S. P., *Silicides for VLSI Applications,* Academic Press, New York, 1983.

NICOLET, M.-A., and S. S. LAU, "Formation and Characterization of Transition-Metal Silicides," in *VLSI Electronics: Microstructure Science,* Vol. 6, ed. Norman G. Einspruch and Graydon B. Larrabee, Academic Press, New York, 1983.

OTTAVIANI, G., and J. W. MAYER, "Mechanisms and Interfacial Layers in Silicide Formation," in *Reliability and Degradation—Semiconductor Devices and Circuits,* ed. M. J. Howes and D. V. Morgan, Wiley, New York, 1981.

RHINES, F. N., *Phase Diagrams in Metallurgy,* McGraw-Hill, New York, 1956.

TU, K. N., J. M. POATE, and J. W. MAYER, eds., *Thin Films—Interdiffusion and Reactions,* Wiley, New York, 1981.

## Specific References

BEYERS, R., *J. Appl. Phys. 56* (1), 147 (1984).

BEYERS, R., K. B. KIM, and R. SINCLAIR, *J. Appl. Phys. 61* (6), 2195 (1987).

D'HEURLE, F. M., A. CROS, R. D. FRAMPTON, and E. A. IRENE, *Philos. Mag. B 55* (2), 291 (1987).

DING, J., J. WASHBURN, T. SANDS, and V. G. KERAMIDAS, *Lawrence Berkeley Laboratory Report 22092,* Aug. 1986.

EL-BORAGY, M., and K. SCHUBERT, *Z. Metallkd. 72,* 279 (1981).

JIANG, H., C. S. PETERSSON, and M.-A. NICOLET, *Thin Solid Films 140,* 115 (1986).

JUAN, T. S., J. L. FREEOUF, P. E. BATSON, and E. L. WILKIE, *J. Appl. Phys, 58,* 1519 (1985).

KÖSTER, P. S. HO, and J. E. LEWIS, *J. Appl. Phys. 53* (11), 7436 (1982).

LEARN, A. J., *J. Electron Mater. 5,* 531 (1974).

McCALDIN, J. O., and H. SANKUR, *Appl. Phys. Lett. 19,* 524 (1971).

PAI, C. S., E. CABREROS, S. S. LAU, T. E. SEIDEL, and I. SUNI, *Appl. Phys. Lett. 47* (7), 652 (1985).

PRETORIUS, R., J. M. HARRIS, and M.-A. NICOLET, *Solid State Electron. 21,* 667 (1978).

PUGH, J. H., and R. S. WILLIAMS, *J. Mat. Res. 1* (2), 343 (1986).

SANDS, T., paper presented at the workshop on III-V semiconductors, *Metal Interface Chemistry and Its Effect on Electrical Properties,* Stanford University, Nov. 1986.

SANDS, T., V. G. KERAMIDAS, A. J. YU, K.-M. YU, R. GRONSKY, and J. WASHBURN, in *Thin Films—Interfaces and Phenomena,* ed. R. J. Nemanich, P. S. Ho, and S. S. Lau, MRS Symposia Proceedings, Vol. 54, 1986, p. 367.

TSAI, C. T., and R. S. WILLIAMS, *J. Mat. Res. 1* (6), 820 (1986).

WANG, S. Q., and J. W. MAYER, *J. Appl. Phys. 64* (9), 4711 (1988).

WILLIAMS, R. S., J. R. LINCE, T. C. TSAI, and J. H. PUGH, in *Thin Films—Interfaces and Phenomena,* MRS Symposia Proceedings, Vol. 54, 1986, p. 335.

## PROBLEMS

**10.1.** An alloy of A and B of overall composition $C_0$ is at a temperature at which it is an equilibrium mixture of two phases $\alpha$ and $\beta$, of composition $C_\alpha$ and $C_\beta$. Make a mass balance to derive the lever rule.

**10.2.** Assume a ternary alloy of overall composition $X_A$, $X_B$, and $X_C$, the sum of which is unity (or 100%). At a certain temperature the system consists of the $\alpha$, $\beta$, and liquid phases. The composition of these phases are denoted by $\chi_A^\alpha$, $\chi_B^\alpha$, $\chi_C^\alpha$, $\chi_A^\beta$, $\chi_B^\beta$, $\chi_C^\beta$, $\chi_A^l$, $\chi_B^l$, and $\chi_C^l$, respectively. Drive the equations to calculate the weight or mole fractions of the phases $\alpha$, $\beta$, and liquid.

**10.3.** Equilibrium phase formations involving zero degrees of freedom are called invariant reactions. An example of the invariant reactions is the eutectic reaction

$$\text{liquid} \xrightarrow{\text{cooling}} \alpha + \beta$$

Can the lever rule be applied at the temperature of an invariant reaction? Explain.

**10.4.** If you have a molten mixture of 20 at % Al and 80 at % Si and start to cool the liquid solution (refer to Fig. 10.6):
**(a)** At what temperature will it start to solidify?
**(b)** What will solidify first (Al, Si, A–Si mixture)?
**(c)** When the molten solution is cooled to a point where it is almost solidified, what is the composition of the liquid?

**10.5.** The solubility of Si in Al is about 1 at % at 500°C (refer to Fig. 10.6). Al spiking is due to Si dissolving in Al at localized areas during contact sintering. Design a process(es) that would eliminate or reduce the Al spiking.

**10.6.** The solid solubility of impurities in Si is shown in the Fig. 7.10a. The solubility of Al in Si is about $5 \times 10^{18}/cm^3$ at about 500°C, increases to about $2 \times 10^{19}/cm^3$ at about 1150°C, and then decreases to about $1 \times 10^{18}/cm^3$ at about 1400°C (the melting point of Si). This behavior is characterized as retrograde solubility. Discuss the dopant concentration of Si islands precipitating out from an Al film containing a few atomic percent Si and the possible electrical contact properties in terms of the processing temperature (refer to Fig. 10.6).

**10.7.** The Au–In–P phase diagram is shown schematically in Fig. 10.28.
**(a)** Is a thin layer of Au stable in contact with InP in an open system? In a closed system? Explain.
**(b)** What are the possible stable metallic layers in contact with InP in a closed system? Explain.
**(c)** What are the stable Au-containing compounds in contact with InP in an open system? Explain.
**(d)** Discuss the significance of the dark bands in the Au–In–P phase diagram.

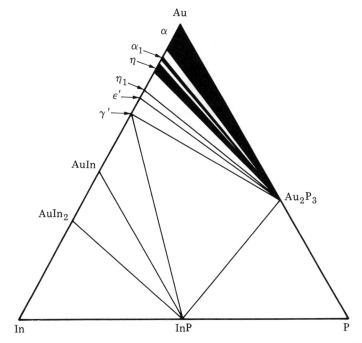

**FIGURE 10.28** Au–In–P phase diagram. [After Tsai and Williams (1986)]

**10.8.** The phase diagrams, for the O–W–Si and O–Ti–Si systems are shown in Fig. 10.18. Based on these diagrams, list the possible reaction products of oxidation of $WSi_2$ and $TiSi_2$. Is elemental W (Ti) a possible oxidation product? Explain in terms of the phase diagrams.

# CHAPTER 11

# Reaction Kinetics: Silicides, Aluminides, and Diffusion Barriers

## 11.1 Thin-Film Reactions

Thin films play a dominant role in modern microelectronics. In integrated-circuit technology almost all vital elements in the active devices, interconnects, and dielectric layers are in thin-film form. An essential criterion of the thin-film structures is that they maintain structural integrity. For large-scale structures, reactions on the 10-nm scale can generally be ignored. This is not the case for thin-film structures, where pronounced reactions or interdiffusion can occur over these distances even at room temperature. The characterization of reaction kinetics in these structures become one of the most important issues in the reliability of integrated circuits. The key to the understanding of thin-film reactions lies in the physical and chemical nature of the interface; at present, we do not have adequate theoretical treatments of this problem to predict accurately thin-film phase formation and reaction kinetics. But we do know that if the thin-film microstructure, the overall impurity content, and the interfacial cleanliness are carefully controlled, thin-film reactions are reproducible. In this chapter we discuss three types of thin-film reactions pertinent to integrated-circuit technology: the formation of silicides and aluminides and the use of diffusion barriers to prevent the reactions between Al and Si (or silicides).

## 11.2 Silicide Formation

Silicide films are used in two ways in integrated devices: (1) as part of the contact that joins an interconnecting line to the substrate, and (2) for interconnections. Silicides are formed usually by means of reactions between a deposited metal layer and the Si substrate. Silicide layers may also be deposited on a semiconductor substrate by means of coevaporation or sputtering. In the latter cases, the deposited layer has the appropriate amount of Si and metal without having to react with the substrate. This type of silicide formation is used on substrates with shallow junctions where penetration is not desirable or on substrates other than Si, such as GaAs in the case of an $WSi_x$ gate for GaAs integrated-circuit technology. Our primary attention here is on the reactions between a thin metal layer and the Si substrate.

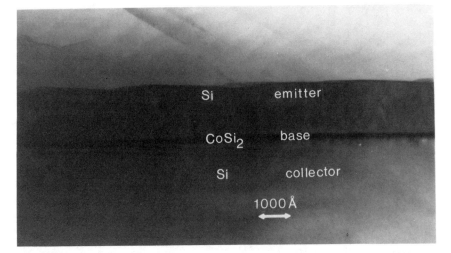

**FIGURE 11.1** Transmission electron micrograph of a CoSi$_2$–Si interface. [After Hensel et al. (1985).]

Under steady annealing conditions (conventional furnace annealing with annealing period of many minutes), transition metals react uniformly with Si single-crystal substrates when no significant amounts of impurities are present. The interface region between the reacted silicide layer and the substrate is flat over many atomic spacings, especially in cases where the silicide layer is epitaxial with respect to the underlying Si substrate. An example of an epitaxial CoSi$_2$ layer on (111) Si is shown in Fig. 11.1. Not all transition metals form uniform layers of silicides on Si substrates. Nonuniform silicides include YSi$_{1.7}$, Rh$_4$Si$_5$, Rh$_3$Si$_4$, PdSi, IrSi$_3$, and HfSi$_2$. These silicides generally form within a narrow temperature range and the formation mechanism is believed to be nucleation controlled. The interfacial non-uniformities are generally much more severe for these silicides than those obtained from non-nucleation-controlled transition-metal silicides.

## 11.2.1 Phase Sequence

The phase diagrams for binary metal–silicide–form systems usually show several equilibrium phases (e.g., Fig. 11.2 shows the Ni–Si system with six equilibrium silicide phases). Only some of these phases are present as detectable growth phases under steady-state annealing conditions. For clean systems only one phase grows at one time in thin-film systems (e.g., Ni$_2$Si grows first in the Ni–Si system). It is generally believed that phase formation in thin films is dictated by kinetics (how fast atoms move) rather than by thermodynamics (how large the driving force is). In a bulk diffusion couple (several centimeters of material A is welded on several centimeters of material B), long-time annealing tends to equilibrate the system and all equilibrium phases would be expected to be present in various amounts. This is definitely *not* the case for thin-film systems (several hundred nanometers of metal deposited on a Si substrate).

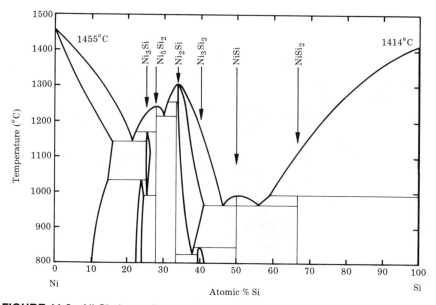

**FIGURE 11.2** Ni-Si phase diagram. [Adapted from the *Bull. Alloy Phase Diagrams 8*(1), 95 (1987).]

The question, then, is: Which of the equilibrium phases will form in thin-film reactions? This question cannot be answered from first principles. There are several phenomenological models that specifically deal with phase sequences in thin films (e.g., Walser and Bené, 1976). The physical origins of these models are not entirely clear at present and will not be discussed.

After the metal layer has been totally consumed in forming the first phase, a second phase, when present in the phase diagram, will form at higher temperatures and/or at longer annealing times. For example, the phase sequence for the Ni–Si system (see Fig. 11.3) is (1) $Ni_2Si$, (2) NiSi, and (3) $NiSi_2$ for samples with a thickness of Si much larger than that of Ni. (The $Ni_3Si_2$ phase is bypassed.) For samples with a thickness of the Ni layer much larger than that of the Si, the phase sequence is (1) $Ni_2Si$, (2) $Ni_5Si_2$, and (3) $Ni_3Si$ (see Fig. 11.3). The experimental observed phase sequence of common silicide phases formed by thermal annealing of a metal film on a Si substrate is summarized in Table 11.1. This table shows that for some metals (V, Cr, Ta, and W) only one silicide, (the disilicide $MSi_2$) is formed by the reaction of a metal film on silicon. For other metals, such as Co and Pt, a series of silicide phases form.

## 11.2.2 Reaction Kinetics

The solid-state reactions for silicide formation under steady-state conditions can be divided into two general categories: (1) systems that exhibit laterally uniform growth with well-defined kinetics and temperature dependence, and (2) systems that show laterally nonuniform growth and exhibit a critical temperature depend-

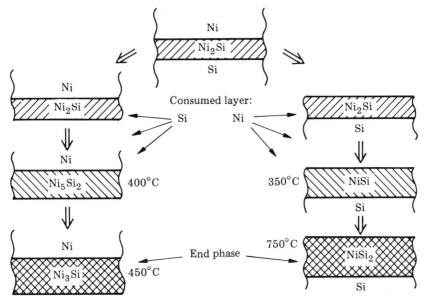

**FIGURE 11.3** Schematic diagrams showing the phase formation in Si–Ni film and silicon crystal–Ni film systems. The indicated temperatures are rough guides for the formation of phases for films a few thousand angstroms thick and an annealing time of 1 hour [After Ottaviani (1979).]

**TABLE 11.1** Experimentally Observed Sequence of Silicide Phases Formed by Thermal Annealing of Metal Films on a Silicon Substrate[a]

| | Ti | V | Cr | Mn | Fe | Co | Ni |
|---|---|---|---|---|---|---|---|
| | $Ti_5Si_3(?)$ | $V\underline{Si}_2$ | $Cr\underline{Si}_2$ | $MnSi$ | $Fe\underline{Si}$ | $Co_2Si$ | $Ni_2Si$ |
| | $TiSi(?)$ | | | $MnSi_2$ | $FeSi_{1.7}$ | $Co\underline{Si}$ | $Ni\underline{Si}$ |
| | $Ti\underline{Si}_2$ | | | | | [†]$CoSi_2$ | [†]$NiSi_2$ |

| Y | Zr | Nb | Mo | | | Rh | Pd |
|---|---|---|---|---|---|---|---|
| [†]$YSi_{1.7}$ | $ZrSi_2$ | $Nb\underline{Si}_2$ | $Mo\underline{Si}_2$ | | | $Rh\underline{Si}$ | $Pd_2Si$ |
| | | | | | | [†]$Rh_4\underline{Si}_5$ | [†]$Pd\underline{Si}$ |
| | | | | | | [†]$Rh_3Si_4$ | |

| Tb/Er | Hf | Ta | W | | | Ir | Pt |
|---|---|---|---|---|---|---|---|
| [†]$TbSi_{1.7}$ | $Hf\underline{Si}$ | $TaSi_2$ | $W\underline{Si}_2$ | | | $Ir\underline{Si}$ | $Pt_2Si$ |
| [†]$ErSi_{1.7}$ | [†]$HfSi_2$ | | | | | $IrSi_{1.75}$ | $Pt\underline{Si}$ |

*Sources:* After Nicolet and Lau (1983), Muraka (1983), and Tu and Mayer (1978).

[a]The silicides observed upon thermal annealing of a thin-metal film on a Si substrate are listed in sequence of their appearance, the uppermost entry listed forming first. A dagger (†) indicates that the compound grows in a laterally nonuniform fashion. The diffusing species during silicide formation is underlined. The diffusing species for the $Pd_2Si$ formation remains to be somewhat unclear, but it is generally believed that both Pd and Si diffuse. It is not clear whether $Ti_5Si_3$ and $TiSi$ form before $TiSi_2$ in a clean system.

ence. For the first category of silicide formation, the reaction takes place over a relatively wide temperature range in a thermally activated manner, usually far below any eutectic or melting temperatures of the system, and the growth proceeds with lateral uniformity and with relatively sharp interfaces. Let us consider a silicide formation reaction:

$$2M(\text{metal}) + Si \longrightarrow M_2Si. \tag{11.1}$$

For this reaction to take place the change in free energy, $\Delta G$, must be negative. Since $\Delta G$ in a solid-state reaction is usually sufficiently large to make $\Delta S$ (the change in entropy due to the reaction) negligible, one may assume that $\Delta G$ is equal to $\Delta H_R$, the heat of reaction. With a negative $\Delta H_R$, we know that this reaction will occur, but we do not know how soon or how fast it will occur. The kinetics of the reaction depend on atomic mobility in general. There are two issues involved in reaction kinetics: (1) nucleation and (2) growth.

## Nucleation

For a new phase to form ($M_2Si$ in this reaction), it is necessary for the atoms of M and Si to get together near the interface and form a nucleus of the appropriate composition and structure. In doing so, one can imagine that the transport of atoms is necessary; in addition, a new surface encompassing the nucleus is also created. This creation of a new surface may offer some resistance to the nucleation of a new phase since there might be an increase in the total surface energy. However, we know that there is a gain in free energy for this reaction to occur; the gain in free energy is proportional to the volume of the newly formed phase. If we assume an initial shape of the new phase to be spherical, the total change in free energy is proportional to $(4/3)\pi r^3$ ($r$ is the radius of the new phase). The increase in surface energy due to the creation of a spherical new phase is proportional to the surface area. At a given temperature $T$, the change in free energy

$$\Delta G = ar^3 \Delta G_v + b \Delta\sigma r^2 \tag{11.2a}$$

where $\Delta G_v$ is the change in free energy per unit volume and $\Delta\sigma$ is the change in surface energy per unit area; $a$ and $b$ are geometric factors [$a = (4/3)\pi$ and $b = 4\pi$ for spherical nucleus]. Figure 11.4 shows schematically the change of $\Delta G$ as a function of $r$. $\Delta G$ goes through a maximum which corresponds to the critical size $r^*$ of the nucleus; nuclei smaller than $r^*$ will tend to disappear, while nuclei larger than $r^*$ will tend to grow. The critical nucleus size is given by maximizing $\Delta G$ with respect to $r$:

$$r^* = \frac{-2b \Delta\sigma}{3a\Delta G_v}. \tag{11.2b}$$

The change in free energy, $\Delta G^*$, at $r^*$ is

$$\Delta G^* = \frac{4b^3 \Delta\sigma^3}{27a^2 \Delta G_v^2}. \tag{11.3a}$$

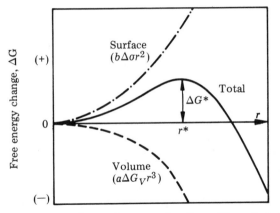

**FIGURE 11.4** Free energy of a nucleus as a function of its radius, showing the volume contribution, the surface energy contribution and their sum. [$a = (4/3)\pi$, $b = 4\pi$ for spherical nucleus.]

For a spherical nucleus, $a = (4/3)\pi$ and $b = 4\pi$,

$$\Delta G^* = \frac{16\pi\ \Delta\sigma^3}{3\ \Delta G_v^2}. \tag{11.3b}$$

The rate of nucleation, the number of nuclei formed per unit volume per unit time, $dn/dt$, is proportional to the density of nuclei, $K\ \exp(-\Delta G^*/kT)$, and to the rate at which nuclei can form [generally proportional to a diffusion term of the type $\exp(-E_A/kT)$]:

$$\frac{dn}{dt} = K\ \exp\left(-\frac{\Delta G^*}{kT}\right)\exp\left(-\frac{E_A}{kT}\right) \tag{11.4}$$

where $K$ is a proportionality constant.

After nucleation, any increase in size of the reacted volume is due to the growth of the nuclei, which requires atomic mobility only. If the growth rate is faster than the nucleation rate, the formation of $M_2Si$ is said to be nucleation controlled (the slower of the two processes). Silicide formation controlled by nucleation often exhibit lateral nonuniformity (type 2 system listed above). If the nucleation rate is faster than the growth rate, the silicide formation is growth controlled (type 1 system listed above) where laterally uniform layers are observed.

## Growth

The growth process proceeds after nucleation and can be limited either by diffusion, where the thickness of the growing silicide, $x$, is proportional to the square root of annealing time, $t^{1/2}$; or by interfacial reactions, where the silicide thickness is proportional to the annealing time, $t$. For diffusion-controlled growth, the transport of atomic species across the growing layer (from one interface to the other) is the limiting step. For interfacial reaction-controlled growth, the reaction rate at the interface to form $M_2Si$ is slower than the transport of atomic species (or the supply of fresh atoms) to the interface for the reaction. For any growth process, the initial stage of the growth is expected to be interfacial reaction controlled. Since the initial concentration gradient of the reacting species across a infinitely thin growing silicide layer is infinitely large, the interfacial reaction must be the limiting step in growth. As the layer increases in thickness, the average gradient and the atomic flux decrease [$F \propto (1/x)$; see eq. (7.3)], and at some critical thickness the diffusion rate becomes slower and may take over as the rate-limiting step in layer growth. This critical thickness for the change over from interfacial reaction limited to diffusion-limited growth depends on the relative magnitudes of the diffusion coefficient and the interfacial reaction constant, and may be negligibly thin.

The situation of silicide growth is analogous to the oxidation process discussed in Chapter 9. Following eq. (9.14), the rate at which the silicide layer grows is

$$\frac{dx}{dt} = \frac{DC_D k_s}{N_D} \frac{1}{D + k_s x(t)}$$

where $x(t)$ is the growing silicide layer thickness, $D$ the average diffusivity of the moving species in the silicide layer, and $C_D$ the metal atomic concentration at the metal–silicide interface assuming that metal is diffusing, $N_D$ the number of the metal atoms incorporated in the silicide, and $k_s$ the silicide reaction rate at the silicide–Si interface. For the case where the interfacial reaction is the limiting step, $D/x(t) > k_s$, $dx/dt = C_D k_s/N_D$, or $x = (C_D k_s/N_D)t$. The silicide thickness therefore increases linearly with time.

For the case where diffusion is the limiting step, $D/x(t) < k_s$, $dx/dt = (DC_D/N_D)(1/x)$, or $x = (2DC_D t/N_D)^{1/2}$. The silicide thickness increases with the square root of time. In the case of diffusion-controlled growth, more reliable experimental data can be obtained for long annealing periods, because the growth regime is far away from the transition region between the two different growth-limiting steps.

The growth of near-noble metal silicides (e.g., $Ni_2Si$, $NiSi$, $Pt_2Si$, $PtSi$, $Pd_2Si$) follow a $t^{1/2}$ relationship. Some transition-metal silicides (e.g., $CrSi_2$) are reported to follow a linear time dependence. For nucleation-controlled silicide formation, the reaction usually occurs within a temperature range of 10 to 30°C. Examples of this type of reaction include the formation of $HfSi_2$ from $HfSi$, $IrSi_3$ from $Ir_4Si_7$, $NiSi_2$ from $NiSi$, $Rh_4Si_5$ from $RhSi$, and certain rare-earth metal silicides (such as $ErSi_{\sim 2}$). The formation characteristics of silicides observed by thermal annealing are listed in Table 11.2.

## 11.2.3 Diffusing Species

To form a compound at the interface between two adjoining elemental solids, there must be relative motion of atoms past each other. The mechanism of atomic motion is by diffusion. One can conceive of a case where the atoms of one element do not alter their relative positions and only atoms of the other species diffuse, or vice versa. The reality usually lies between these two limiting cases. It is of interest to

**TABLE 11.2**  Characteristics of Transition-Metal Silicides Formed by Thermal Annealing of Metal/Si Substrate Samples[a]

| Element | Silicide | Formation Temperature (°C) | Activation Energy (eV) | Growth Rate |
|---------|----------|----------------------------|------------------------|-------------|
| $^{22}$Ti | TiSi | 500 | — | — |
|  | TiSi$_2$ | 550 | — | — |
| $^{23}$V | VSi$_2$ | 600 | 2.9, 1.8 | $t, t^{1/2}$ |
| $^{24}$Cr | CrSi$_2$ | 450 | 1.7 | $t$ |
| $^{25}$Mn | MnSi | 400–500 | — | — |
|  | MnSi$_2$ | 800 | — | — |
| $^{26}$Fe | FeSi | 450–550 | 1.7 | $t^{1/2}$ |
|  | FeSi$_2$ | 550 | — | — |
| $^{27}$Co | Co$_2$Si | 350–500 | 1.5 | $t^{1/2}$ |
|  | CoSi | 375–500 | 1.9 | $t^{1/2}$ |
|  | CoSi$_2$ | 550 | — | — |
| $^{28}$Ni | Ni$_2$Si | 200–350 | 1.5 | $t^{1/2}$ |
|  | NiSi | 350–750 | 1.4 | $t^{1/2}, t$ |
|  | NiSi$_2$ | ≥750 | — | — |
| $^{40}$Zr | ZrSi$_2$ | 700 | — | — |
| $^{41}$Nb | NbSi$_2$ | 650 | — | — |
| $^{42}$Mo | MoSi$_2$ | 525 | 3.2 | $t$ |
| $^{45}$Rh | RhSi | 350–425 | 1.95 | $t^{1/2}$ |
| $^{46}$Pd | Pd$_2$Si | 100–300 | 1.5 | $t^{1/2}$ |
|  | PdSi | 850 | — | — |
| $^{72}$Hf | HfSi | 550–700 | 2.5 | $t^{1/2}$ |
|  | HfSi$_2$ | 750 | — | — |
| $^{73}$Ta | TaSi$_2$ | 650 | 3.7 | $t$ |
| $^{74}$W | WSi$_2$ | 650 | 3.0 | $t, t^{1/2}$ |
| $^{77}$Ir | IrSi | 400–500 | 1.9 | $t^{1/2}$ |
|  | IrSi$_{1.7}$ | 500–1000 | — | — |
|  | IrSi$_3$ | 1000 | — | — |
| $^{78}$Pt | Pt$_2$Si | 200–500 | 1.5 | $t^{1/2}$ |
|  | PtSi | 300 | 1.6 | $t^{1/2}$ |

*Source:* After Tu and Mayer (1978) and Nicolet and Lau (1983).

[a]The silicides listed are those observed upon thermal annealing of a thin-metal film on a silicon substrate. Cases characterized by a laterally uniform growth with well-defined kinetics are identified by their mode of growth [linear ($t$) or parabolic ($t^{1/2}$) in time] and the activation energy of the process. When growth modes are not given, the kinetics are either not known or the growth is not laterally uniform. Typical formation temperatures are also given.

know when predominantly one or the other type of atoms diffuse. This knowledge sheds light on the microscopic processes that control the reactions. Also, some practically relevant properties of the grown layer can depend very directly on the identity of the diffusing species.

Several types of marker experiments have been developed and used to obtain information on the motion of atoms during a reaction. Atoms of one element (or both) can be marked by substituting a radioactive isotope for the stable atom. These isotopes are chemically indistinguishable from the stable ones, but their net movement can be monitored through their radioactivity, for example, in the form of concentration profiles before and after the reaction. The decay of $^{31}$Si into $^{31}$P by the emission of a beta particle (maximum $e^-$ energy $= 1.48$ MeV) with a half-life of 2.62 hours is a convenient way of marking Si atoms for silicide reactions, as $^{31}$Si can be formed readily in a nuclear reactor from conventional high-purity Si single crystals.

A different marker type is obtained if a small, yet finite-size object quite distinct from the two reacting elements is placed within the growing layer. It is assumed that the marker object (1) is chemically inert with respect to its surrounding, (2) does not interfere significantly with the overall development of the reaction (although it will necessarily do that locally), and (3) is not selectively coupled to the diffusive motion of either element involved in the reaction. The marker serves as an object against which the overall flux of atoms of either kind may be measured (inert marker). The displacement of an inert marker depends on the net cumulative flow of atoms past it. Objects that have been used as inert markers to study silicide formation of thin films are implanted inert gases (e.g., Xe) that form microscopic (about 10 nm) bubbles during implantation or after some thermal annealing, a discontinuous, thin (about 3 nm) tungsten film with a surface coverage of a few percent, an oxide layer of similar thickness, and ZrO$_2$ in thick-film studies (Nicolet and Lau 1983, Muraka 1983, Tu and Mayer 1978).

The results obtained by these various techniques are compiled in Table 11.1. Every technique that was applied to the refractory metals that react uniformly with Si to form MSi$_2$ indicates that Si is the diffusing species. This consistency contrasts with the poor reproducibility of kinetics measurements for these same silicides. The parameters responsible for the variations in the kinetics apparently do not alter the diffusing species. For the near-noble metals, the metal is the moving species, with Pd standing out as an unclarified special case. With that exception only, the table reveals an obvious correlation: When the first phase that forms is rich in one element (e.g., CrSi$_2$), the richer element (Si) is also the diffusing species. It is tempting to ascribe a causal relationship to this observation by inverting the statement. Following this idea, the composition of the first compound to grow would then depend in an essential way on the abundant presence of the diffusing element at the interface where the compound grows. This thought is consistent with the observation made by Ottaviani et al. (1979) that Ni$_2$Si or Pt$_2$Si grows always and only when unreacted metal is present on the opposite side of the silicide layer. Corresponding facts hold for Co$_2$Si and presumably, for all dimetal silicides. According to this view, then, the composition of the first phase formed derives in an essential way from peculiarities of the dynamics of the growth process (the identity of the moving species), not only from considerations of equilibrium (e.g., phase diagram information). However, knowledge of the diffusing species alone does not suffice to explain which compound rich in this element should form. How that choice is made, at an atomistic level, is a question that has resisted clarification so far and may do so for some time, because tools presently available to investigate an interface buried some 100 nm below the surface of a solid without perturbing that interface are falling far short of the needs. To develop such techniques is a major requirement if the understanding of solid-phase reactions is to progress beyond the present stage of hypotheses.

Considerable effort has been devoted to understanding the sequence of phase formation in thin films. From such work it is now established that arguments based only on thermodynamic driving forces, or based only on the rate of growth of compounds, generally fail to explain the outcome of a reaction. It is clear that the identity of the moving species and its presence at the reaction interface in smaller or larger numbers constitutes an essential part of the process. That being so, an understanding of what determines the dominant moving species in any given silicide becomes a major issue which has largely been neglected so far.

## 11.2.4 Thin-Film Diffusion versus Bulk Diffusion

It was mentioned previously that the major observed difference between thin-film samples and bulk samples is that in thin-film couples only one single phase is formed at one time until one of the elements of the binary system is completely consumed, whereas all or nearly all equilibrium phases are present in a bulk diffusion couple. Due to the dimensions of the samples, the diffusion distance in thin films is of the order of tens to hundreds of nanometers, and the diffusion distance in bulk samples is of the order of millimeters.

The case of thin film "versus" bulk phase growth was analyzed by Gösele and Tu (1982). Their model is briefly summarized here to illustrate the main features. The phase formation in a diffusion couple in their model is assumed to be controlled by growth, not nucleation. The driving force for the phase to grow is due to the gain in free energy from the reaction. Therefore,

$$\phi_\mu = -\frac{\partial \bar{\mu}}{\partial x} \simeq -\frac{\Delta H_R}{x} \tag{11.5}$$

where $\Delta H_R$ is the heat of reaction (we assume that $\Delta H_R \simeq \Delta G$ here) and $x$ the layer thickness of the growing phase. From eqs. (7.31) and (7.34), we have the flux

$$F = vC = \frac{k°\lambda^2 C}{kT}\left(-\frac{\Delta H_R}{x}\right)$$
$$= \frac{DC}{kT}\left(-\frac{\Delta H_R}{x}\right) \tag{11.6}$$

where $D$ is the diffusion coefficient.

The diffusional flux $F$ is therefore decreasing as $1/x$. The velocity of the diffusing atoms, $v$, also decreases as $1/x$,

$$v = \frac{D}{kT}\left(-\frac{\Delta H_R}{x}\right). \tag{11.7}$$

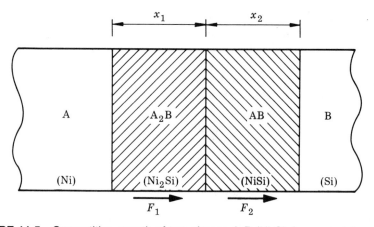

**FIGURE 11.5** Competitive growth of two phases $A_2B$ ($Ni_2Si$, for example) and $AB$ ($NiSi$, for example), assuming A atoms (Ni) are mobile in $A_2B$ and $AB$. [After d'Heurle and Gas (1986).]

Let us consider the situation where two phases are growing in a diffusion couple as sketched in Fig. 11.5. The growth rate of the $A_2B$ phase is

$$\frac{dx_1}{dt} = F_1 - F_2 \tag{11.8}$$

where $F_1$ is the flux of A atoms across $A_2B$ and $F_2$ is the flux across $AB$. The growth rate of the $AB$ phase is

$$\begin{aligned}
\frac{dx_2}{dt} &= F_2 - \frac{dx_1}{dt} \\
&= F_2 - (F_1 - F_2) = 2F_2 - F_1.
\end{aligned} \tag{11.9}$$

From eq. (11.6),

$$F_1 = \frac{\alpha}{x_1}$$
$$F_2 = \frac{\beta}{x_2} \tag{11.10}$$

where $\alpha$ and $\beta$ are constants pertaining to the diffusivity, $kT$, heat of reaction, and so on. Therefore,

$$\frac{dx_1}{dt} = \frac{\alpha}{x_1} - \frac{\beta}{x_2}$$
$$\frac{dx_2}{dt} = \frac{2\beta}{x_2} - \frac{\alpha}{x_1}. \tag{11.11}$$

Inspection of eq. (11.11) shows that neither $x_1$ nor $x_2$ can become zero, since whenever these two quantities become small, $dx_1/dt$ and $dx_2/dt$ tend toward infinity. This means that once the $A_2B$ phase and the AB phase are formed, they will grow together; no single phase will disappear as long as the growth of these two phrases are diffusion limited.

The result is different if the growth of the AB phase is interfacial reaction limited initially (the thickness of the AB phase is below the critical thickness, $x^c$, for diffusion-controlled growth to take over). One may write $F_2 = \gamma$ (a constant). Then

$$\frac{dx_2}{dt} = 2\gamma - \frac{\alpha}{x_1}. \tag{11.12}$$

For $dx_2/dt > 0$,

$$x_1^c \geq \frac{\alpha}{2\gamma}. \tag{11.13}$$

The AB phase *will not* grow until the $A_2B$ phase grows to a critical thickness $x_1^c$. For the AB phase and the $A_2B$ phase to grow together, all one has to do is to wait until the $A_2B$ phase has grown to a sufficient thickness ($> x_1^c$).

In thin-film samples, the diffusion distance is very limited. One of the elements is usually consumed before the critical thickness to form the second phase can be reached. In bulk samples where the distances are many times longer than the thin-film case before one element runs out, all equilibrium phases will then form since the growth of these phases will become diffusion controlled eventually (d'Heurle and Gas, 1986).

The changeover of phase growth from the thin-film case to the bulk case may be probed experimentally using lateral diffusion couples (Zheng et al., 1982; Chen et al., 1984). Figure 11.6 shows schematically a Ni–Si lateral diffusion sample before and after annealing. The diffusion distance in such samples is of the order of 1 to 100 μm. It was found that the phase $Ni_2Si$ grows proportional to (time)$^{1/2}$ until a length of 25 to 30 μm (at 600°C), where multiple-phase formation is observed. The critical length, therefore, appears to be between 25 and 30 μm for

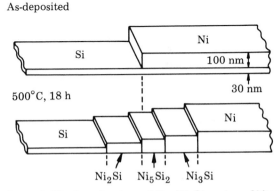

**FIGURE 11.6** Lateral diffusion structure of the Ni–Si system. [After Chen et al. (1984).]

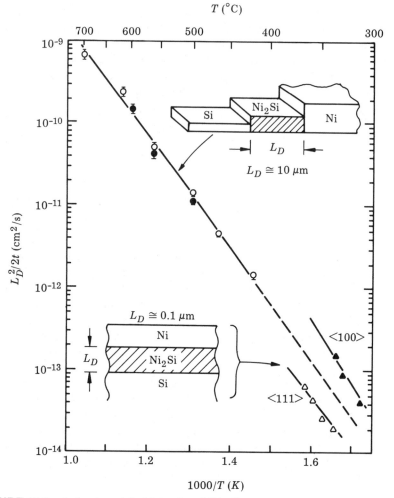

**FIGURE 11.7** Arrhenius plot of lateral and thin-film structures. The straight line and the dashed lines represent an activation energy of 1.4 eV. [After Zheng et al. (1982).]

other phases to form for the Ni–Si system. The lateral growth rate of $Ni_2Si$ is consistent with those for thin-film samples extrapolated to the appropriate temperature regime (Fig. 11.7).

## 11.2.5 Epitaxial Silicides

A number of silicides have been found to grow epitaxially on Si substrates (Tung, 1988). The better known epitaxial silicides are $NiSi_2$, $CoSi_2$, $Pd_2Si$, and $PtSi$ (Nicolet and Lau, 1983; Chen et al., 1986). Table 11.3 lists the commonly observed epitaxial silicides on Si. These silicide layers normally consist of well-oriented polycrystals. The morphology and perfection of the layer depend on the deposition

**TABLE 11.3** Epitaxial Silicides on Silicon where a = 0.54307 nm and Lattice Parameters Given in nm[a]

| Ti | V | Cr | Fe | Co | Ni |
|---|---|---|---|---|---|
| $TiSi_2$(C 54)<br>a = .8253<br>b = .4783<br>c = .8540<br><br>$TiSi_2$(C 49)<br>a = .362<br>b = 1.376<br>c = .360 | $VSi_2$(C 40)<br>a = .4571<br>c = .6372 | $CrSi_2$(C 40)<br>a = .4428<br>c = .6363 | $FeSi_2$(Tet)<br>a = .2684<br>c = .5128<br><br>$FeSi_2$(A 11)<br>a = .9863<br>b = .7791<br>c = .7833 | $CoSi_2$(C 1)<br>a = .5365 | $NiSi_2$(C 1)<br>a = .5406 |
| **Zr** | **Nb** | **Mo** | **Ru** | **Rh** | **Pd** |
| $ZrSi_2$(C 49)<br>a = .3721<br>b = 1.468<br>c = .3683 | $NbSi_2$(C 40)<br>a = .47971<br>b = .6592 | $MoSi_2$(C 40)<br>a = .4613<br>c = .6424<br><br>$MoSi_2$(C 11b)<br>a = .3203<br>c = .7855 | | | $Pd_2Si$(C 22)<br>a = .649<br>c = .343 |
| **Hf** | **Ta** | **W** | **Os** | **Ir** | **Pt** |
| | $TaSi_2$(C 40)<br>a = .47821<br>c = .65695 | $WSi_2$(C 40)<br>a = .4614<br>c = .6414<br><br>$WSi_2$(C 11b)<br>a = .3211<br>c = .7868 | | | $PtSi$(B 31)<br>a = .593<br>b = .560<br>c = .360 |

*Source:* After Chen et al., 1986

[a] A 11 = orthorhombic, B 31 = orthorhombic
C 11 = cubic, C 11b = tetragonal, C 22 = hexagonal
C 40 = hexagonal, C 49 = orthorhombic
C 54 = orthorhombic, Tet = tetragonal

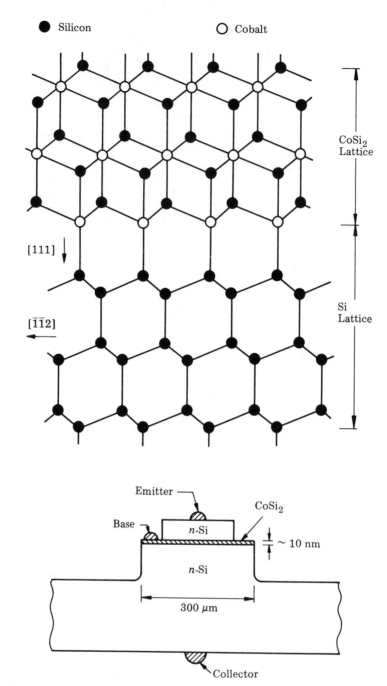

**FIGURE 11.8** (a) A ball-and-stick model of the crystal structure of silicon and cobalt disilicide shows how well the lattices match, allowing the silicide base to be grown epitaxially on top of silicon; [1$\bar{1}$0] projection. Cobalt disilicide has a conductivity comparable to that of gold or other metals used in integrated circuits. [After Poate and Dynes. © 1986 IEEE.] (b) Schematic cross section of a single mesa device fabricated from an epitaxial Si–CoSi$_2$–Si heterostructure. [After Hensel, et al. (1985).]

process, the layer thickness, and the annealing treatment. In addition to those listed in Table 11.3, rare earth silicides (R.E. $Si_{\sim 2}$) and $YSi_{\sim 2}$ have also been observed to form epitaxial layers on Si under special annealing conditions (Chen et al., 1986; Knapp and Picraux, 1986; Gurvitch et al., 1987).

Nearly perfect quality epitaxial layers of $CoSi_2$ and $NiSi_2$ can be grown on Si substrates. In particular, the $CoSi_2$–Si interface can be smooth and sharp without the presence of growth steps over many micrometers in lateral dimensions (Gibson et al., 1987). Figure 11.8a shows a ball-and-stick model of lattice matching between the Si lattice ($a = 0.5431$ nm at 300 K) and the $CoSi_2$ lattice ($a = 0.5365$ nm at 300 K) at the interface. This model shows how well the lattices match, allowing the silicide to grow epitaxially on Si without excessive strain. On top of the epitaxial $CoSi_2$ layer, a single-crystal Si layer can again be grown, thus forming a Si–$CoSi_2$–Si heterostructure. This heterostructure provides the possibility of fabricating a metal–base transistor (Fig. 11.8b). The idea of a metal-base transistor is to replace the base region (about 100 nm) of a bipolar transistor by a metal layer ($CoSi_2$ in this case), in which electrons move with a Fermi velocity $v_F$ of about $10^8$ cm/s (velocity of the electrons at the Fermi energy $E_F$; see Chapter 12). A drastic shortening of the base transit time may then be realized (see Chapter 5) (Hensel et al., 1985). An understanding of the operational mechanisms and the perfection of the transistor performance still require more research effort. At present, however, the integration of epitaxial silicide and semiconductor offers the possibilities for an advanced class of ultrafast devices.

## 11.2.6 Impurity Effects

Impurities can affect the properties of silicides by virtue of their presence. Impurities can also influence the processes by which the silicides are formed. Electrical conductivity and the contact characteristics (see Section 11.6.2) of the silicide are, perhaps, two of the most relevant properties of silicides that can be affected by impurities. Figure 11.9 shows the effect of nitrogen in sputtered $TiSi_2$ films on film resistivity. The impurity ($N_2$) was added to the sputtering gas to introduce the impurity in the sputtered film ($TiSi_2$). The films were then annealed in vacuum. The as-deposited films which are amorphous have high sheet resistance. After annealing the $TiSi_2$ crystallizes and the sheet resistance of the film goes down; however, the final resistance is higher for films with a higher nitrogen content (Lien and Nicolet, 1984). The increased resistivity of films is a typical example of the influence of impurities due to their presence.

For silicides that are formed by reaction between a deposited metal film and the substrate, impurities often redistribute during silicide growth, leading to different reaction kinetics (often slower) with altered contact properties. Take the Pt–Si reaction as an example (Ottaviani and Mayer, 1981). The reaction kinetics to form Pt silicides are found to be essentially unchanged when a thin native oxide layer (often containing pinholes) is present at the Pt–Si interface. Thus the native oxide layer that is formed in normal device processing does not have a major influence on Pt silicide formation.

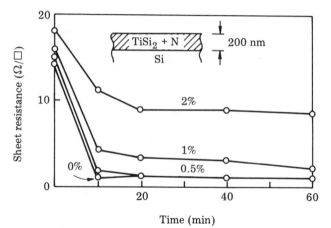

**FIGURE 11.9** Sheet resistance of $TiSi_2$ films which contain about 5, 7.5, 10, and 15 at % of N (formed by adding 0, 0.5, 1, and 2% of $N_2$ to the sputtering Ar gas). [After Lien and Nicolet (1984).]

When impurities such as oxygen are incorporated in the Pt film, the reaction kinetics to form Pt silicide can be changed significantly. Figure 11.10 shows the (thickness)$^2$ versus annealing time relationship for the formation of $Pt_2Si$ at 315°C for three types of samples. The reaction rate (the slope of the straight lines) is fastest for Pt films deposited in ultrahigh-vacuum conditions. The sputter-deposited Pt films (contains less than 0.1 at % oxygen) react with Si with a slower rate, but faster than that for Pt films implanted with 3 at % oxygen.

The oxygen in the Pt films tends to accumulate at the $Pt_2Si$–Pt interface after reaction. If the accumulation of oxygen is sufficiently large, the phase formation

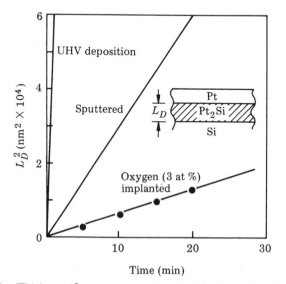

**FIGURE 11.10** (Thickness)$^2$ versus annealing time for formation of $Pt_2Si$ at 315°C for three types of samples. [After Ottaviani and Mayer (1981).]

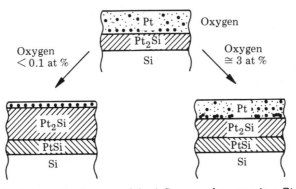

**FIGURE 11.11** Schematic diagram of the influence of oxygen in a Pt film on Si when the structure is annealed at 300°C. Oxygen accumulates at the Pt–Pt₂Si interface in the initial stages of silicide growth. For long anneals of the sample containing 3 at % oxygen, the accumulation of oxygen in Pt stops silicide growth. [After Ottaviani and Mayer (1981).]

sequence will be affected. Figure 11.11 shows a schematic diagram of the influence of oxygen in Pt–Si reactions. When the oxygen concentration in Pt is low (<0.1 at %), the accumulation of oxygen at the $Pt_2Si$–Pt interface only slows the formation kinetics. The oxygen eventually accumulates at the PtSi surface without any changes in phase sequence. When the oxygen content is large (about 3 at %) in the Pt films, the accumulation of oxygen at the $Pt_2Si$–Si interface is high enough to limit the supply of Pt at the $Pt_2Si$–Si interface, thus causing the formation of PtSi at the $Pt_2Si$–Si interface without the total consumption of the Pt layer. A general model of impurity redistribution has been developed by Lien and Nicolet (1984).

Many researchers have investigated the redistribution of dopants (mainly As and P) in Si during near-noble metal (Ni, Pd, and Pt) silicide formation (annealing temperatures are about 300°C). They found that although the dopant has very low diffusivity at these low annealing temperatures (see Table 7.5), part of the dopant is accumulated at the Si–silicide interface and another part incorporated in the silicide during silicide formation. This phenomenon can be explained by the large number of point defects generated in the Si neighboring the Si–silicide interface during the formation of these silicides. These defects enhance the diffusivity of substitutional dopants at low temperatures (Wittmer and Tu, 1984). The accumulation of dopants can reduce the contact resistance for ohmic contacts and possibly be used to adjust the barrier height for Schottky diodes.

## 11.3 Aluminide Formation

Aluminum and some of its alloys are commonly used as metallization for silicon integrated circuits. Aluminum has high electrical conductivity and good adhesion to Si, $SiO_2$, and other metallizations. In Chapter 10 we have seen the spiking phenomenon of Al into Si substrates at integrated-circuit processing temperatures.

To prevent this from happening, Al is kept from making direct contact with the Si substrate or with silicides layers by use of a diffusion barrier layer. Most of the actual or candidate diffusion barriers are refractory metals such as W and Ti. The characteristics of aluminide formation, therefore, plays an important role in the effectiveness of diffusion barriers.

## 11.3.1 Initial Phase Growth

For Al–metal bilayer interactions, an Al-rich phase grows initially. For the Al–Ti, Ta, Cr, Mo, W, Co, and Ni systems, the initial phase to grow is the most Al-rich phase in the phase diagram. For the Al–V, -Pd, -Pt systems, the initial phase is Al-rich, but not the most Al-rich phase in the phase diagram. The initial phase formation is summarized in Table 11.4, where the phase composition, reaction temperature, interface uniformity, kinetics dependence, and dominant diffusing species are shown. In general, there is no observable trend in the reaction temperature; the low end is about 225°C for $Pt_2Al_3$ formation and the high end is about 525°C for $TaAl_3$, $MoAl_{12}$, and $WAl_{12}$ formation. Most of the reactions take place in a layer-by-layer manner ($TiAl_3$, $VAl_3$, $TaAl_3$, $Cr_2Al_{13}$, $Co_2Al_9$, $NiAl_3$, and $Pt_2Al_3$). Other aluminide-reacted layers are not as uniform. For a few systems, the Al films develop large voids at or slightly below the initial reaction temperature. The formation of voids and the thickening are especially pronounced with Al films on Pt and Pd, and are less pronounced with Ta, Cr, Mo, and Co (Colgan, 1987).

**TABLE 11.4** Summary of Initial Phase Formation of Aluminides

| IVA | VA | VIA | VIIA | |
|---|---|---|---|---|
| Ti[a] <br> $TiAl_3$ <br> 350°C <br> planar <br> $\sqrt{t}$, Al | V <br> $VAl_3$ <br> 425°C <br> planar <br> layer by layer | Cr[a] <br> $Cr_2Al_{13}$ <br> 375°C <br> planar <br> $\sqrt{t}$ | Co[a] <br> $Co_2Al_9$ <br> 350°C <br> planar <br> $\sqrt{t}$ | Ni[a] <br> $NiAl_3$ <br> 300°C <br> planar <br> $\sqrt{t}$, Al |
| Zr[a] <br> $ZrAl_3$ <br> 350°C <br> irregular <br> nonparabolic | Nb[a] <br> $NbAl_3$ | Mo[a] <br> $MoAl_{12}$ <br> 525°C <br> irregular <br> Al | Element, <br> First phase, <br> Reaction T, <br> Interface, <br> Kinetics & <br> diffusing <br> species <br> in the <br> initial phase. | Pd <br> $Pd_2Al_3$ <br> 250°C <br> irregular <br> $\sqrt{t}$, Al |
| Hf[a] <br> $HfAl_3$ <br> 350°C <br> irregular <br> nonparabolic | Ta[a] <br> $TaAl_3$ <br> 525°C <br> planar <br> nonparabolic | W[a] <br> $WAl_{12}$ <br> 525°C <br> planar? | | Pt <br> $Pt_2Al_3$ <br> 225°C <br> planar <br> $\sqrt{t}$, Al |

*Source:* After Colgan (1987), Bower (1973), Howard et al. (1976), and Colgan and Mayer (1986 and 1989).

[a]The initial phase is the most Al-rich.

### 11.3.2 Reaction Kinetics

The reaction kinetics of aluminide formation have not been investigated as thoroughly as those for silicide formation. Of the aluminides included in Table 11.4, the formation of $TiAl_3$ is among the few that have been relatively well characterized. The $TiAl_3$ phase appears to grow in a layer-by-layer manner. The reaction rate of $TiAl_3$ is diffusion controlled and can be expressed as $x^2 = Dt$, where $x$ is the thickness of the growing $TiAl_3$ layer, $D$ an effective diffusion constant, and $t$ the reaction time. The constant $D$ may be further expressed as

$$D = D_0 e^{-E_A/kT}$$

where $D_0$ is a preexponential constant, $E_A$ an activation energy, and $kT$ has the usual meaning. For the formation of $TiAl_3$, $D = 0.15 \exp(-1.85/kT)$ (Bower, 1973). While the experimental data for the formation of $TiAl_3$ were obtained in 1973, the kinetics results still appear to be valid at present in spite of the possibly cleaner thin-film samples. Table 11.4 also shows the reaction kinetics dependence on annealing time. It is possible to correlate the planar interface (usually, a result of layer-by-layer growth) with diffusion-controlled kinetics ($x \propto \sqrt{t}$). For irregular interfaces, meaningful kinetics data become difficult to obtain.

### 11.3.3 Diffusing Species

The dominant diffusing species during aluminide formation has been investigated for a few Al–metal systems. The results are included in Table 11.4. It was found that Al is always the dominant diffusing species in the initial phase formation. For example, the diffusing flux ratio Al/M is about 13:1 for $TiAl_3$ formation; and is about 30:1 for $NiAl_3$ formation (Colgan and Mayer, 1986). Comparing aluminide formation with silicide formation, we see a correlation that the initial phase to form is enriched with the dominant diffusing species in both silicides and aluminides (see Tables 11.1 and 11.4). Since the initial phase of aluminides listed in Table 11.4 is always Al-rich, it is tempting to speculate that Al is always the dominant diffusing species for those systems that have not been investigated experimentally. For silicide formation, the initial phase is sometimes Si-rich ($CrSi_2$, for example); therefore, the metal is not always the dominant diffusing species.

# 11.4 Al–Pd₂Si Interactions

Palladium silicide ($Pd_2Si$) is often used in Si integrated-circuit technology because it can be formed reproducibly at a relatively low temperature of about 200°C. As mentioned previously, Al is often deposited on top of $Pd_2Si$ for interconnection purposes. Upon thermal annealing above 300°C, the Al–$Pd_2Si$/Si junction is found to be unstable. In particular, the Schottky barrier height $\phi_B$ and the ideality factor $n$ vary as a function of annealing. This junction instability is found to be a result of Al interacting with $Pd_2Si$ and eventually causing junction degradation (Köster

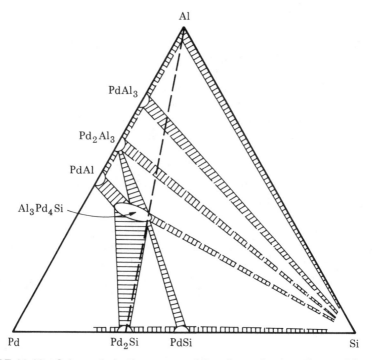

**FIGURE 11.12** Schematic isotherm map of the phase diagram observed for the Al–Pd–Si ternary system. The temperature is about 400 to 500°C. [After Köster et al. (1982).]

et al., 1982). The ternary phase diagram of the Al–Pd–Si system is shown in Fig. 11.12. This phase diagram is valid between about 400° and 500°C. To understand the interactions of this system, we first look at the initial stage (e.g., 200 to 250°C) annealing, where an intermediate layer of $Al_3Pd_4Si$ is formed at the Al–$Pd_2Si$ interface and accompanied by the dissolution of the excess Si and some Pd into Al (see Fig. 11.13). In the intermediate stage, a continuous layer of $Al_3Pd_4Si$ is formed. According to the phase diagram in Fig. 11.12, $Al_3Pd_4Si$ is stable against $Pd_2Si$. Further reaction requires the diffusion of Al (most probably through the structural defects, e.g., pinholes and grain boundaries) to the interface in order to sustain the conversion of $Pd_2Si$ to $Al_3Pd_4Si$.

In contrast, $Al_3Pd_4Si$ is not stable against Al at the $Al_3Pd_4Si$–Al interface. The dissociation of $Al_3Pd_4Si$ leads to the formation of Al–Pd compounds with high Al concentration (e.g., $Al_3Pd_2$ and/or $Al_3Pd$) and the excess Si has to precipitate out because there is only a small solubility of Si in the Al–Pd binary alloys. In the final stage, $Pd_2Si$ is completely converted into $Al_3Pd_4Si$, which in turn dissociates into Al–Pd compounds and Si precipitates (now doped with Al) epitaxially grown on the $n$-type Si substrate, as shown in Fig. 11.13c. The electrical characteristics can be altered significantly since the junction is more of a $p$-$n$ junction rather than a Schottky barrier. The presence of $Al_3Pd_4Si$ actually slows down the Al penetration

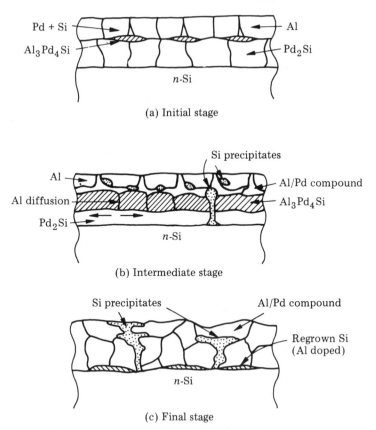

(a) Initial stage

(b) Intermediate stage

(c) Final stage

**FIGURE 11.13**   Schematic presentation of Al–$Pd_2Si$–Si reactions. [After Köster et al. (1982).]

into the Si substrate. To prevent Al–$Pd_2Si$ interaction, a diffusion barrier such as TiN may be placed between $Pd_2Si$ and Al to slow the junction degradation further.

## 11.5 Al–Au interactions

The Al–Au system is an important binary system for microelectronics; the equilibrium phase diagram is shown in Fig. 11.14. The connection from a chip to the outside is made by bonding Au or Al wires to the conductor pads on the chip. Al wire may fracture just beyond the bond, but Au wires can be bonded on to Al bonding pads easily due to the ductility of the Au wire. However, Al and Au interact rather easily during thermal-compression bonding. Brittle intermetallic compounds form between Al and Au and the bonds eventually break under even a slight stress condition. Bonding is described in Section 15.3.

The interaction between Al and Au has been the subject of considerable attention. Figure 11.15 summarizes the interactions for the Al–Au thin-film system

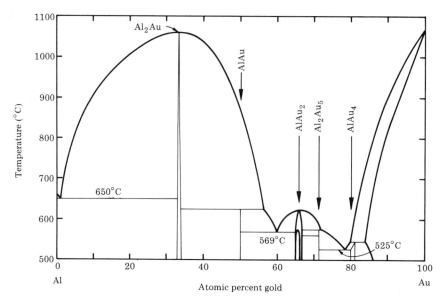

**FIGURE 11.14** Phase diagram of the Au–Al system. [Adapted from the *Bull. Alloy Phase Diagrams* 8(1), 71 (1987).]

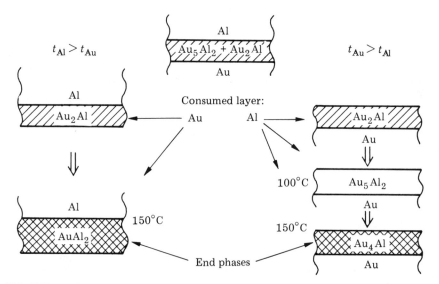

**FIGURE 11.15** Schematic diagrams showing the phase formation in Au–Al thin-film couples. The indicated temperatures are those typically required for the formation of phases in films several thousand angstroms thick after an anneal time on the order of 1 hour. Initially, $Au_5Al_2$ is formed, followed by the formation of $Au_2Al$ with a transition to $Au_2Al$ at higher temperatures. When the Al layer is consumed, $Au_5Al_2$ is formed at temperatures $\geq 100°C$, and at higher temperatures ($\geq 150°C$), $Au_4Al$ is formed. When Au layer is consumed, $AuAl_2$ is formed at 150°C. [After Campisano et al. (1975).]

(Majni et al., 1981). If the interface between Al and Au is relatively clean, a reaction takes place at room temperatures. The initial phase to form is always the $Au_5Al_2$ compound. At slightly higher temperature the $Au_2Al$ phase forms. The coexistence of $Au_5Al_2$ and $Au_2Al$ is a peculiar case in thin-film reactions, since in almost all thin-film couples only one phase forms at one time until the total consumption of one of the initial elements in the couple. Further phase transformation depends on the relative thickness of Al and Au. For thickness of Al ($t_{Al}$) larger than the thickness of Au ($t_{Au}$), the end phase is $AuAl_2$ in equilibrium with Al. For $t_{Au} > t_{Al}$, the end phase is $Au_4Al$ in equilibrium with Au. The reaction kinetics for the $Au_2Al$ phase and the $AuAl_2$ phase have been found to follow a (anneal time)$^{1/2}$ dependence. For $Au_2Al$, $D_0 = 1.4$ cm$^2$/s, $E_A = 1.03$ eV; for $AuAl_2$, $D_0 = 2.0$ cm$^2$/s, $E_A = 1.2$ eV (Campisano et al., 1975).

## 11.6 Diffusion Barriers

Diffusion barriers are an important part of integrated-circuit technology. As we have discussed previously, Al spiking into the Si substrate during annealing leads to electrical shorts and other problems at the junctions. One of the solutions to improve contact reliability is to place a diffusion barrier between Al and Si to prevent the dissolution of Si in the Al layer. A diffusion barrier therefore separates material A and material B physically by interposing a barrier layer of material X chosen so that the undesirable intermixing of A and B will be suppressed (Fig. 11.16). An ideal diffusion barrier should meet the following conditions (Nicolet, 1978):

1. The transport rate of A across X and of B across X should be small.
2. The loss rate of X into A and of X into B should be small.
3. X should be thermodynamically stable against A and against B.
4. There should be strong adhesion of X with A and with B.

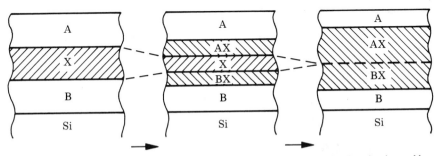

**FIGURE 11.16** Sacrificial diffusion barrier. If the reactions of the barrier layer X with A and B are each fully characterized and laterally uniform, the point at which the layer X has been fully consumed is predictable. As long as this point is not reached, a separation of A and B is in fact accomplished. For any prescribed time–temperature cycle there is a minimum thickness of X that will assure separation, and this amount can be predicted. [After Nicolet (1978).]

**5.** The specific contact resistance of A to X and of X to B should be small.

**6.** X should be laterally uniform in thickness and structure.

**7.** X should be resistant to mechanical and thermal stresses.

**8.** X should be highly conductive (thermal and electrical).

In practice, not all conditions can be met for one single diffusion barrier and compromises will have to be made. Practical diffusion barriers are generally divided into (1) sacrificial barriers, (2) stuffed barriers, and (3) amorphous diffusion barriers (Nicolet, 1978).

## 11.6.1 Sacrificial Barriers

The idea of a sacrificial barrier is illustrated in Fig. 11.16, where the barrier layer X reacts with A or B or both in a laterally uniform manner with fully characterized rates. As long as X is not completely consumed in the reactions with A or B or both, the separation between A and B is still in effect. An example of the sacrificial barrier is the use of a Ti layer between Al and Si to delay Al spiking (Fig. 11.17). At 450°C, Ti reacts negligibly with Si but reacts with Al to form a continuous and laterally uniform layer of TiAl$_3$ (Fig. 11.17b). When Ti is consumed, the lifetime of the sacrificial barrier is used up. Further heat treatment leads to contact degradation (Fig. 11.17d). The lifetime of the Ti sacrificial layer can be estimated from $t = L^2/D$, where $L$ is the thickness of the sacrificial Ti layer and diffusion is the

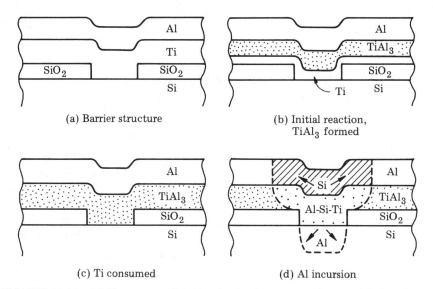

(a) Barrier structure

(b) Initial reaction, TiAl$_3$ formed

(c) Ti consumed

(d) Al incursion

**FIGURE 11.17**   (a) Ti as a sacrificial barrier basic contact structure. (b) TiAl$_3$ begins to form over the whole structure. (c) All the titanium has been consumed. At this point, the Ti–Si–Al ternary forms over the contact region (d), and silicon begins migrating into the aluminum while aluminum displaces the migrating silicon in the contact hole. The majority of the silicon displaced from the contact hole is diverted into the TiAl$_3$ adjacent to the contact area to extend the ternary region [After Bower (1973).]

effective constant $[D = 0.15 \exp(-1.85/kT) \text{ cm}^2/\text{s}]$ for the formation of $TiAl_3$. For example, a layer of 100 nm of Ti is consumed to form $TiAl_3$ in about 1.5 hours at 450°C. It can be seen from Table 11.4 that V and Ta have higher reaction temperatures with Al; these barriers have longer lifetimes against reactions with Al compared with that of Ti (e.g., Al reacts with V about 20 times slower than with Ti). Sacrificial barriers therefore delay the metallization failure with predictable lifetimes when the sacrificial reactions are well characterized.

## 11.6.2 Stuffed Barriers

While sacrificial barriers may provide adequate protection for the metallizations, they have a definite lifetime after which the barriers no longer function. For a more permanent protection, the barrier layer X should be stable against A and B (i.e., there are no driving forces for X to react with either A or B). The lack of thermodynamical driving force to react is not enough to make X a barrier; it is also necessary to stop or reduce diffusion of A and B across X via short-circuit paths. There are two ways to achieve this: (1) eliminate the short-circuit paths (i.e., the grain boundaries), and (2) plug the easy paths with appropriate atoms or molecules (''stuffing'' the barrier; see Fig. 11.18). An example of an elemental stuffed barrier is the Ti–Mo–Au system. Molybdenum and gold are mutually insoluble; there is no driving force for them to react. Gold atoms migrate into a Mo layer due to structural defects in the Mo film. When the Ti and Mo are deposited under a good vacuum conditions, the two metals intermix via grain boundaries. When Au is added on top of the film, Au and Ti readily combine at 600°C across the virtually unperturbed Mo layer. However, if Ti is exposed to oxygen before the Mo layer is deposited, intermixing between Ti and Mo is not observed. The implication is that the Ti–Mo–Au barrier system owes its blocking capacity to the presence of oxygen, which plugs the easy paths in the Mo and the Ti layers.

Reactive RF sputtering is one convenient method of stuffing a barrier layer. It was found that reactive sputtering of Mo in the presence of nitrogen can produce a nearly stoichiometric $Mo_2N$ layer which works well as a stuffed barrier between

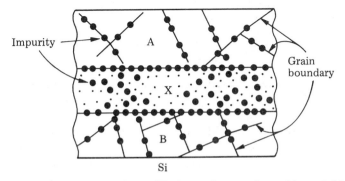

**FIGURE 11.18** "Stuffed" barrier. Easy diffusion paths are plugged by suitable impurities; the stuffed barrier can successfully withstand the heat treatment. [After Nicolet (1978).]

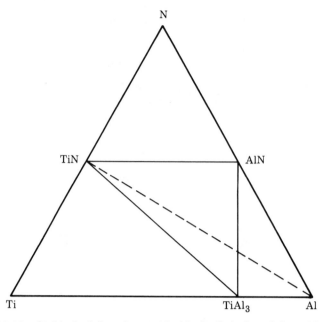

**FIGURE 11.19**  Stable (solid) and unstable (dashed) tie lines inferred from the TiN–Al and TiN–TiAl$_3$ reactions; $T$ = 600°C. [After Beyers et al. (1984).]

Au and Ti. In this case the barrier layer Mo$_2$N is a compound that is stable against Ti and Au, and at the same time the easy paths in the compound barrier layer are plugged (Nowicki and Wang, 1978). The use of electrically conductive nitrides (e.g., TiN and Mo$_2$N), borides (TiB$_2$), and carbides (TiC and NbC) as diffusion barriers has met with reasonable success. While TiN has been applied successfully to the gate metallization of insulated-gate field-effect transistors (IGFETs), where TiN is used as a barrier between Ti and Pt, TiN is *not* stable against Al. Figure 11.19 shows the ternary phase diagram for the N–Ti–Al system (Chapter 10). There is no tie line between Al and TiN; therefore, Al reacts with TiN. On the other hand, TiN is stable against TiAl$_3$. A diffusion barrier system with a sequence of TiN–TiAl$_3$–Al should provide better contact stability.

## 11.6.3 Amorphous Barriers

In Section 11.6.2 we showed how easy paths for diffusion may be plugged by impurities. Another way to stop easy path diffusion is to eliminate the grain boundaries altogether by making the diffusion layer either single crystalline or amorphous in structure. Single-crystalline barrier layers are impractical at present; the alternative solution of depositing amorphous layers appears to be a feasible approach. It should be recognized that amorphous layers are metastable in nature. When the amorphous layer crystallizes into a polycrystalline film, easy diffusion paths are present again. It is also possible for the amorphous layer, X, to react with layers A and/or B; in doing so the amorphous diffusion barrier becomes a sacrificial barrier. During reaction the composition of the amorphous layer may change and may lead to a lowering of the crystallization temperature.

The next question is: How can one make an amorphous metallic material (often called a metallic glass)? Pure elemental metals do not form amorphous phases at room temperature. There are a number of empirical rules on how to select the elements to obtain amorphous metallic alloys. Generally, the larger the differences between the constituents in terms of atomic size, crystalline structure, and electro-negativity, the easier it is for the two constituents to form a metallic amorphous alloy. In addition, fast quenching the liquid phase near a deep eutectic point in the equilibrium phase diagram also favors amorphous phase formation. For diffusion barrier application in microelectronics, amorphous metallic alloys usually consist of a near-noble metal and a refractory metal such as $Ni_xW_{1-x}$ and $Cu_xTa_{1-x}$ with $x$ centering around 0.5. These amorphous films are often deposited by coevaporating and/or cosputtering onto a substrate held at low or room temperatures (Nicolet et al., 1983).

There are two parameters that characterize an amorphous metallic alloy: the crystallization temperature $T_c$ and the reaction temperature $T_r$ at which reaction occurs with the overlayer A or with the underlayer B. The lower of the two temperatures determines how reliable the amorphous layer is as a diffusion barrier. Most of the amorphous diffusion barrier layers are used between an Al top layer and a Si substrate. The phase transformations of interest to us for an amorphous layer in this situation are (1) crystallization, (2) reaction with Si, and (3) reaction with Al. Figure 11.20 shows a plot of the reaction temperature of commonly used amorphous alloys in microelectronics versus their crystallization temperature. There is a general increase in $T_r(Si)$ for the Si reaction with increasing $T_c$, whereas $T_r(Al)$ for the Al reaction remains between 400 and 500°C. A more definite correlation is found between the reaction temperature of the amorphous alloy with the reaction

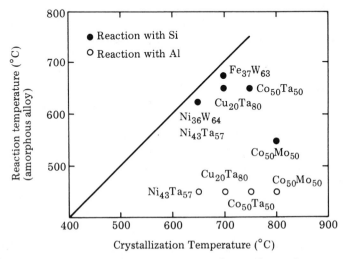

**FIGURE 11.20** Plot of the reaction temperature of amorphous alloys versus their crystallization temperature. [After Saris et al. (1986).]

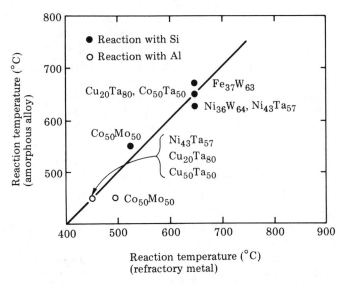

**FIGURE 11.21**  Plot of the reaction temperature of amorphous alloys as a function of the reaction temperature of their refractory components. [After Saris et al. (1986).]

temperature of their refractory components (see Fig. 11.21). This correlation shows that the onset of the reactions between the amorphous alloy and the Si (or Al) is largely determined by the reaction between the refractory constituent (referred to as $M_2$ in the $M_1M_2$ alloy) and the Si(or Al). Referring to Tables 11.2 and 11.4, one can see that the reaction temperatures between Si (or Al) and refractory metals are much higher than those between Si (or Al) and near-noble metals. This behavior may be understood in terms of the low atomic mobility in an amorphous matrix and the different reaction characteristics for individual constituents. In reaction with Si, near-noble metals (referred to as $M_1$ in the $M_1M_2$ alloy) are known to be the moving species in forming silicides at 250 to 300°C (Table 11.1). The absence of a detectable reaction with the amorphous layer at low temperatures is apparently due to the fact that $M_1$ is locked in the amorphous matrix. At about 650°C, the temperature is sufficiently high for Si atoms to be dislodged from the lattice by breaking covalent bonds. Reaction between Si and the amorphous alloy can therefore take place due to sufficient Si mobility at the interface. Then the amorphous alloy is progressively converted into silicides.

   In reaction between Al and metal alloys, Al is expected to be the moving species (see Table 11.4). At about 300 to 350°C, Al begins to react with $M_2$ (the near-noble metal constituent), resulting in a $M_2$-rich alloy near the reaction front. Further reaction with a $M_2$-rich alloy cannot proceed until the characteristic reaction temperature between $M_2$ and Al is reached (i.e., about 450 to 500°C). The protection provided by the amorphous barriers shown in Fig. 11.21 is good up to about 450°C against Al overlayers. Once the onset of the reaction is reached, the mode of protection of the amorphous layer switches to, perhaps, a sacrificial barrier.

# 11.7  Alloy–Si Interactions

One crucial step in achieving large-scale integration in electronic devices is the dimensional reduction of the contacts. The shrinking is necessary in both lateral direction as well as in depth. Since the depth of a silicide contact is measured by the amount of Si consumed in forming the silicide, the shallow silicide contact requires a limited but uniform interaction between the Si substrate and the contact materials without sacrificing reliable coverage of the contact areas. One way to limit the Si consumption from the substrate during formation is to form shallow silicide layers by using a codeposited $M_{1x}M_{2y}$ alloy film, such as a $Pd_{80}W_{20}$ layer. This idea came from the observation that near-noble metals (i.e., Pt, and Pd) react with Si at much lower temperatures than those for refractory metals (i.e., Mo, W, and V). An alloy layer of a near-noble metal and a refractory metal, such as $Pd_xV_y$, codeposited on Si may react with Si selectively such that a near-noble metal silicide forms alone in contact with the Si substrate, with an outer refractory-metal-rich region (Nicolet and Lau, 1983).

It was found that upon annealing at low temperature (200 to 300°C), an alloy layer consisting of a near-noble metal (Pt, Pd, and Ni) and a refractory metal (W, V, and Cr) deposited on Si forms a near-noble metal silicide and a mixed layer of both metals in the outer region of the deposited film. This mixed layer is enriched in the refractory-metal component and can serve as a diffusion barrier against further metallization (such as Al metallization) and forms a reliable coverage for the contact areas. Annealing at higher temperature ($\geq$500°C) causes the formation

**TABLE 11.5**  Phases After Annealing for Alloy Films About 200 nm Thick on Si

| Alloy Layer on Si | Phases After Annealing at 800°C | | Percent of Near-Noble Metal in Outer Layer |
| --- | --- | --- | --- |
| | Inner Layer (Next to Si) | Outer Layer | |
| $Pt_{50}W_{50}$ | PtSi | $WSi_2$ | — |
| $Pt_{80}W_{20}$ | PtSi | $WSi_2$ | ~60 |
| $Pt_{78}V_{22}$ | PtSi | $VSi_2$ | 4 |
| $Pt_{11}Cr_{89}$ | PtSi | $CrSi_2$ | 15 |
| $Pt_{80}Cr_{20}$ | PtSi | $CrSi_2$ | 20 |
| $Pd_{78}W_{22}$ | $Pd_2Si$ | $WSi_2$ | 6 |
| $Pd_{77}V_{23}$ | $Pd_2Si$ | $VSi_2$ | 5 |
| $Pd_{11}V_{89}$ | $Pd_2Si$ | $VSi_2$ | 20 |
| $Pd_{15}Cr_{85}$ | $Pd_2Si$ | $CrSi_2$ | 7 |
| $Ni_{75}Cr_{25}$ | $NiSi_2$ | $CrSi_2$ | |
| $Ni_{50}Cr_{50}$ | $NiSi_2$ | $CrSi_2$ | |
| $Ni_{60}Cr_{40}$ | $NiSi_2$ | $CrSi_2$ | |

*Source:* After Ottaviani et al. (1980).

of a refractory-metal silicide in the outer region. No mixing or ternary phases have been observed even after annealing at 800°C. Table 11.5 lists the final composition of the layers of some of the alloy films on Si after annealing at 800°C for 1 hour. In every case, a near-noble metal silicide forms next to the Si substrate.

The uniformity of coverage of the near-noble silicide next to the Si substrate and the silicide formation depends on the composition of the alloy. Generally speaking, alloying a refractory metal to a near-noble metal raises the formation temperature of the near-noble metal silicide. The presence of near-noble metals in the alloy lowers the formation temperature of the refractory-metal silicide. For example, a $Pd_{80}W_{20}$ alloy layer (about 200 nm) on Si requires an annealing of about 200°C to form a uniform layer of $Pd_2Si$ next to the Si substrate with an outer region enriched in W. The outer region transforms into a $WSi_2$ layer at about 500°C. At this point, the alloy layer may cease to serve as shallow silicide contact, since the formation of $WSi_2$ requires further consumption of Si from the substrate. On the other hand, for a $Pd_{30}W_{70}$ alloy, an annealing temperature of about 350°C is required to form the $Pd_2Si$ phase, and the coverage of $Pd_2Si$ on the Si substrate is not necessarily uniform and pinhole-free. The outer region requires a temperature of about 700°C to form the $WSi_2$ phase.

# REFERENCES

### General References

COLGAN, E. G., "Aluminum–Transition Metal Thin Film Reactions," Ph.D. thesis, Cornell University, 1987. See also E. G. Colgan, *J. Appl. Phys,* *62*(4), 1224 (1987).

COLGAN, E. G., and J. W. MAYER, "Thin Film Aluminide Formation," *J. Materials Research* (in press) (1989).

MURAKA, S. P., *Silicides for VLSI Applications,* Academic Press, New York, 1983, Chapter 4.

NICOLET, M.-A., and S. S. LAU, "Formation and Characterization of Transition-Metal Silicides," in *VLSI Electronics: Microstructure Science,* Vol. 6, ed. N. G. Einspruch and G. B. Larrabee, Academic Press, New York, 1983.

OTTAVIANI, G., and J. W. MAYER, "Mechanisms and Interfacial Layers in Silicide Formation," in *Reliability and Degradation—Semiconductor Devices and Circuits,* ed. M. J. Howes and D. V. Morgan, Wiley, New York, 1981, Chapter 2.

TU, K. N., and J. W. MAYER, "Silicide Formation," in *Thin Films—Interdiffusion and Reactions,* ed. J. M. Poate, K. N. Tu, and T. W. Mayer, Wiley, New York, 1978.

### Specific References

BEYERS, R., R. SINCLAIR, and M. E. THOMAS, *J. Vac. Sci. Technol. B2*, 781 (1984).

BOWER, R., *Appl. Phys. Lett. 23*, 99 (1973).

CAMPISANO, S. U., G. FOTI, E. RIMMINI, S. S. LAU, and J. W. MAYER, *Philos. Mag. 31,*903 (1975).

CHEN, S. H., L. R. ZHENG, J. C. BARBOUR, E. C. ZINGU, L. S. HUNG, C. B. CARTER, and J. W. MAYER, *Mater. Lett., 2,*469 (1984).

CHEN, L. J., H. C. CHENG, and W. T. LIN, in *Thin Films—Interfaces and Phenomena,* ed. by R. J. Nemanich, P. S. Ho, and S. S. Lau, MRS Symposia Proceedings, Vol. 54, 1986, p. 245.

COLGAN, E. G., and J. W. MAYER, *Nucl. Instrum. Methods B17,*242 (1986).

D'HEURLE, F. M., and P. GAS, *J. Mat. Res. 1,*205 (1986).

GIBSON, J. M., J. L. BATSTONE, and R. T. TUNG, *Appl. Phys. Lett. 51,*45 (1987).

GÖSELE, U., and K. N. TU, *J. Appl. Phys. 53*(4), 3252 (1982).

GURVITCH, M., A. F. J. LEVI, R. T. TUNG, and S. NAKAHARA, *Appl. Phys. Lett. 51,*311 (1987).

HENSEL, J. C., A. F. J. LEVI, R. T. TUNG, and J. M. GIBSON, *Appl. Phys. Lett. 47,*151 (1985).

HOWARD, J. K., R. F. LEVER, P. J. SMITH, and P. S. HO, *J. Vac. Sci. Technol. 13*(11), 68 (1976).

KNAPP, J. A., and S. T. PICRAUX, *Appl. Phys. Lett. 48,* (1986).

KÖSTER, U., P. S. HO, and J. E. LEWIS, *J. Appl. Phys. 53*(11), 7436 (1982).

LIEN, C. D., and M.-A. NICOLET, *J. Vac. Sci. Technol. B24,*738 (1984).

MAJNI, J. G., C. NOBILI, G. OTTAVIANI, M. COSTATO, and E. GALLI, *J. Appl. Phys. 52*(6), 4047 (1981).

NICOLET, M.-A., *Thin Solid Films 52,*415 (1978).

NICOLET, M.-A., I. SUNI, and M. FINETTI, *Solid State Technol.,* Dec. 1983, p. 129.

NOWICKI, R. S., and I. WANG, *J. Vac. Sci. Technol. 15,*235 (1978).

OTTAVIANI, G., *J. Vac. Sci Technol. 116,*1112 (1979).

OTTAVIANI, G., J. G. MAGNI, and C. CANALI, *Appl. Phys. 18,*285 (1979).

OTTAVIANI, G., K. N. TU, J. W. MAYER, and B. Y. TSAUR, *Appl. Phys. Lett. 36,*331 (1980).

POATE, J. M., and R. C. DYNES, *IEEE Spectrum* Feb. 1986, p. 39.

SARIS, F. W., L. S. HUNG, M. NASTASI, and J. W. MAYER, in *Thin Films—Interfaces and Phenomena,* ed. R. J. Nemanich, P. S. Ho, and S. S. Lau, MRS Symposia Proceedings, Vol. 54, 1986, p. 81.

TSAUR, B. Y., S. S. LAU, M.-A. NICOLET, and J. W. MAYER, *Appl. Phys. Lett, 38,*922 (1981).

TUNG, R. T., "Epitaxial Silicides," Chapter 10 in *Silicon-Molecular Beam Epitaxy;* ed. E. Kaspar and J. C. Bean, (CRC Press, Boca Raton, Florida, 1988), pp. 13–64.

WALSER, R., and R. BENÉ, *Appl. Phys. Lett. 28,*624 (1976).

WITTMER, M., and K. N. TU, *Phys. Rev. B29,*2010 (1984).

ZHENG, L. R., L. S. HUNG, J. W. MAYER, G. MAJNI, and G. OTTAVIANI, *Appl. Phys. Lett. 41,*646 (1982).

## PROBLEMS

**11.1.** You want to make contact to Si and deposit a layer of palladium (Pd) that is 1000 Å thick ($10^{-5}$ cm) on Si. The density of Pd is $6.8 \times 10^{22}$ atoms/cm$^3$ as compared to that of silicon of $5 \times 10^{22}$ atoms/cm$^3$. You want to form the compound Pd$_2$Si (a silicide with two Pd atoms per Si atom) and the equilibrium phase diagram shows a Pd–Si eutectic at 750°C and the melting point of Pd$_2$Si at 1200°C. For integrated circuit processing for contact formation:

(a) Would you heat the sample to temperatures $T < 750°C$, $T = 750°C$, $T$ between 750 and 1200°C, $T = 1200°C$, $T > 1200°C$?

(b) What thickness of *silicon* would be consumed in forming Pd$_2$Si?

(c) The Schottky barrier height of Pd$_2$Si on *n*-type Si is 0.75 eV. For forward bias of 0.3 V, what *polarity* of voltage (positive or negative) would be applied?

(d) For the same forward applied voltage, if you replace the Schottky barrier with an ohmic contact, will the current increase, decrease, or remain unchanged?

**11.2.** You are measuring the diffusion of Si in an Al film from a source of Si that establishes a surface concentration $C_0$ of $5 \times 10^{20}$ atoms/cm$^3$. At 327°C you measure a diffusion length of 40 μm for a diffusion time of 30 min.

(a) What is the value of the diffusion coefficient $D$?

(b) If the surface concentration $C_0$ decreases by a factor of 20, will the values listed below increase, decrease, or remain unchanged?

(1) Diffusion coefficient $D$

(2) Diffusion length $\lambda$

(3) Flux $F$

(c) If the diffusion length is a factor of 400 larger when the sample is heated to 527°C for 30 min as compared to 327°C, what is the value of the activation energy $E_A$ in eV?

**11.3.** Nickel has a density of $9 \times 10^{22}$ atoms/cm$^3$. If a 1000-Å layer of Ni is deposited on (100) Si, how much Si is consumed in the formation of Ni$_2$Si, NiSi, and NiSi$_2$?

**11.4.** You are asked to compare the following elements as sacrificial diffusion barriers against Al: (1) Ti, (2) V, (3) Cr, and (4) Hf. Which element(s) would be your choice, and why.

**11.5.** An alloy film of Pd$_{77}$V$_{23}$ deposited on Si would phase separate into an inner layer of Pd$_2$Si and an outer layer of VSi$_2$ after annealing, as shown in Table 11.5. Predict the layer sequence of samples with the following initial configuration after annealing: (a) Si/Pd/V and (b) Si/V/Pd.

**11.6.** According to Fig. 11.19, Al is not stable in contact with TiN. In practice, however, TiN is often used in conjunction with Al. Explain why and how may TiN act as diffusion barrier in contact with Al.

**11.7.** Gold metallization is commonly used in GaAs technology. What barrier systems would you use to improve the stability of Au on GaAs?

**11.8.** For shallow junction contacts in Si technology, alloy films such as PtW and PdW can be used to form an inner layer of silicide without an excessive consumption of Si from the substrate. Design another scheme of metallization for shallow contacts using only one metallic element.

# CHAPTER 12

# Energy Bands and Wave Behavior

## 12.1 Introduction

We have treated the electron as a particle in our discussion of $p$-$n$ junctions and transistors. Even though we relied on the results of quantum mechanics for energy levels of atoms and the formation of energy bands, we discussed the bands with a physical model of the semiconductor. We associated the valence band with the covalently bound electrons and the conduction band with the free electrons. The existence of an energy gap between the valence and conduction bands followed naturally from the concept that a certain amount of energy—equivalent to the energy gap—was required to break a covalent bond.

Again we relied on the results of quantum mechanics for the effective density of states in the conduction band and the valence band. However, in discussing the Fermi level and the number of electrons in the conduction band, we treated the electrons as classical particles that obeyed Maxwell–Boltzmann statistics. Of course, the classical description broke down when discussing heavily doped semiconductors, where the number of electrons in the conduction band was greater than the effective density of states.

Despite these exceptions, charge transport in the components of integrated circuits can by and large be treated on the basis of electrons as particles. However, when treating the light-emitting properties of GaAs, one realizes that a more detailed treatment of electrons as waves in crystals is required. This wave aspect of electrons is also required when one explores the differences between the band structure of Si and that of GaAs.

## 12.2 The Electron as a Wave

Electrons have wavelike properties with a wavelength inversely proportional to their particle momentum,

$$\lambda = \frac{h}{mv} \tag{12.1}$$

where Planck's constant $h = 6.626 \times 10^{-34}$ J · s. For free electrons with a kinetic energy of 0.026 eV and thermal velocity of $10^7$ cm/s ($10^5$ m/s), the wavelength corresponds to

$$\lambda = \frac{h}{mv} = \frac{6.6 \times 10^{-34}}{9.1 \times 10^{-31} \times 10^5} \frac{J \cdot s}{kg} = 7.3 \times 10^{-9} \text{ m} = 7.3 \text{ nm}$$

a value that is equivalent to many lattice spacings. With light waves or photons where $E = hv$, the wavelength is given by

$$\lambda = \frac{hc}{E} \tag{12.2}$$

where the velocity of light $c = 2.998 \times 10^8$ m/s. For photons with a wavelength of $7.3 \times 10^{-9}$ m (for comparison with electrons), the energy

$$E = \frac{hc}{\lambda} = \frac{6.6 \times 10^{-34} \times 3 \times 10^8}{7.3 \times 10^{-9}} \frac{J \cdot m \cdot s}{m \cdot s} = 2.7 \times 10^{-17} \text{ J} = 170 \text{ eV}$$

where 1 eV $= 1.602 \times 10^{-19}$ J. The energy of the photon is orders of magnitude greater than that of an electron of comparable wavelength. A 1-eV photon has a wavelength of 1.24 μm.

For discussions of wave properties it is convenient to use the wave vector **k** or in one dimension the wave number $k$, where

$$k = \frac{2\pi}{\lambda}. \tag{12.3}$$

Although this is the same symbol as that used with Boltzmann's constant, there should be no confusion, since Boltzmann's constant is used in this book with the temperature $T$ (i.e., $kT$). For a thermal electron with a wavelength of $7.3 \times 10^{-9}$ m, the wave number is

$$k = \frac{2\pi}{\lambda} = \frac{2\pi}{7.3 \times 10^{-9}} \frac{1}{m} = 8.7 \times 10^8 \text{ m}^{-1} = 8.7 \times 10^6 \text{ cm}^{-1}.$$

We can introduce the general statement that for any particle there is associated a wave whose wave vector **k** is parallel to the momentum **p** of the particle such that

$$\mathbf{p} = \hbar\mathbf{k} \tag{12.4}$$

where $\hbar = h/2\pi$. The $\hbar$ notation is standard in quantum mechanics. From classical mechanics where $E = mv^2/2 = p^2/2m$, we can express the energy of a free electron as

$$E = \frac{\hbar^2 k^2}{2m}. \tag{12.5}$$

A plot of energy versus wave number is shown in Fig. 12.1 and is referred to as an E–k diagram. A free electron can occupy any point on the continuous line in

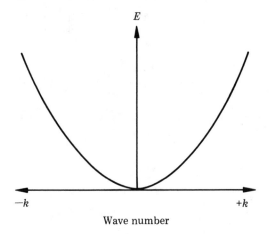

**FIGURE 12.1** Energy versus wave number (the $E$–$k$ diagram) for a free electron.

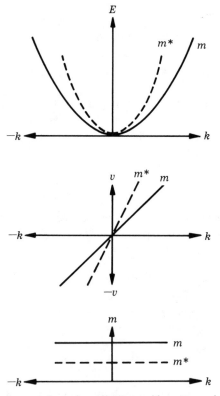

**FIGURE 12.2** $E$–$k$ diagram for a free electron with mass $m$ (solid line) and a smaller mass, $m^*$. The parabolic $E$–$k$ diagram leads to a linear $v$ versus $k$ relation and a constant mass.

the $E$–$k$ diagram. For those familiar with quantum-mechanical treatment, eq. (12.5) is the solution of the time-independent Schrödinger equation with the potential $V = 0$.

The same treatment can be carried over to electrons in semiconductors, where for low values of $k$, $E$ is proportional to $k^2$. However, to account for the differences in mobilities between electrons in Si and those in GaAs, we introduce the concept of an effective mass $m^*$ which changes the curvature of the $E$–$k$ diagram as shown in Fig. 12.2. In analogy with $E = p^2/2m$ and $v = dE/dp$, we can write the electron velocity as

$$v = \frac{1}{\hbar} \frac{dE}{dk} \tag{12.6}$$

(the straight line in Fig. 12.2) and the effective mass as

$$m^* = \hbar^2 \left(\frac{d^2E}{dk^2}\right)^{-1}. \tag{12.7}$$

The dashed line in Fig. 12.2 shows the case where the effective mass $m^*$ is less than that of the free electron mass. In silicon the effective mass of the electrons $m_e^*$ is close to that of the free electron, but in GaAs, $m_e^* = 0.067$ m. This difference accounts for the higher mobility $\mu$ of electrons ($\mu = e\tau/m^*$, Chapter 2) and the smaller density of states [$N_c = 2(2\pi m^* kT/h^2)^{3/2}$, Chapter 3] in GaAs as compared to those in Si.

## 12.3  Standing Waves and Electron in a Box

When electrons are confined to a semiconductor, there are only certain allowed values of $E$ and $k$. The continuous curve in the $E$–$k$ diagram is now replaced by a series of points, each specifying a particular $E$ and associated $k$ value. In reference to the Schrödinger equation, the solutions are quantized.

We can gain insight into the problem by considering the behavior of standing waves built up from superimposition of sine waves. The expression for a sine wave of amplitude $A$ and wavelength $\lambda$ is

$$y = A \sin \frac{2\pi x}{\lambda} = A \sin kx \tag{12.8a}$$

where $k = 2\pi/\lambda$ as before. For a sine wave traveling toward positive values of $x$ (Fig. 12.3) with a period $T$ for the wave to travel one wavelength $\lambda$,

$$y = A \sin \left(kx - \frac{2\pi t}{T}\right) = A \sin (kx - \omega t) \tag{12.8b}$$

where the angular frequency $\omega = 2\pi/T$. We can construct a standing wave by superimposing two waves with the same amplitude $A$, wave number $k$, and frequency $\omega$, but one $y_1$ traveling toward positive values of $x$ (Fig. 12.4):

$$y_1 = A \sin (kx - \omega t)$$

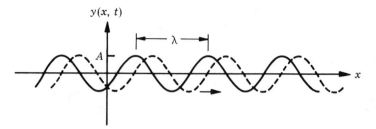

**FIGURE 12.3** Sine wave of amplitude $A$ and wavelength $\lambda$ moving toward positive values of $x$.

and the other $y_2$ toward negative values of $x$:

$$y_2 = A \sin (kx + \omega t)$$

so that

$$y = y_1 + y_2 = A[\sin (kx - \omega t) + \sin (kx + \omega t)]. \qquad (12.9)$$

From geometrical relations for the sum of two sine curves, we obtain:

$$y = 2A \sin kx \cos \omega t \qquad (12.10)$$

which is the relation for a standing wave where all points vibrate with the same frequency but the amplitude depends on $x$ at a particular time $t$.

We can apply boundary conditions to the standing wave and specify that the amplitude be zero at points $x = 0$ and $x = L$, that is,

$$y(x = 0, t) = 0 \qquad (12.11)$$
$$y(x = L, t) = 0$$

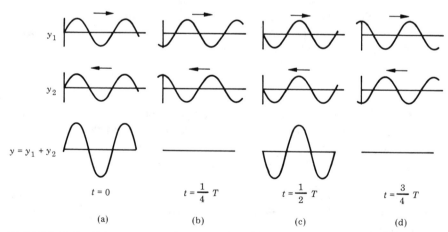

**FIGURE 12.4** Sine waves moving toward positive, $y_1$, and negative, $y_2$, values of $x$ are superimposed, $y = y_1 + y_2$: (a) $t = 0$; (b) $t = \frac{1}{4} T$; (c) $t = \frac{1}{2} T$; (d) $t = \frac{3}{4} T$.

**FIGURE 12.5** Boundary conditions such that $y = 0$ at $x = 0$ and $x = L$.

as shown in Fig. 12.5. These boundary conditions can be met by

$$\sin kx = 0 \qquad \text{at } x = 0, L. \tag{12.12}$$

The solution for $x = 0$ allows any value of $k$, but that for $x = L$ can only be met for

$$k_n = n \frac{\pi}{L} \tag{12.13}$$

where $n$ is an integer and the subscript on the wave number identifies the particular value of $n$. The wave shapes are shown in Fig. 12.6.

This solution can be carried over to the case of an electron confined in a region of length $L$. It is the solution of the one-dimensional Schrödinger equation for an electron where the potential $V = \infty$ at $x = 0$ and $x = L$. Since the electron cannot penetrate outside the potential well, the wave functions are zero at $x = 0, L$. The energy of the electron is written

$$E_n = \frac{\hbar^2 k_n^2}{2m} = \frac{h^2 n^2}{8mL^2} \tag{12.14}$$

where $n = 1, 2, 3. \ldots$ . This quantized solution for the energy of the electron leads to an $E$–$k$ diagram composed of discrete, allowed solutions of $E$ as shown in Fig. 12.7. This figure is a gross exaggeration of the energy scale because the separation between the energy values is exceedingly small. The allowed energy values are multiples of $h^2/8mL^2$, which for $L = 10^{-2}$ m and $n = 1$ has a value

$$E_1 = \frac{(6.626 \times 10^{-34})^2}{8 \times 9.1 \times 10^{-31} \times 10^{-4}} = 6.03 \times 10^{-34} \text{ J} = 3.77 \times 10^{-15} \text{ eV}.$$

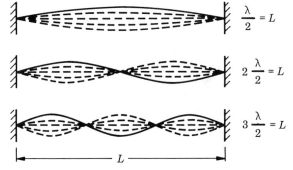

**FIGURE 12.6** Standing waves where $\lambda = 2L$, $\lambda = L$, and $\lambda = 2L/3$.

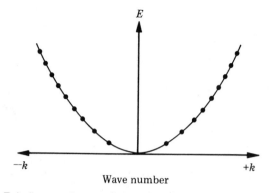

**FIGURE 12.7** *E–k* diagram for an electron confined in a region of length *L* in which the dots represent allowed solutions.

Even to achieve a value near $kT = 0.025$ eV, the value of $n$ would exceed $2 \times 10^6$. It is these large number of states in an energy interval that allows us to treat electrons as particles in descriptions of current flow and transistor action.

In three dimensions, with a cube of side $L$ containing the electrons, the energy can be written

$$E = \frac{\hbar^2}{2m} (k_x^2 + k_y^2 + k_z^2)$$

$$= \frac{h^2}{8mL^2} (n_x^2 + n_y^2 + n_z^2) \tag{12.15}$$

where $n_x$, $n_y$, and $n_z$ are integers.

## 12.4 Density of States in a Semiconductor

In Chapter 3 we introduced without derivation the concept of an effective density of states in the conduction band. We are now in a position to carry out the derivation. For electrons in a cube of side $L$, the energy $E$ [eq. (12.15)] can be written

$$E = E_1(n_x^2 + n_y^2 + n_z^2) \tag{12.16}$$

where $E_1 = h^2/8mL^2$. Each set of values for $n_x$, $n_y$, and $n_z$ corresponds to one energy level, and we construct an $n$-space (Fig. 12.8) formed by the axes $n_x$, $n_y$, and $n_z$. Only an octant is shown, because the integer values are positive. In this space every integer specifies a state with a unit cube containing exactly one state. The number of states in any volume is just equal to the numerical value of the volume. The energy is

$$E = R^2 E_1 \tag{12.17}$$

where $R^2 = (n_x^2 + n_y^2 + n_z^2)$.

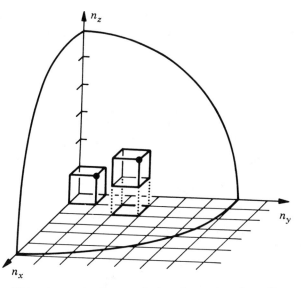

**FIGURE 12.8** Allowed energies for an electron in a cube where $E = E_1 (n_x^2 + n_y^2 + n_z^2)$.

To find the number of electrons that could occupy these states, we must introduce another concept from quantum mechanics: Electrons have two spin states (a fact that requires a fourth quantum number to specify the state an electron occupies), and there are two spin states per $n$-space state. The number $N$ of states within radius $R$ is

$$N = 2 \frac{1}{8} \frac{4}{3} \pi R^3 = \frac{1}{3} \pi \left( \frac{E}{E_1} \right)^{3/2} \tag{12.18}$$

where the factor of 2 is due to the spin states and the $\frac{1}{8}$ is required for an octant (positive $n$ values).

To determine how many states there are between the energy levels $E$ and $E + dE$, we differentiate eq. (12.18):

$$dN = \frac{\pi}{2} E_1^{-3/2} E^{1/2} \, dE \tag{12.19}$$

$$= \frac{\pi}{2} \left( \frac{8mL^2}{h^2} \right)^{3/2} E^{1/2} \, dE.$$

The density of states per unit volume $g(E) = (V)^{-1} dN/dE$ is

$$g(E) = \frac{\pi}{2} \left( \frac{8m}{h^2} \right)^{3/2} E^{1/2} = CE^{1/2} \tag{12.20}$$

with the volume $V = L^3$ and $C = 4\pi (2m/h^2)^{3/2}$.

In Chapter 3 we used an effective density of state $N_S$ rather than $g(E)$ to find the number $n$ of electrons per unit volume in the conduction band. To obtain $n$ we must take the density of states $g(E)$, multiply by the probability of occupation $F(E)$, and integrate from the bottom to the top of the conduction band, which we set at infinity,

$$n = \int_{E_G}^{\infty} g(E)F(E)\, dE \qquad (12.21)$$

using the energy ordinate shown in Fig. 12.9. For the case of lightly doped semiconductors where the Fermi level lies in the middle of the gap, we can use Maxwell–Boltzmann statistics and write the probability of occupation as

$$F_{\mathrm{MB}}(E) = \exp\left(-\frac{E - E_F}{kT}\right) \qquad (12.22)$$

where $E - E_F \gg kT$. With the zero of energy at the top of the valence band, we write the density of states as

$$g(E) = C(E - E_G)^{1/2} \qquad (12.23)$$

so that eq. (12.21) becomes

$$n = C \int_{E_G}^{\infty} (E - E_G)^{1/2} \exp\left(-\frac{E - E_F}{kT}\right) dE. \qquad (12.24)$$

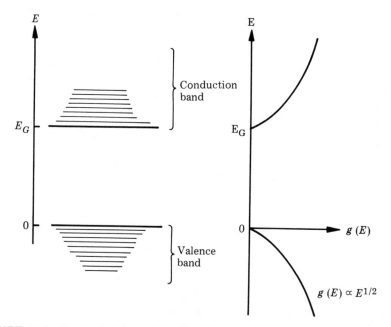

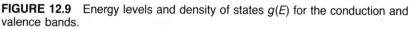

**FIGURE 12.9** Energy levels and density of states $g(E)$ for the conduction and valence bands.

With a new variable,

$$x = \frac{E - E_G}{kT}$$

eq. (12.24) takes the form

$$n = C(kT)^{3/2} \exp\left(-\frac{E_G - E_F}{kT}\right) \int_0^\infty x^{1/2} e^{-x} \, dx. \qquad (12.25)$$

The integral now has a numerical value, $(\pi)^{1/2}/2$, so that the number of electrons per unit volume is

$$
\begin{aligned}
n &= \frac{C}{2} (kT)^{3/2} \pi^{1/2} \exp\left(-\frac{E_G - E_F}{kT}\right) \\
&= 2 \left(\frac{2\pi mkT}{h^2}\right)^{3/2} \exp\left(-\frac{E_G - E_F}{kT}\right)
\end{aligned}
\qquad (12.26)
$$

using the expression for $C$ in eq. (12.20). We can treat holes in a similar fashion and obtain for the number $p$ of holes per unit volume in the valence band

$$p = 2 \left(\frac{2\pi mkT}{h^2}\right)^{3/2} \exp\left(-\frac{E_F}{kT}\right). \qquad (12.27)$$

In this treatment we have neglected the effective mass $m^*$ (Section 12.2), which should be included in eqs. (12.26) and (12.27). We denote $m_e^*$ as the electron effective mass and $m_h^*$ as the hole effective mass, so that

$$
\begin{aligned}
n &= N_C \exp\left(-\frac{E_G - E_F}{kT}\right) \\
p &= N_V \exp\left(-\frac{E_F}{kT}\right)
\end{aligned}
\qquad (12.28)
$$

where the effective densities of states $N_C$ and $N_V$ in the conduction and valence are given by

$$N_C = 2 \left(\frac{2\pi m_e^* kT}{h^2}\right)^{3/2}$$

and

$$N_V = 2 \left(\frac{2\pi m_h^* kT}{h^2}\right)^{3/2}$$

$$(12.29)$$

Note that we could substitute the conduction band energy $E_C$ for $E_G$ in eq. (12.28) since $E_C = E_G$ in the notation of Fig. 12.9. Equation (12.28) is, of course, the basis of our use of lines at $E_C$ and $E_V$ to denote the conduction and valence bands in schematic diagrams of p-n junctions and transistors. As we show in Section 12.5, eq. (12.28) holds only for $n \ll N_C$ and $p \ll N_V$ or, alternatively, $E_G - E_F \gg kT$ and $E_F \gg kT$.

**TABLE 12.1**  Values for $N_C$, $N_V$, and $n_i$ at 300 K

|  | Si | GaAs | Ge |
|---|---|---|---|
| $E_G$ (eV) | 1.12 | 1.42 | 0.66 |
| $N_C$ (cm$^{-3}$) | $2.8 \times 10^{19}$ | $4.7 \times 10^{17}$ | $1.04 \times 10^{19}$ |
| $N_V$ (cm$^{-3}$) | $1.04 \times 10^{19}$ | $7.0 \times 10^{18}$ | $6 \times 10^{18}$ |
| $n_i$ (cm$^{-3}$) | $1.45 \times 10^{10}$[a] | $1.8 \times 10^6$ | $2.4 \times 10^{13}$ |

*Source:* Sze (1981).
[a]Value for $n_i$ calculated from eq. (12.31) differs from that in Table 12.1.

The intrinsic carrier concentration, $n_i$, can be found from the mass-action relation (Chapter 3),

$$n_i^2 = np$$
$$= N_C N_V \exp\left(-\frac{E_G}{kT}\right) \tag{12.30}$$

and

$$n_i = (N_C N_V)^{1/2} \exp\left(-\frac{E_G}{2\,kT}\right) = N_C \exp\left(-\frac{E_C - E_F}{kT}\right)$$
$$= 4.9 \times 10^{15} \left(\frac{m_e^* m_h^*}{m^2}\right)^{3/4} T^{3/2} \exp\left(-\frac{E_G}{2kT}\right). \tag{12.31a}$$

The location of the Fermi level in intrinsic material is

$$E_F = \frac{E_G}{2} + \frac{3}{4} kT \ln \frac{m_h^*}{m_e^*}. \tag{12.31b}$$

Values for $n_i$ and the densities of states are given in Table 12.1. The density of states in Si is substantially greater than in GaAs, reflecting differences in the effective masses. The differences in $n_i$ values are due to both the $N_C N_V$ product and the values of $E_G$.

## 12.5 Electrons in Metals and Degenerate Semiconductors

For lightly doped semiconductors the conduction band is mostly empty of electrons and we can use Maxwell–Boltzmann statistics. The electrons are treated as a classical free-particle gas. However, the energy distribution of an electron in a metal is not even approximately Maxwellian. In an ideal metal such as Li with a valence of 1, the conduction band is half-full of electrons.

The density of states $g(E)$ is unchanged from that in eq. (12.20), but the probability of occupation $F(E)$ is given by the Fermi–Dirac distribution

$$F_{\text{FD}}(E) = \frac{1}{\exp\left[(E - E_F)/kT\right] + 1} \tag{12.32}$$

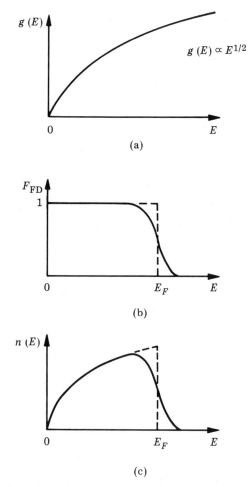

**FIGURE 12.10** For electrons in metals or degenerate semiconductors with $n > N_c$: (a) density of states, $g(E)$; (b) Fermi–Dirac distribution $F_{FD}$ with Fermi energy $E_F$; (c) concentration of electrons, $n(E) = g(E)F_{FD}$.

which has the property that $F_{FD} = \frac{1}{2}$ at $E = E_F$. As shown in Fig. 12.10, $F_{FD}(E)$ has a nearly square shape with $F_{FD}$ nearly unity for energies less than the Fermi level $E_F$. This means that all the states are occupied (one electron per available state according to the Pauli exclusion principle) up to energies near $E_F$. At the Fermi level the probability of occupation is one-half.

The properties of the distribution function can be more easily seen at $T = 0$, where $F_{FD}(E) = 1$ for $E < E_F$ and $F_{FD} = 0$ for $E > E_F$ (Fig. 12.10b, dashed line). The expression [eq. (12.21)] for the number $n$ of electrons per unit volume is

$$n = \int_0^\infty g(E)F_{FD}(E)\, dE = \int_0^{E_F} g(E)\, dE \tag{12.33}$$

where we have chosen the energy at the bottom of the conduction band as zero. Then

$$
\begin{aligned}
n &= 4\pi \left(\frac{2m}{h^2}\right)^{3/2} \int_0^{E_F} E^{1/2}\, dE \\
&= \frac{\pi}{3} \left(\frac{8m}{h^2}\right)^{3/2} E_F^{3/2}.
\end{aligned}
\tag{12.34}
$$

The Fermi energy is

$$
E_F = \frac{h^2}{8m}\left(\frac{3n}{\pi}\right)^{2/3}.
\tag{12.35}
$$

If we consider Al, which has an atomic density $N = 6.02 \times 10^{22}$ atoms/cm$^3$ and assume 3 valence electrons/atom, the value of $E_F$ is

$$
E_F = \frac{(6.6 \times 10^{-34})^2}{8 \times 9.1 \times 10^{-31}} \left(\frac{3 \times 18.06 \times 10^{28}}{\pi}\right)^{2/3} = 18.5 \times 10^{-19}\ \text{J} = 11.6\ \text{eV}.
$$

For Al, the electron energies extend 11.6 eV above the bottom of the band. Table 12.2 lists electron concentrations and values of $E_F$ for several metals.

The wave numbers $k_F$ at the Fermi energy $E_F$ are related to $E_F$ by eq. (12.5),

$$
E_F = \frac{\hbar^2}{2m} k_F^2
\tag{12.36}
$$

so that from eq. (12.35),

$$
k_F = (3\pi^2 n)^{1/3}.
\tag{12.37}
$$

The velocity $v_F$ of the electrons at energy $E_F$ can be found from the momentum $p = \hbar k$,

$$
v_F = \frac{\hbar k_F}{m} = \frac{\hbar}{m} (3\pi^2 n)^{1/3}.
\tag{12.38}
$$

For Al, using values from Table 12.2,

**TABLE 12.2**  Calculated Parameters for Metals at Room Temperature

| Metal | Atomic Concentration, $N$ ($\times 10^{22}$ cm$^{-3}$) | Electron Concentration, $n$ ($\times 10^{22}$ cm$^{-3}$) | Fermi Energy (eV) | Fermi Velocity ($\times 10^8$ cm/s) |
|---|---|---|---|---|
| Ag | 5.85 | 5.85 | 5.48 | 1.39 |
| Al | 6.02 | 18.06 | 11.63 | 2.02 |
| Au | 5.9 | 5.9 | 5.51 | 1.39 |
| Cu | 8.45 | 8.45 | 7.0 | 1.57 |

*Source:* Kittel (1976).

$$v_F = \frac{6.6 \times 10^{-34}}{2\pi \times 9.1 \times 10^{-31}} (3\pi^2 \times 18.06 \times 10^{28})^{1/3}$$

$$= 2 \times 10^6 \text{ m/s} = 2 \times 10^8 \text{ cm/s}.$$

This is a value an order of magnitude greater than the classical free-electron velocity at room temperature.

At elevated temperatures, $T = 300$ to $400$ K, the distribution function $F_{FD}(E)$ becomes rounded near $E_F$ (Fig. 12.10b, solid line), indicating a gradual change in the occupation of states from mostly full to mostly empty. Around $E_F$, the distribution function changes from $F_{FD}(E) = 0.9$ to $F_{FD}(E) = 0.1$ over an energy range equivalent to $4.4kT$. This energy range, about $0.1$ eV at $300$ K, is two orders of magnitude less than $E_F$ in Al.

For electron energies well above the Fermi energy, $E - E_F \gg kT$, the unity term in eq. (12.32) may be neglected, so that

$$F_{FD}(E) \simeq \exp\left(-\frac{E - E_F}{kT}\right) \tag{12.39}$$

which is the same form as the classical Maxwell–Boltzmann distribution function [eq. (12.22)]. We took advantage of this approximation in deriving the expressions [eqs. (12.28) and (12.29)] for the carrier concentrations in lightly doped semiconductors.

For more heavily doped semiconductors where $n \geq N_C$, the effective density of states in the conduction band [eq. (12.29)], the number $n$ of electrons per unit volume is given by

$$n = \int_{E_G}^{\infty} g(E)F_{FD}(E) \, dE$$

$$= N_C \frac{2}{(\pi)^{1/2}} F_{1/2}(\eta_F) \tag{12.40}$$

where the Fermi–Dirac integral $F_{1/2}$ is

$$F_{1/2}(\eta_f) = \int_0^{\infty} \frac{\eta^{1/2} \, d\eta}{1 + e^{\eta - \eta_f}} \tag{12.41}$$

and

$$\eta_f = \frac{E_F - E_G}{kT} \quad \text{and} \quad \eta = \frac{E - E_G}{kT}. \tag{12.42}$$

Values of $F_{1/2}$ are given in Fig. 12.11.

When the Fermi level is greater than $5kT$ above the conduction band edge where $E_C = E_G$,

$$F_{1/2} \simeq \frac{2}{3}\left(\frac{E_F - E_G}{kT}\right)^{3/2} \tag{12.43}$$

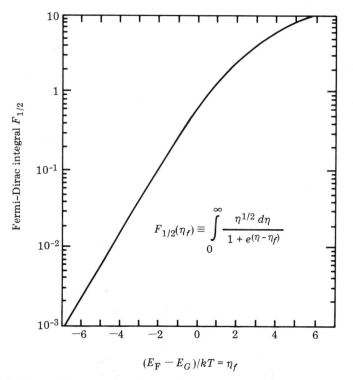

$$F_{1/2}(\eta_f) \equiv \int_0^\infty \frac{\eta^{1/2}\, d\eta}{1 + e^{(\eta - \eta_f)}}$$

$$(E_F - E_G)/kT = \eta_f$$

**FIGURE 12.11**  Values of the Fermi–Dirac integral $F_{1/2}$ versus $(E_F - E_G)/kT$.

and from eqs. (12.29) and (12.40),

$$
\begin{aligned}
n &= 2 \left( \frac{2\pi m_e^* kT}{h^2} \right)^{3/2} \cdot \frac{2}{\pi^{1/2}} \cdot \frac{2}{3} \left( \frac{E_F - E_G}{kT} \right)^{3/2} \\
&= \frac{\pi}{3} \left( \frac{8m_e^*}{h^2} \right)^{3/2} (E_F - E_G)^{3/2}.
\end{aligned}
\tag{12.44}
$$

Equation (12.44) is the same as derived for metals [eq. (12.34)] except for the use of an effective mass and the choice of the zero of energy at $E_V$ ($E_C - E_V = E_G$). Under these conditions the semiconductor is said to be degenerate, and often the symbol $n^+$ is used instead of $n$ to denote a heavily doped $n$-type semiconductor.

When the Fermi level is several $kT$ below the conduction band,

$$
F_{1/2} = \frac{(\pi)^{1/2}}{2} \exp \left( \frac{E_F - E_G}{kT} \right)
\tag{12.45}
$$

and the carrier concentration $n$ returns to that given in eq. (12.28) for the non-degenerate lightly doped semiconductor. At $E_F - E_G = 6\,kT$, $F_{1/2} = 10$ and $n = N_C(2/\sqrt{\pi})F_{1/2} \approx 10N_C$. For $E_G - E_F = 2kT$, $F_{1/2} = 0.1$, and $n \approx 0.1N_C$. Within this range, the graphical values for the Fermi–Dirac integral given in Fig. 12.11 should be used.

## 12.6 Electrons, Holes, and Electric Fields

We have shown in Section 12.2 that we can represent the energy $E$ and momentum $p$ of an electron in a crystal by an $E$–$k$ diagram (where $\mathbf{p} = \hbar\mathbf{k}$, with $\mathbf{k}$ the wave vector). If we have an electron in a crystal subject to the force $\mathbf{F}$ caused by an applied electric field $\mathscr{E}$ ($\mathbf{F} = -e\mathscr{E}$), then

$$\mathbf{F} = \hbar\frac{d\mathbf{k}}{dt}. \qquad (12.46)$$

The electron in the crystal is subject to forces from the crystal lattice as well as from the external field, so that $\hbar d\mathbf{k}/dt$ is used rather than $d(m\mathbf{v})/dt$ as in free space.

In one dimension, Fig. 12.12 shows an electron subject to a force $F$ due to an electric field. During the time $\tau$ between collisions, the value of $k$ changes from $k(0)$ to $k(\tau)$,

$$\Delta k = k(\tau) - k(0) = -e\frac{\tau}{\hbar}. \qquad (12.47)$$

If we apply an electric field of $10^4$ V/m directed toward negative values of $k$ and assume a mean time $\tau = 10^{-12}$ s, for an electron at $k = 0$ and $t = 0$, the wave number $k$ is

$$k = \frac{1.6 \times 10^{-19} \times 10^4 \times 10^{-12} \times 2\pi}{6.6 \times 10^{-34}} = 1.52 \times 10^7/\text{m}$$

which leads to an energy $E$,

$$E = \frac{\hbar^2 k^2}{2m} = \frac{(6.6 \times 10^{-34})^2 \times (1.52 \times 10^7)^2}{2(2\pi)^2 \times 9.1 \times 10^{-31}}$$
$$= 1.4 \times 10^{-24} \text{ J} = 8.8 \times 10^{-6} \text{ eV}.$$

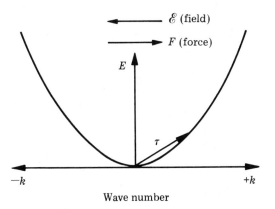

**FIGURE 12.12** *E–k* diagram for an electron subject to a force *F* at *k* = 0 which changes to *k*(τ), where τ is the time between collisions.

As expected, the application of modest values of the electric field does not make a major change in the energy distribution of electrons in an $E-k$ diagram (the scale in Fig. 12.12 is blown up for clarity) and the electrons reside near the minimum in the $E-k$ diagram. This is not true under high-field conditions, where the electrons can gain appreciable energy between collisions.

The valence band is associated with the covalently bound outer electrons. If there are no vacant bonds, the valence band is completely filled. In an $E-k$ diagram (Fig. 12.13a), for every electron with a given $k$, for example $k_1$, there is an electron with an equal but oppositely directed value of $k$, $-k_1$. The sum of the $k$ values = 0 and there are no vacant states. If an electric field is applied, the current is zero (assuming no free electrons in the conduction band) because there are no available empty states to make a transition into to gain momentum. This conclusion could also be reached from the simple model of a pure semiconductor with no broken bonds discussed in Chapter 3.

If there is a broken bond, so that there is an electron missing (or hole present) at $k_1$, for example, the band has a total momentum value of $-k_1$. The hole has momentum $k_h = -k_1$ which is the $k$ value equal and opposite to that of the missing electron.

In an energy-band diagram of the valence band, the states are filled with electrons from the bottom up. An electron near the top of the valence band is less tightly bound than those farther down in the band; that is, it takes less work to remove an electron from the top of the band than from lower in the band. The energy of the hole is opposite to that of the electrons, so that the lowest energy of the hole is at the top of the valence band. In analogy to water, holes just like bubbles (absent water molecules) float to the top. In a semiconductor with both

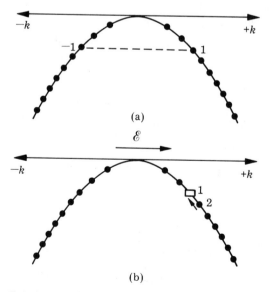

FIGURE 12.13   $E-k$ diagram for electrons and missing electrons (holes) in the valence band.

holes and electrons, the electrons would occupy states at the bottom of the conduction band and the holes would occupy states at the top of the valence band.

The effective mass [eq. (12.7)] is inversely proportional to the curvature of the band:

$$m^* = \hbar^2 \left(\frac{d^2E}{dk^2}\right)^{-1}$$

so that for electrons in the valence band (where $E$ decreases with an increase in $k$) the effective mass is negative. In the upward-curving bands in the conduction band the effective mass of the electrons is positive. Since the energy of the hole is opposite in sign to that of the missing electron, the effective mass of the hole is positive; that is, the hole energy increases with an increase in $k$.

If we consider an electric field $\mathscr{E}$ applied in the positive $x$ direction, the force on the electrons is in the negative $k$ direction. An electron in state $k_2$ (Fig. 12.13b) would make a transition to the empty state at $k_1$, moving the hole to state $k_2$. Thus the hole velocity increases in the direction of the electric field and responds as if the hole had a positive charge.

In summary, we can treat the missing electron in the valence band as either the ensemble of all the remaining electrons or as a hole that has a positive charge $e$ and positive mass with the magnitude of the mass, the effective mass $m_h^*$, given by the curvature of the bands. That is,

$$\text{hole} \longrightarrow \text{positive charge and positive mass.} \tag{12.48}$$

# 12.7 Direct and Indirect Gap Semiconductors, GaAs and Si

We have treated electrons in crystals on the basis of $E$–$k$ relationships,

$$E = \frac{\hbar^2 k^2}{2m^*}$$

where we assumed spherical energy surfaces ($k^2 = k_x^2 + k_y^2 + k_z^2$) with the curvature of the conduction bands determined by $m^*$, which differs from one material to the next. The effective masses are then different in different semiconductors.

There are further complications when we treat real crystals. The band structure for GaAs is shown in Fig. 12.14 in the region near $k = 0$. There is no problem with the conduction band, which has a curvature equivalent to $m^* = 0.07$ m; however, the valence bands are threefold. The curvatures of the upper two are different and are denoted the heavy hole (hh) and light hole (lh) bands. There is a lower band displaced in energy (split off in energy from the hh and lh bands) due to spin-orbit interactions. This is denoted the split-off hole (soh) band. Normally, holes only occupy the light and heavy bands so that the effective mass of the valence band can be expressed for density of state $N_V$ calculations as

$$m_h^* = [(m_{lh}^*)^{3/2} + (m_{hh}^*)^{3/2}]^{2/3}. \tag{12.49}$$

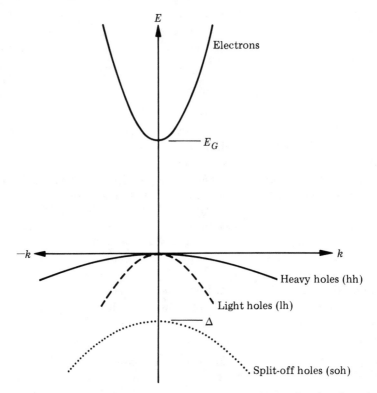

**FIGURE 12.14** *E–k* diagram for a direct-gap semiconductor showing three hole masses: heavy holes, light holes, and split-off holes.

Values of the light-hole and heavy-hole effective masses $m_{lh}^*$ and $m_{hh}^*$ are given in Table 12.3 for several III–IV semiconductors.

Gallium arsenide as well as the semiconductors listed in Table 12.3 are referred to as direct-gap semiconductors, because the minimum in the conduction band and the maximum in the valence band are located at $k = 0$. When holes and electrons

**TABLE 12.3** Effective Mass Ratios $m^*/m$ of Electrons and Holes

| Semiconductor | Electron, $m_e^*$ | Heavy Hole, $m_{hh}^*$ | Light Hole, $m_{lh}^*$ |
|---|---|---|---|
| GaAs | 0.07 | 0.45[a] | 0.08[a] |
| GaSb | 0.047 | 0.3 | 0.06 |
| InP | 0.073 | 0.4 | 0.08 |
| InAs | 0.026 | 0.41 | 0.025 |
| InSb | 0.015 | 0.39 | 0.021 |

*Source:* Kittel (1976).

[a]From Sze (1981).

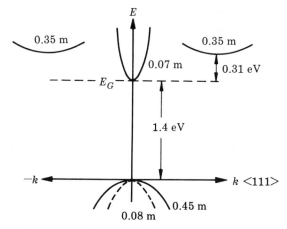

**FIGURE 12.15** *E–k* diagram for gallium arsenide showing the subsidiary valley in the conduction band.

are present, the electrons can make a direct transition to a hole state without a change in momentum.

The conduction band of GaAs has another minimum located away from $k = 0$ along the <111> direction, as shown in Fig. 12.15. The energy separation between the conduction band minimum and the upper band (or upper valley) minimum is about 0.31 eV. The band curvature in the upper valley is greater than that in the lower valley (at $k = 0$). The effective mass is 0.35 m, compared to that, 0.07 m, around $k = 0$.

Under low to moderate electric field conditions, electron conduction in GaAs is due to electron transitions in the low-effective-mass, high-mobility valley. Under high electric fields greater than about $3 \times 10^3$ V/cm, electrons can be transferred from the high mobility-valley to the upper valley, where the electron mobility is lower, because of the field-induced increase in kinetic energy. This high-field transferred electron effect produces a decrease in the drift velocity, as shown in Fig. 12.16 for both GaAs and InP (the valley separation is 0.53 eV for InP). The decrease in velocity with increase in electric field gives rise to a negative differential resistivity in GaAs which has been exploited in the design of microwave devices.

In Si as well as Ge and some of the III–V semiconductors, such as GaP and AlSb, the minimum in the conduction band is located away from $k = 0$ (in Si along the <100> axes in $k$ space), as shown in Fig. 12.17. In thermal equilibrium, electrons occupy states at the minima of the conduction band. There is no net current because for every electron with a positive $k$ value in one minimum there is an electron with a negative $k$ value in another minimum.

In Si, constant-energy surfaces in the conduction band are ellipsoids, with the six ellipsoids located along the <100> axes. With an ellipsoid the effective mass is characterized by the curvature along the long axis, the longitudinal effective mass $m_l^*$, and the transverse effective mass $m_t^*$. For calculation of the effective

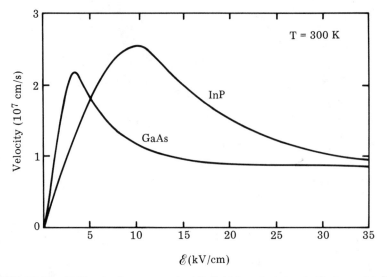

**FIGURE 12.16**   Drift velocity versus electric field for electrons in GaAs and InP. [From Sze (1981).]

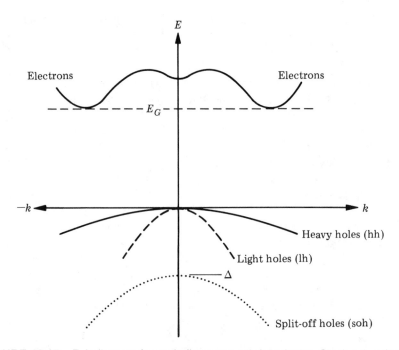

**FIGURE 12.17**   $E$–$k$ diagram for an indirect-gap semiconductor, Si, showing that the minimum in the conduction band occurs at $k$ values away from $k = 0$.

**TABLE 12.4** Effective Mass Ratios $m^*/m$ for Electrons and Holes in Si and Ge[a]

|  | Electrons | | Holes | |
|---|---|---|---|---|
|  | **Longitudinal** | **Transverse** | **Heavy** | **Light** |
| Si | 0.98 | 0.19 | 0.49 | 0.16 |
| Ge | 1.64 | 0.082 | 0.28 | 0.04 |

*Source:* Sze (1981).

density of states $N_C$ in the conduction band, the effective mass of the electron $m_e^*$ is

$$m_e^* = (m_l^* m_t^{*2})^{1/3}. \qquad (12.50)$$

The effective mass ratios ($m^*/m$) for electrons and holes are given in Table 12.4 for Si and Ge. In calculating the density of states in Si, for example (Table 12.1), we accounted for the six ellipsoids by multiplying $N_C$ by 6; for Ge we multiply $N_C$ by 4 because of the eight half-ellipsoids along the <111> axes.

Silicon is an example of an indirect-gap semiconductor (along with Ge, GaP, and AlSb). The electrons at the conduction band minimum are displaced in $k$ space from the holes at the maximum in the valence band. Electrons cannot make a direct transition to the valence band without interacting with lattice atoms in order to conserve momentum.

## 12.8 Optical Transitions and Phonons

If light is incident on a high-purity semiconductor crystal at photon energies, $E = h\nu$, that are less than the band gap, the light passes through without attenuation except for reflection at the surfaces. When the energy is greater than the band gap, the photons can be absorbed by the promotion of an electron from the valence band to the conduction band, as shown in Fig. 12.18b for the direct-band-gap semiconductor GaAs. The transition is indicated by a vertical line because there is essentially no change in the wave number $k$ of the electron. The photon momentum is negligible compared to that of the electron at photon energies around 1 eV (see Section 12.2). Consequently, the optical absorption increases rapidly with photon energy above the threshold at $h\nu = E_G$ where the direct photon transitions set in.

The situation is not as simple for indirect-gap semiconductors such as Si, as shown in Fig. 12.19. At photon energies $h\nu = E_G$ there are no conduction band states available at $k = 0$, so that a valence-band electron cannot make a direct transition to the conduction band. Both energy and momentum must be conserved in the transition involving the absorption of a photon. Momentum can be conserved by interaction with the lattice atoms so that a lattice vibrational wave or phonon is generated or absorbed.

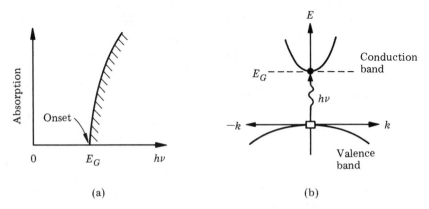

**FIGURE 12.18** For a direct-gap semiconductor: (a) absorption of photons versus photon energy showing a steep onset at $h\nu = E_G$; (b) absorption process in an $E$–$k$ diagram.

Phonons represent periodic variations in the lattice spacings and as traveling waves can move through the crystal at velocities equal to the velocity of sound (about $10^5$ cm/s). There are a range of phonon energies and wave numbers that can be generated in crystals. Since the mass of the lattice atoms is large, the phonon wave numbers $K$ can be large, but the energies $E_p$ small ($\approx 0.01$ to $0.03$ eV). Thus the phonon energy is much less than the band gap $E_G$.

In the indirect-gap material at or just above the threshold of absorption at $h\nu = E_G$, for photon absorption we must conserve momentum,

$$k(\text{photon}) = k_e + K \tag{12.51}$$

and energy,

$$h\nu(\text{photon}) = E_G + E_p \tag{12.52}$$

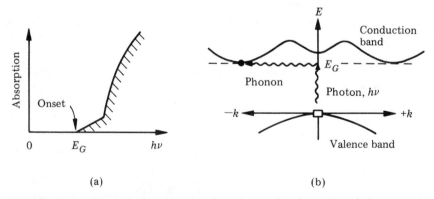

**FIGURE 12.19** For an indirect-gap semiconductor: (a) absorption of photons versus photon energy showing a shallow onset at $h\nu = E_G$ followed by a steep direct absorption; (b) two-step absorption process showing the change in $E$ due to the photon and change in $k$ due to the phonon.

where $k_e$ is the electron wave number near the minimum in the conduction band. The wave number of the photon is negligible, as is the phonon energy, so that optical absorption can set in at $h\nu = E_G$. However, as shown in Fig. 12.19, the absorption is not very strong, as it involves a three body interaction (electron, photon, and phonon).

The absorption coefficient per unit length $\alpha$ is given in Fig. 12.20 for Si and GaAs. The band-gap energy of Si is less than that of GaAs, so the onset of optical absorption is at lower energies in Si than in GaAs. However, since GaAs is a direct-gap material, the absorption coefficient is nearly an order of magnitude greater than that of Si up to photon energies near 3 eV where direct transitions are allowed in Si. The strong optical absorption of GaAs makes it a good candidate for solar cell applications.

Some of the III–V semiconductors such as GaP and AlAs are indirect-gap semiconductors. The GaAs–AlAs system shows a transition from direct gap (GaAs) to indirect gap (AlAs) as the amount of Al in $Al_xGa_{1-x}As$ is increased. Figure 12.21a shows the $E$–$k$ diagram for the conduction band in GaAs along the $<100>$ and $<111>$ directions (Fig. 12.15 showed the $E$–$k$ diagram only along $<111>$ direction). In GaAs the direct transitions occur at $k = 0$ where the energy minima is denoted by $\Gamma$. There are two subsidiary values along the $<111>$ and $<100>$ directions, denoted $L$ and $X$, respectively. As the fraction $x$ of Al in $Al_xGa_{1-x}As$

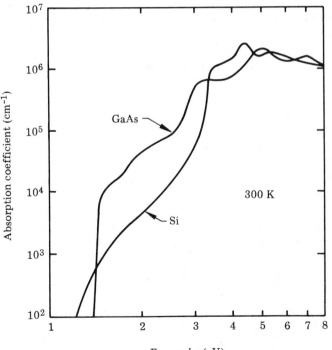

**FIGURE 12.20**  Absorption coefficient versus photon energy for Si and GaAs at 300 K. [Adapted from Sze (1981).]

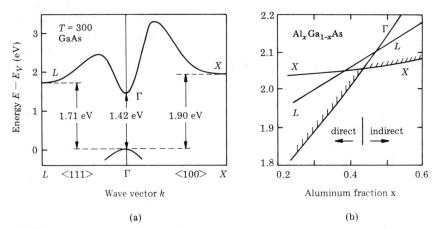

**FIGURE 12.21** (a) E–k diagram for GaAs along ⟨111⟩ and ⟨100⟩ directions. [From Sze, (1981).] (b) Energies of conduction band minima versus Al fraction in Al$_x$Ga$_{1-x}$As. [From Dingle et al. (1977).]

increases, the energies of these minima at $\Gamma$, $X$, and $L$ increase as shown in Fig. 12.21b for $x$ values between 0.2 and 0.6. The value of the energy difference between $E_V$ and the minima increases faster for $\Gamma$ (at $k = 0$) than for $X$ (along the ⟨100⟩ direction). At Al fractions of about 0.4, the $\Gamma$ and $X$ lines cross and AlGaAs becomes an indirect-gap material.

At this point it is appropriate to give the dispersion relation $\omega$ versus $K$ for phonons, where we have used $K$ for the wave number of the phonons ($K = 2\pi/\lambda$). In Chapter 2 we discussed the continuum model, which we gave a linear dispersion relation,

$$\omega = v_s K$$

where $v_s$ is the velocity of sound. The acoustic wave was for the longitudinal mode, where the atom displacement is parallel to the propagation direction. This is denoted LA (longitudinal acoustic). There are also transverse, or shear, waves, where the atom displacement is transverse to the wave velocity. These are known as TA (transverse acoustic) modes.

In the diamond lattice, Si and GaAs, with two atoms per primitive cell, there are two branches in the dispersion relation, acoustical and optical, as shown in Fig. 12.22. In the optical mode the atoms vibrate against one another so that the center of mass is fixed (their motion is out of phase by $\pi$), whereas for the acoustic motion, the atoms move nearly in phase in the same direction. Again, there are longitudinal and transverse optical modes (LO and TO).

A feature of the optical branch, both for transverse and longitudinally polarized modes, is that there is a phonon frequency $\nu_{ph}$ at $K = 0$. The energies are larger than $kT$ (0.026 eV at 300 K), and for the longitudinal optical (LO) phonon in Si,

$$h\nu_{ph} = 0.063 \text{ eV}$$

which corresponds to a frequency of $1.5 \times 10^{13}/\text{s}$.

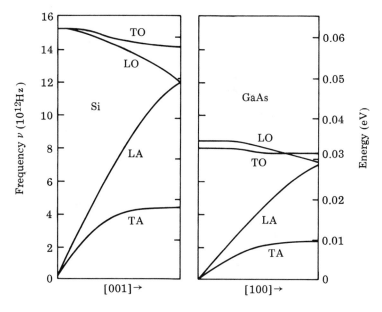

Phonon wave number, $K$

**FIGURE 12.22** Phonon spectrum in Si and GaAs. [From Sze (1981).]

## 12.9 Ohmic Contacts

Ideally, the contact to a *p-n* junction or bipolar transistor should impose no barrier to the flow of charge carriers. Such contacts are called ohmic contacts because when they are applied to a uniform semiconductor sample, the measured current–voltage characteristics show linear current–voltage behavior, with the measured resistance directly related to the sample resistivity, with no resistance attributed to the metal–semiconductor contact. The ohmic contact, by definition, offers negligible resistance to current flow compared to the bulk.

The performance of the integrated circuit chip with thousands of metal–semiconductor connections depends crucially on well-behaved ohmic contacts. There is little room for error with the reduced dimensions in large-scale integrated circuits, where the contact layer can only extend a few tens of nanometers into the semiconductor. The conventional ohmic contact structure to *n*-type semiconductors in integrated circuits is shown in Fig. 12.23. It consists of a metal layer in contact with a heavily doped, denoted $n^+$, *n*-type layer. The metal–semiconductor contact itself may be a silicide layer (Chapter 10), and there may also be a diffusion barrier interposed between the metal and silicide to prevent the metal from penetrating the silicide layer.

As described in Chapter 4, metal films deposited on semiconductors produce rectifying contacts. The resistance of the contact at zero-applied voltage, the contact resistance $R_C$, can be defined as

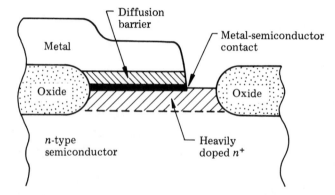

**FIGURE 12.23**  Ohmic contact to an integrated circuit formed by a metallic layer in contact with a heavily doped, $n^+$, layer with a diffusion barrier separating the metallic contact layer from the metal interconnect.

$$R_C = \left( \frac{dJ}{dV} \right)^{-1}_{V=0} \tag{12.53}$$

where $R_C$ is in units of $\Omega \cdot cm^2$. For the Schottky barrier diode current–voltage relation given in Chapter 4,

$$R_C = \frac{kT}{eJ_s} = \frac{k}{eAT} \exp\left( \frac{e\phi_B}{kT} \right). \tag{12.54}$$

The value of $A$, the Richardson constant, is on the order of 100 A/cm$^2$, so that for a barrier height of about 0.6 eV, the contact resistance will be about 10 $\Omega \cdot cm^2$. This value is orders of magnitude too large for integrated-circuit applications.

The contact resistance can be reduced by choosing a metal–semiconductor system with a lower barrier height. This option may not be available because the metal–semiconductor contact is chosen due to other constraints imposed by the metallization system. In that case, rather than use a contact where current transport is determined by thermionic emission over the barrier, the metal is placed on a heavily doped semiconductor where current transport is determined by penetrating, or tunneling through, the barrier.

In heavily doped $n$-type semiconductors the number of electrons is greater than the effective density of states $N_C$ in the conduction band. In contrast to the case where $n \ll N_C$ and the Fermi level $E_F$ lies below the conduction band, for the degenerate case the Fermi level lies above the bottom of the conduction band (Fig. 12.24), and Fermi–Dirac statistics must be used to describe the occupancy of the states in the conduction band. As a first approximation we consider that all donors are ionized so that $n = N_D$. In practice, this condition is not met at high doping concentrations: in As-doped Si, the maximum carrier concentration is about $2 \times 10^{20}/cm^3$, and in Si-doped GaAs, the maximum is about $2 \times 10^{18}/cm^3$, despite the fact that much higher concentrations of donors are present. In both semiconductors the maximum carrier concentration is well in excess of the value of $N_C$: $N_C(Si) = 2.8 \times 10^{19}/cm^3$ and $N_C(GaAs) = 4.7 \times 10^{17}/cm^3$ (Table 12.1).

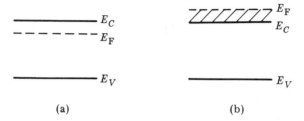

(a)                                   (b)

**FIGURE 12.24**   Relative positions of the Fermi level $E_F$ and conduction band edges $E_C$: (a) in a nondegenerate semiconductor, $N_D \ll N_C$, with Maxwell–Boltzmann statistics; (b) in a degenerate semiconductor, $N_D > N_C$, with Fermi–Dirac statistics.

When a metal film is deposited on a degenerate semiconductor, the potential barrier in the semiconductor is approximately equal to the barrier height $\phi_B$. In thermal equilibrium (Fig. 12.25), the width $d_n$ of the depletion region is (Chapter 4)

$$d_n = \left(\frac{2\epsilon V_0}{eN_D}\right)^{1/2} \simeq \left(\frac{2\epsilon\phi_B}{eN_D}\right)^{1/2}. \tag{12.55}$$

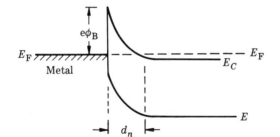

(a) Thermal equilibrium

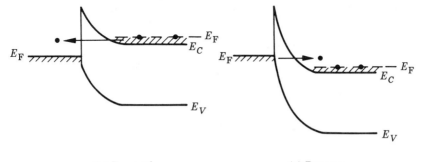

(b) Forward                         (c) Reverse

**FIGURE 12.25**   Energy-level diagrams for a metal–Schottky barrier contact on a heavily doped, degenerate, n-type semiconductor in (a) equilibrium showing the barrier height $e\phi_B$, (b) forward bias, and (c) reverse-bias, showing electron tunneling through the barrier.

For a permittivity $\epsilon = 10^{-12}$ F/cm, $\phi_B = 0.6$ V, and $N_D = 10^{20}$/cm$^3$, the width of the depletion region is about 2.5 nm. For potential barriers this thin ($<5$ to 10 nm) electrons have a finite probability of transmission through, rather than over, the barrier. This quantum-mechanical phenomenon is called tunneling.

Under forward bias, the states in the semiconductor are shifted upward in energy compared to those in the metal (Fig. 12.25b). Electrons in the filled states above $E_C$ in the semiconductor are shifted above the Fermi level in the metal. Consequently, these electrons can tunnel through the barrier into the unoccupied states in the metal. The current density can be expressed

$$J_T = envT_t \tag{12.56}$$

where $v$ is the thermal velocity of the electrons and $T_t$ is the tunneling transition probability. Under reverse bias (Fig. 12.25c), the states in the semiconductor are shifted downward and the electrons in the metal can tunnel into the empty states in the conduction band of the semiconductor.

This procedure of forming a metallic layer on a heavily doped (degenerate) layer produces ohmic contacts with contact resistance values of $10^{-6} \, \Omega \cdot$ cm$^2$ if $N_D = 10^{20}$/cm$^3$. This is the standard procedure for forming contacts in silicon integrated circuits. In GaAs it is difficult to make heavily doped $n$-type layers, and conventional metal to $n$-type GaAs contacts do not have low contact resistance. Graded-band gap structures can be used as described in Section 14.5.

## 12.10 Tunnel Diodes

The tunneling mechanism is also illustrated by the current–voltage characteristics of a $p$-$n$ junction formed between degenerate $p$-type and $n$-type semiconductors. As shown in Fig. 12.26, the Fermi level $E_F$ lies below the valence-band edge $E_V$, so that there is a region of unoccupied or empty states between $E_F$ and $E_V$. In the $n$-type material, the valence band is fully occupied and the states between the conduction-band edge $E_C$ and the Fermi level are also occupied. The occupancy of the states in both $p$- and $n$-type material are described by Fermi–Dirac statistics.

When the degenerate $p$-type and $n$-type regions are joined (Fig. 12.27a), the Fermi level is constant throughout the junction, and a potential barrier is formed

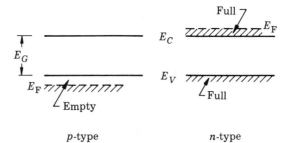

$p$-type $\qquad\qquad$ $n$-type

**FIGURE 12.26** Degenerate $p$-type and $n$-type semiconductors showing the position of the Fermi level $E_F$ and the occupation of the bands.

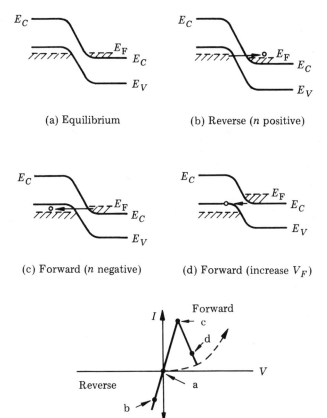

(a) Equilibrium        (b) Reverse ($n$ positive)

(c) Forward ($n$ negative)        (d) Forward (increase $V_F$)

(e) $I$-$V$ curve

**FIGURE 12.27** Tunnel diode junction formed between degenerate $n$-type and $p$-type material showing the energy levels for (a) equilibrium, (b) reverse bias with the $n$-type material with positive applied voltage, (c) small forward voltage, $n$-type material with negative applied voltage, and (d) an increase in forward voltage along with (e) the associated $I$–$V$ characteristic.

just as in the case of the $p$-$n$ junction formed with nondegenerate material. For an abrupt junction with a sharply defined transition between the $n$- and $p$-type material, the depletion layer will be narrow, $d < 10$ nm. With a small forward or reverse bias, the shift in the bands will lead to filled states opposite empty states, as shown in the figure. Since the filled and empty states are separated by only a narrow potential barrier, the electrons can tunnel from filled to empty states.

Under reverse bias ($n$-type semiconductor positive; Fig. 12.27b), electrons from the full valence states below $E_F$ in the $p$-type material can tunnel to the empty conduction-band states above the Fermi level in the $n$-type material. As the reverse bias is increased, the shift in the bands place more filled states in the $p$-type material opposite empty states. Thus the current due to tunneling of electrons increases with increasing reverse-bias voltage.

Under forward bias the situation is different because there is only a narrow band of empty states in the *p*-region opposite a narrow band of filled states in the *n*-region. When the forward bias is initially increased (Fig. 12.27c), these empty and filled state are aligned in energy and the forward current increases as shown in the *I–V* curve in Fig. 12.27e. With further increase in forward bias the current decreases because there are a fewer filled states aligned with empty states (Fig. 12.27d). The decrease of tunneling current with increased forward-bias voltage produces a region of negative slope—a negative-resistance region.

If the forward-bias voltage is increased further, the tunneling component is negligible when the filled and empty bands pass each other, but the current increases, as shown by the dashed line in the *I–V* curve. In this case the current is dominated by electrons in the *n*-region surmounting the potential barrier and being injected in the *p*-region while the holes in the *p*-region are injected into the *n*-region.

## REFERENCES

Ashcroft, N. W., and N. D. Mermin, *Solid State Physics*, Holt, Rinehart and Winston, New York, 1976.

Blakemore, J. S., *Solid State Physics*, 2nd ed., Cambridge University Press, Cambridge, 1985.

Dingle, R., R. A. Logan, and J. R. Arthur, Jr., *GaAs and Related Compounds (Edinburgh)*, Institute of Physics Conference Series 33a, 1977.

Howes, M. J., and D. V. Morgan, eds., *Gallium Arsenide*, Wiley, Chichester, West Sussex, England, 1985.

Kittel, C., *Introduction to Solid State Physics*, 5th ed., Wiley, New York, 1976.

Pierret, R. F., *Advanced Semiconductor Fundamentals*, Addison-Wesley, Reading, Mass., 1987.

Smith, R. A., *Semiconductors*, 2nd ed., Cambridge University Press, Cambridge, 1979.

Solymar, L., and D. Walsh, *Lectures on the Electrical Properties of Materials*, 3rd ed., Oxford University Press, Oxford, 1984.

Sze, S. M., *Physics of Semiconductor Devices*, 2nd ed. Wiley, New York, 1981.

Tipler, P. A., *Modern Physics*, Worth, New York, 1978.

## PROBLEMS

**12.1.** (a) An electron is confined in a one-dimensional potential well where $L = 2 \times 10^{-10}$ m ($L = 2$ Å). Calculate the energies $E_1$ for $n = 1$ and $E_2$ for $n = 2$ in eV.

(b) The energy of free electrons in a cube of length $L = 10^{-2}$ m is given by $E = h^2/8$ mL$^2$ ($n_x^2 + n_y^2 + n_z^2$). If $n_x^2 = n_y^2 = n_z^2$, calculate the value of $n_x$ for electrons with energy $= 1.0$ eV.

**12.2.** (a) Calculate the Fermi energy $E_F$ in eV at $T = 0$ in the free-electron model for a metal with $4 \times 10^{22}$ atoms/cm$^3$ for the case where there is (1) one electron/atom and (2) two free electrons/atom.

(b) What are the wave numbers, $k_F$?

(c) What are the velocities, $v_F$?

**12.3.** (a) Use the Fermi function $F_{FD}$ to find the temperature at which there is a 1% probability that a state with an energy 0.15 eV above the Fermi energy $E_F$ (i.e., $E - E_F = 0.15$ eV) will be occupied by an electron.

(b) For Si ($N_C = 2.8 \times 10^{19}$/cm$^3$) and GaAs ($N_C = 4.7 \times 10^{17}$/cm$^3$) at 300 K, at what electron concentrations will the Fermi level lie $6kT$ and $1kT$ below the conduction-band edge?

**12.4.** Consider conduction in metals

(a) For an electric field $\mathscr{E}$ of 10 V/cm and a mean time $\tau$ between collisions of $10^{-13}$ s, what is the change $\Delta k = e\mathscr{E}\tau/\hbar$?

(b) For a value of the Fermi energy $E_F = 4$ eV, what is the value of the wave number $k_F$ using the relation $E_F = \hbar^2 k_F^2/2m$? (Note that $\Delta k \ll k_F$ for only a small change in the electron energy distribution.)

(c) What is the mobility of the electrons?

(d) What is the number of electrons/cm$^3$?

(e) What is the value of the resistivity $\rho$?

**12.5.** Consider two semiconductors with $E$–$k$ diagrams as shown. Both have $10^{17}$ electrons/cm$^3$ in the conduction band and have the same mean time $\tau$ between collisions.

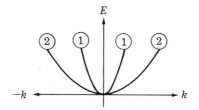

(a) In which case (1, 2, or both the same) will the electrons have the highest mobility?

(b) In which case will the effective density of states $N_C$ in the conduction band be greatest?

(c) In which case will the Fermi level be located closest to the conduction band?

**12.6.** (a) For a semiconductor, the density of states $N_V$ in the valence band is given by $N_V = 2\,(2\pi mkT/h^2)^{3/2}$. Evaluate $N_V$ at 500 K for $m$ equal to the free-electron mass.

(b) A semiconductor at $T = 500$ K has an effective mass for electrons of $0.2m$, where $m$ is the free-electron mass (effective mass $m^* = m$ for holes).

 (1) Calculate the effective density of states $N_C$ for electrons in the conduction band.

 (2) If the number of electrons equals the number of holes, will the position of the Fermi level be at the middle of the energy gap, closer to the conduction band, or closer to the valence band?

 (c) A semiconductor with an effective density of states $N_C$ in the conduction band of $10^{19}/cm^3$ has an electron concentration of $10^{16}/cm^3$. If the effective mass of the electrons increases, will the Fermi level shift toward or away from the conduction band or remain unchanged if the electron concentration is unchanged?

**12.7.** In a semiconductor with an $E–k$ diagram as shown, and with $10^{17}$ electrons/cm$^3$ and $10^6$ holes/cm$^3$:

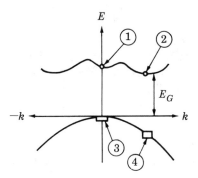

 (a) Which position (1 or 2) would be occupied by electrons, and which (3 or 4) by holes?

 (b) If electrons occupied position 1, what would be the most likely transition for recombination with holes, 1 to 3 or 1 to 4?

 (c) If electrons occupied position 2, is the most likely transition for recombination with holes 2 to 3 or 2 to 4 (if there were holes at both 3 and 4)?

 (d) For part (c), what would be the most likely transition if hole occupation of 3 and 4 followed Fermi–Dirac statistics?

**12.8.** In a one-dimensional lattice with spacing $a$ between atoms, electrons with wave number $k = \pi/a$ are reflected and cannot propagate. If $a = 1$ Å $(10^{-8}$ cm), compare the value of $k = \pi/a$ with the value of $k$ for electrons with velocities $v$ of $10^3$ cm/s.

**12.9.** What is the probability $F$ that a state $2kT$ and a state $4kT$ above the Fermi level will be occupied
 (a) using Fermi–Dirac statistics?
 (b) using Maxwell–Boltzmann statistics?

**12.10.** Consider a semiconductor with equal hole and electron effective masses equal to $0.1m$ (where $m$ is the free-electron mass) and with equal mobilities $= 100$ cm$^2$/V $\cdot$ s. The band gap $= 0.9$ eV and $kT = 0.03$ eV.

(a) For the intrinsic case (no dopants):
  (1) What is the intrinsic carrier concentration?
  (2) What is the resistivity?
  (3) Does the Fermi level $E_F$ lie below, above, or at the middle of the gap?
  (4) If the temperature is increased:
    (i) Does $n_i$ increase, decrease, or remain unchanged?
    (ii) Does $E_F$ increase, decrease, or remain unchanged?
(b) Acceptors are added so that the Fermi level is $3kT$ above the valence band.
  (1) What is the hole concentration?
  (2) What is the intrinsic carrier concentration?
  (3) What is the electron concentration?
  (4) What is the resistivity?

**12.11.** Consider real semiconductors with light and heavy holes and spherical or ellipsoidal conduction band minima, as in the text.
  (a) Calculate $N_C$ and $N_V$ for Si and GaAs at 300 K and compare with tabulated values.
  (b) If there are $4 \times 10^{16}$ electrons/cm$^3$, what is the position of the Fermi level below $E_C$ for the two semiconductors?
  (c) If the semiconductors are intrinsic, does the Fermi level lie above or below $E_G/2$, and by how much?

**12.12.** Compare Ge, an indirect-band-gap material, with energy gap $E_G = 0.66$ eV, density of states $N_C$ in the conduction band $= 1 \times 10^{19}$/cm$^3$, with GaAs—a direct-band-gap material with $E_G = 1.4$ eV and $N_C = 5 \times 10^{17}$/cm$^3$. The samples are at 300 K.
  (a) Which makes the most efficient light-emitting diode?
  (b) Which has the lowest intrinsic carrier concentration?
  (c) Which has the lowest effective mass for electrons?
  (d) Which has the greatest absorption of light at a wavelength of 1.2 μm?
  (e) Which has the highest mobility for electrons?

**12.13.** (a) A semiconductor has an effective density of states $N_C$ of $10^{19}$/cm$^3$, a donor concentration of $2 \times 10^{17}$/cm$^3$, a carrier concentration of $10^{17}$ electrons/cm$^3$, and a hole concentration of $2 \times 10^5$/cm$^3$. How far below the conduction band in eV will the Fermi level $E_F$ be at 300 K?
  (b) Will the energy level $E_D$ lie above, just at, or below the Fermi level?
  (c) What is the value of the intrinsic carrier concentration $n_i$?
  (d) Will the number of holes increase, decrease, or remain fixed if
    (1) the temperature decreases?
    (2) the donor concentration decreases?
    (3) the energy gap $E_G$ decreases?

**12.14.** You have a sample of germanium (Ge) with a band gap $E_G = 0.66$ eV and an effective density of states for electrons and holes $= 10^{19}$/cm$^3$ (assume that all are temperature independent). You add $10^{16}$ donor atoms which have an energy level at a position 0.1 eV below the conduction band $(E_C - E_D = 0.1 \text{ eV})$.

**(a)** At what temperature will the donor levels become half-filled?

**(b)** At what temperature will the intrinsic carrier concentration equal the donor concentration?

**12.15. (a)** You have a semiconductor at 300 K with a band gap of 0.8 eV and equal effective masses for electrons and holes of $0.1m$, where $m$ is the free-electron mass. Calculate the intrinsic carrier concentration.

**(b)** If you know that the intrinsic carrier concentration $n_i$ is $10^{11}/\text{cm}^3$ and you have a dopant concentration which provides $10^{16}$ electrons/$\text{cm}^3$, what is the hole concentration?

**(c)** If you know that the effective density of states for electrons is $10^{18}/\text{cm}^3$ and there are $10^{16}$ electrons/$\text{cm}^3$ at 300 K, what is the energy separation in eV between the conduction band and the Fermi level?

**12.16.** You are given a III–V semiconductor, AlP, and are told that the energy band gap $E_G$ is 1.5 eV, that it is an indirect gap material, and that the effective density of states in the conduction band $N_C$ is $6 \times 10^{19}/\text{cm}^3$ and in the valence band $N_V$ is $2 \times 10^{18}/\text{cm}^3$. At 300 K:

**(a)** Calculate the intrinsic carrier concentration $n_i$.

**(b)** Is the Fermi level closer to the conduction band $E_C$ or to the valence band $E_V$?

**(c)** Calculate the effective mass for electrons relative to the free-electron mass $m$.

**(d)** Which would have the higher mobility, electrons or holes?

**(e)** For what photon energies would you expect the biggest photoconductivity changes: energies less than or greater than $E_G$?

**(f)** Which of the following dopants would be a donor, and what site (Al or P) would it occupy: B, Al, Ga, Si, As, Cd, Te?

**12.17.** Consider the data at 300 K for Si $E_G = 1.1$ eV, $N_C = 2.8 \times 10^{19}/\text{cm}^3$, $N_V = 1.04 \times 10^{19}/\text{cm}^3$, $\mu_n = 1500 \text{ cm}^2/\text{V}$, $\mu_p = 450 \text{ cm}^2/\text{V} \cdot \text{s}$, and $n_i = 1.45 \times 10^{10}/\text{cm}^3$.

**(a)** What is the resistivity of intrinsic Si?

**(b)** If you add donors so that there are $10^{15}$ electrons/$\text{cm}^3$:

    **(1)** What is the energy separation in eV between the conduction band $E_C$ and the Fermi level?

    **(2)** What is the hole concentration?

    **(3)** What is the resistivity?

**(c)** Indicate the changes (increase, decrease, or remain the same) in the carrier concentrations ($n$, $p$, $n_i$) if dopants are added so that the $E_F$ level moves toward the conduction band or if the temperature increases.

**(d)** From measurement of the photoconductive response to a flash of light on $n$-type material, you measure the hole lifetime $\tau$ to be $10^{-4}$ s. What is the hole diffusion length $L = (D\tau)^{1/2}$?

# CHAPTER 13

# Heterostructures: Transistors and Lasers

## 13.1 Introduction

In the previous chapters we have been concerned with transistors made from either silicon or gallium arsenide. In the remaining chapters we consider heterostructures formed by epitaxial growth of a layer of one semiconductor on another. Hetero-epitaxy (described in Chapter 14) usually involves the growth of a ternary semiconductor such as $Al_xGa_{1-x}As$ on a binary semiconductor such as GaAs. In this case the band gap of AlGaAs is usually in the range 1.6 to 1.9 eV and is greater than that of GaAs (1.45 eV). In general, the composition is chosen to obtain close matching between the two lattices as well as the desired energy-band differences. For the AlGaAs–GaAs system, however, the lattices are matched at all compositions. Lattice-matched heterojunction structures allow us to join two different band-gap materials without degrading the properties of either semiconductor. The epitaxial growth processes are sufficiently controlled to allow us to introduce large carrier concentrations in narrow regions 10 to 100 nm thick in the epitaxial structure. Use of epitaxial growth allows us to fabricate high-speed devices with ultrasmall dimensions exhibiting good charge transport.

Heterojunction devices have also been the key to obtaining efficient laser operation. When we consider optical fibers for transmission of information over long distances, the wavelength of interest is at 1.3 or 1.55 μm, where the optical fibers have the lowest absorption losses and also the lowest dispersion, which allows high-fidelity transmission over long distance. At this wavelength the loss is less than 1 dB/km, where 1 dB (decibel) is a loss of intensity of a factor of $10^{0.1}$ or 1.26. It is necessary to choose semiconductors with not only the correct band gap, but also a direct gap for efficient optical emission. Quaternary semiconductors such as $Ga_xIn_{1-x}As_yP_{1-y}$, which has an emission wavelength of 1.3 μm at $x = 0.26$ and $y = 0.42$, are epitaxially grown on InP to achieve lattice matching.

It is also possible to relax the constraint imposed by lattice matching to produce strained-layer superlattices. These are structures with a series of thin, non-lattice-matched layers; their thinness allows them to be grown epitaxially without generating misfit dislocations. This whole range of lattice-matched and non-lattice-matched structures allows us to do band-gap engineering for specific device applications.

In this chapter we consider some of the advantages that can be gained by use of heterostructures. We start with bipolar transistors with heavily doped base regions, discuss FETs and lasers where the electrons are tightly confined in a narrow layer, and conclude with optoelectronic integrated circuits. In this chapter we cover applications of heterostructures; materials and growth are described in Chapter 14.

## 13.2 Heterojunction Bipolar Transistors (HBT)

The key feature of the bipolar transistor is that the emitter current is overwhelmingly carried by the injection of carriers from the emitter into the base (Chapter 5). For an *npn* transistor, this condition means that electrons are injected into the base with negligible hole injection into the emitter from the base. In conventional, homojunction silicon transistors this is achieved by making the donor concentration $N_D$ in the emitter much greater than the acceptor concentration $N_A$ in the base. In heterojunction bipolar transistors, this is achieved by growing a wider band-gap emitter on a narrower band-gap base (Kroemer, 1982). In silicon structures, the base can be formed by a GeSi alloy ($E_G < E_G(\text{Si})$) and the emitter by an epitaxial Si layer (Bean, 1988).

From the relations given in Chapters 4 and 5, the ratio of injected electrons $n_p$ to injected holes $p_n$ is given for an abrupt homojunction by the ratio of dopant concentrations in the emitter and base:

$$\frac{n_p}{p_n} = \frac{N_D}{N_A}. \tag{13.1}$$

The only limitation to the lightly doped based approach is that the series resistance of the base is high and must be considered in high-current and also high-frequency ($RC$ limited) applications. It would be desirable to increase the dopant concentration in the base without degrading the injection criterion.

The dopant concentration in the base can be increased if the emitter has a wider band gap $E_G$ than the base. Figure 13.1 shows an energy-band diagram assuming that there are no interface states for a wide-band material, such as AlGaAs ($Al_xGa_{1-x}As$), grown epitaxially on GaAs. The flat-band energy diagram (Fig. 13.1a) for lightly *n*-doped materials shows that the difference in the energy gap, $\Delta E_G$, is given by the sum of the discontinuities at the conduction band $\Delta E_C$ and valence band $\Delta E_V$:

$$\Delta E_G = \Delta E_C + \Delta E_V. \tag{13.2}$$

Values of $E_G$ and the discontinuities are given in Chapter 14.

The energy diagram for the emitter–base configuration is shown in Fig. 13.1b for heavily doped *n*- and *p*-type regions. In this case the valence-band discontinuity, $\Delta E_V$, adds directly to the barrier to hole injection. If all the dopants are ionized and there are $N_D$ donors in the emitter and $N_A$ acceptors in the base, the electron concentration $n_p$ just inside the *p*-type base is

$$n_p = N_D e^{-(eV_n)/kT} \tag{13.3}$$

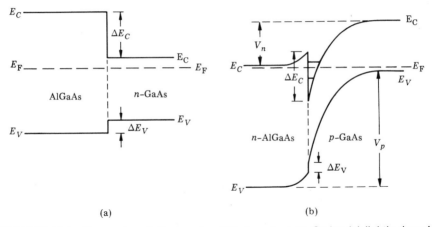

(a)                                                      (b)

**FIGURE 13.1**   Energy-band diagram for AlGaAs grown on GaAs: (a) lightly doped
*n*-type material showing the conduction- and valence-band discontinuity; (b) *n*-type
AlGaAs on *p*-type GaAs showing the potential barriers.

and the injected hole concentration $p_n$ just inside the *n*-type emitter is

$$p_n = N_A e^{-(eV_p)/kT} \tag{13.4}$$

where $eV_n$ and $eV_p$ are the energy barriers for electrons in the emitter and holes in
the base, respectively.

For homojunctions (junctions on the same material), $eV_n = eV_p = eV_0$, as
shown in Chapter 4. However, for heterojunctions, $eV_n \neq eV_p$ and

$$eV_p - eV_n = \Delta E_C + \Delta E_V = \Delta E_G. \tag{13.5}$$

The ratio of electrons to that of injected holes is

$$\frac{n_p}{p_n} = \frac{N_D}{N_A} e^{\frac{(eV_n - eV_p)}{kT}} = \frac{N_D}{N_A} e^{\Delta E_G/kT}. \tag{13.6}$$

At room temperature, $kT = 0.026$ eV, so that even a small change of 0.15 eV in
$\Delta E_G$ leads to a value of exp $(\Delta E_G/kT) = 320$. Thus the dopant concentration in
the base of a heterojunction can exceed that in the emitter by orders of magnitude
without degrading the injection ratio.

# 13.3 Heterojunction Field-Effect Transistors

The concept of a field-effect transistor (Fig. 13.2) is to establish a current-carrying
channel between two contacts, source and drain, and to have a third electrode—
the gate—to modulate the current flow. For high-speed operation, high carrier
mobilities and velocities are required as well as a high sheet carrier concentration,
$n_s$, in the channel under the gate. The value of $n_s$ should lie above $10^{12}$ carriers/cm$^2$
for good values of transconductance.

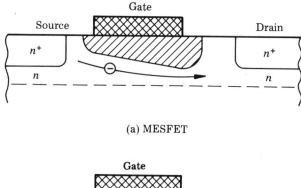

(a) MESFET

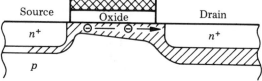

(b) MOSFET

**FIGURE 13.2** $n$-channel Field-effect transistors: (a) MESFET to show the $n$-channel containing $n_s$ carriers/cm²; (b) MOSFET structure where application of a positive gate voltage causes accumulation of electrons in the $n$-channel.

The natural choice for such a high-speed FET is to fabricate an $n$-channel structure on GaAs to take advantage of the high electron mobility. In conventional, metal–semiconductor (Schottky barrier) FETs (MESFETs), the channel is doped with donors and the mobility is determined by scattering from ionized donors. In silicon metal–oxide–semiconductor FETs (MOSFET), the channel is induced by the electric field, which causes electron accumulation at the interface between the gate oxide and $p$-type Si (Fig. 13.2b).

In some heterojunction FETs, a heavily doped, wide-band gap semiconductor, AlGaAs, is grown on an epitaxial layer of high-purity, high-mobility GaAs. The electrons associated with the donors in the AlGaAs transfer to the GaAs establishing a channel in the high-mobility GaAs. The electrons in the undoped GaAs channel are separated from the donors in the heavily doped wide-band AlGaAs leading to the term "modulation doping" (MODFET). The electrons in the channel form a high-electron mobility transistor (HEMT). The thin sheet of electrons form a two-dimensional electron gas field-effect transistor (TEGFET). Acronyms are listed in Table 13.1. We will use the generic term "heterojunction FET" or, if required, MODFET for this modulation-doped structure.

The energy-band discontinuity, $\Delta E_C$, is the key to the spatial separation of the electrons and donors. The structure is shown in Fig. 13.3 with the band gap of $Al_xGa_{1-x}As = 1.80$ eV ($x = 0.3$) and that of GaAs $= 1.40$ eV. On the undoped GaAs, a thin (10 nm) layer of undoped AlGaAs is grown followed by a heavily doped, $N_D = 1.5 \times 10^{18}/cm^3$, layer of AlGaAs. The electrons depleted from the AlGaAs accumulate in the GaAs in the thin layer defined by the potential "notch."

**TABLE 13.1** Acronyms and Structures Used in Heterojunctions

| | 1. Structure |
|---|---|
| APD | Avalanche photodiode |
| BT | Bipolar transistor |
| DH laser | Double-heterojunction laser |
| FET | Field-effect transistor |
| HBT | Heterojunction bipolar transistor |
| HEMT[a] | High-electron-mobility transistor |
| LED | Light-emitting diode |
| MESFET | Metal–semiconductor FET |
| MODFET[a] | Modulation-doped FET |
| MOSFET | Metal–oxide–semiconductor FET |
| MQW | Multiple (multi-)-quantum well |
| NDR | Negative differential resistance |
| SDHT[a] | Selectively doped heterojunction transistor |
| SL | Strained layer |
| TEGFET[a] | Two-dimensional electron-gas FET |
| QW | Quantum well |
| | |
| | 2. Growth Processes |
| CVD | Chemical vapor deposition |
| LPE | Liquid-phase epitaxy |
| MBE | Molecular-beam epitaxy |
| MOCVD | Metal–organic chemical vapor deposition |
| MOMBE | Metal–organic molecular-beam epitaxy |
| MOVPE | Metal–organic vapor-phase epitaxy |
| VPE | Vapor-phase epitaxy |

[a]HEMT, MODFET, SDHT, and TEGFET are different acronyms for the same structure (see Section 13.3).

This establishes a plane or two-dimensional electron gas of carriers with a sheet concentration of about $10^{12}$ carriers/cm$^2$. Although similar in concept to the electron gas in the $n$-channel of a silicon MOSFET, there are three advantages in the GaAs MODFET:

1. The mobilities of electrons in GaAs are higher than those in silicon.
2. The heterojunction is nearly structurally perfect (as compared to the amorphous $SiO_2$/crystal Si interface), so that electrons traveling parallel to the heterojunction are not scattered by structural defects at the interface.
3. Electron scattering from ionized impurities is minimal for the electron gas in the channel in the undoped GaAs.

The electron mobilities in a modulation-doped heterojunction as well as in doped GaAs and in $Al_{0.3}Ga_{0.7}As$ are shown in Fig. 13.4. The striking features are the high electron mobilities and the strong temperature dependence of the mobility in the heterostructure. In the absence of ionized dopant scattering centers, the mobility

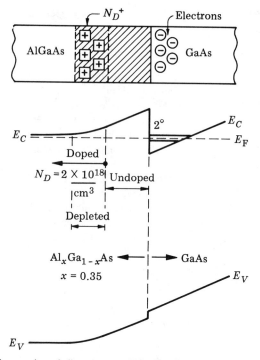

**FIGURE 13.3** Energy-band diagram on AlGaAs–GaAs heterojunction, showing the formation of an electron gas at the interface. The outer layer of AlGaAs is doped to $2 \times 10^{18}$ electrons/cm$^3$ and is separated from the GaAs by an undoped AlGaAs spacer layer 10 nm thick.

is dominated by lattice vibrations (phonon scattering). At room temperature, the modulation-doped structure exhibits a mobility that is a factor of 1.5 to 2 higher than in impurity-doped GaAs. At liquid nitrogen temperatures (77 K), the electron mobilities can exceed $10^5$ cm$^2$/V $\cdot$ s.

In a short-channel structure operated at high voltages such that the carrier drift velocity is at its peak value, $v_s$, the maximum drain-to-source current, $I_D(\text{max})$, for a sheet carrier concentration $n_s$ is

$$I_D(\text{max}) = e n_s v_s W \tag{13.7}$$

where $W$ is the width of the device. There is an improvement in $v_s$ in heterojunction devices, although not as dramatic as for the mobility. The peak velocity at 300 K has a value of $2.1 \times 10^7$ cm/s in undoped GaAs and $1.8 \times 10^7$ cm/s for a doping level in GaAs of $10^{17}$/cm$^3$. As another comparison, the value of $v_s$ in silicon is about $10^7$ cm/s for lightly doped material.

In device fabrication, a Schottky barrier is formed on the AlGaAs for the gate contact (Fig. 13.5). The requirement for low-leakage current across the gate Schottky junction gives an upper limit of about $10^{18}$ donors/cm$^3$ in the AlGaAs which has a total thickness less than 70 nm. The thickness of the AlGaAs must be

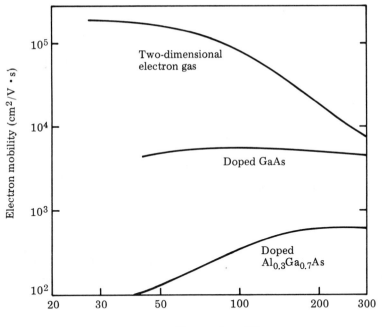

**FIGURE 13.4** Electron mobilities as a function of temperature for the electron gas in a MODFET and the two uniformly doped ($n = 10^{17}$ to $10^{18}/cm^3$) materials used to form a heterojunction.

controlled so that the entire thickness of the AlGaAs is depleted by the gate and heterointerface fields. As shown in Fig. 13.4, the electron mobilities in AlGaAs are so low that it is necessary to deplete the AlGaAs so that the source–drain current $I_D$ is carried mostly by the electron gas in the GaAs channel.

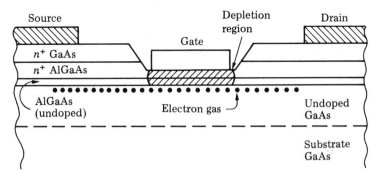

**FIGURE 13.5** Schematic cross section of a heterojunction FET made from AlGaAs–GaAs forming an electron gas at the interface. The source and drain are formed on heavily doped GaAs for low-contact resistance. The depletion region under the gate extends through the AlGaAs layer.

As discussed in Section 5.10, the channel conductance $g_d$ can be defined

$$g_d = \left. \frac{\partial I_D}{\partial V_D} \right|_{V_G} \tag{13.8}$$

where $I_D$ and $V_D$ are the current and applied voltage, respectively, at the drain. In the low-voltage region the drift velocity $v_d$ is proportional to the mobility, so that

$$I_D = \frac{e n_s \mu V_D W}{L} \tag{13.9}$$

where $L$ is the channel length and $W$ is the channel width normal to the current flow in the channel. The channel conductance is

$$g_d = \frac{e n_s \mu W}{L} \tag{13.10}$$

where $n_s$ is the number of electrons per cm$^2$ in the channel. For an electron mobility of 7000 cm$^2$/V · s and with $n_s = 10^{12}$ electrons/cm$^2$ in a channel 1 μm long and 10 μm wide,

$$g_d = \frac{1.6 \times 10^{-19} \times 10^{12} \times 7000 \times 10}{1} = 1.12 \times 10^{-2} \text{ S.}$$

This channel conductance is a factor of 10 higher than that in a Si MOSFET.

For the MODFET at a fixed drain voltage, the drain current is controlled by the gate voltage $V_G$. The thickness of the AlGaAs layer between the gate and the electron-gas channel is critical. For small thicknesses the gate Schottky barrier can completely deplete both the AlGaAs layer and the electron gas even at zero gate bias, thus leading to an enhancement mode or normally off device. A positive gate voltage is required to turn the device on. The terminology is the same as that used for the Si MOSFET (Chapter 5). The threshold voltage $V_T$ is adjusted by control of the AlGaAs thickness and doping. Thicker AlGaAs layers do not lead to depletion of the electron gas at zero gate voltage, and the devices are normally-on or depletion-mode in operation. A negative gate bias voltage is used to deplete the electron gas and pinch the device off, as shown in Fig. 13.6.

In a normally-off device, the gate capacitance $C_G$ is given by

$$C_G = \frac{\epsilon_{\text{AlGaAs}}}{d} \times WL \tag{13.11}$$

where $d$ is the thickness of the total AlGaAs layer, $\epsilon_{\text{AlGaAs}}$ is the permittivity of the AlGaAs layer under the gate, and $W$ and $L$ are the device channel width and length, respectively. Here we have neglected the contribution of correction terms in $d$, which can be equivalent to thicknesses of 8 to 10 nm. As in the case of the Si MOSFET, the charge $Q_s$ in the channel is then

$$Q_s = e n_s = C_G(V_G - V_T) \tag{13.12}$$

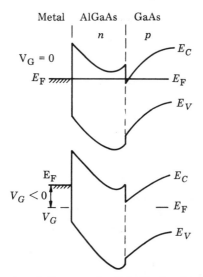

**FIGURE 13.6** Energy-band diagram of a Schottky barrier on a AlGaAs–GaAs, normally-on FET. Application of a negative gate bias voltage pinches off the channel. [From N. T. Linh, Ch. 3 in Dingle (1987).]

where $V_G - V_T$ is the effective voltage controlling the charge in the channel, and we assume that the voltage drop due to the resistance in the channel is negligible.

The device conductance or mutual transconductance $g_m$ is defined as

$$g_m = \frac{\partial I_D}{\partial V_G}\bigg|_{V_D} \tag{13.13}$$

which in the region of constant velocity $v_s$ (or "saturated" velocity) leads to

$$g_m(\text{sat}) = \epsilon_{\text{AlGaAs}} \times v_s \times \frac{W}{d} \tag{13.14}$$

$$= C_G \frac{v_s}{L} \tag{13.15}$$

by use of eq. (13.11). Values of $v_s$ at 300 K are $\leq 2 \times 10^7$ cm/s.

To a first-order approximation, the maximum transconductance $g_m(\text{max})$ can be expressed as

$$g_m(\text{max}) \approx \frac{en_s\mu W}{L}. \tag{13.15a}$$

This equation is identical to eq. (13.10). For the 10-μm-wide gate device, the maximum transconductance is 11.2 mS or 1120 mS/mm gate width at 300 K. More realistic calculations show $g_m(\text{max})$ to be about 600 mS/mm, and experimental values of $g_m$ of about 400 mS/mm have been observed (Morkoç and Unlu, 1987).

In this constant-velocity regime, the transit time $T_R$ is given by

$$T_R = \frac{C_G}{g_m(\text{sat})} = \frac{L}{v_s}.$$ (13.16)

For a channel length of 1 $\mu$m and a velocity of $2 \times 10^7$ cm/s, the transit time would be $5 \times 10^{-12}$ s or 5 ps.

These MODFETs are high-speed devices with current-gain cutoff frequencies $f_T$ of 50 to 100 GHz and switching times less than 10 ps. We have used AlGaAs–GaAs structures to illustrate charge transfer across a heterojunction for the placement of electrons in a high-purity semiconductor without the presence of mobility-limiting donor impurities. To optimize performance there are other combinations of lattice-matched and strained lattice combinations used to obtain the highest velocities and sheet carrier conductances. All rely on the control of composition, thickness, and doping provided by the epitaxial growth process.

## 13.4 Quantum Well Structures

The modulation-doped heterojunction FET is an example of a quantum well with a triangular potential well. Figure 13.7 shows the conduction band and Fermi level for an AlGaAs–GaAs heterostructure. The level $E_1$ is the ground level—or lowest allowed energy level—of electrons in the potential well. The shaded region indicates the energy region occupied by the two-dimensional gas of electrons.

A simple system to visualize is multiple layers, about 10 nm thick, of undoped AlGaAs and GaAs, a multiple-quantum well (MQW). Figure 13.8 shows the conduction and valence-band edges for a set of square wells. The carriers are confined in each of the layers (for reasonably wide spacings and values of the energy discontinuities $\Delta E_C$ and $\Delta E_V$) and will behave almost independently of the carriers in the other wells. In these wells the electrons do not move with their usual three degrees of freedom. There is two-dimensional behavior in the planes of the layers ($x$ and $y$ directions) and one-dimensional behavior normal ($z$ direction) to the layers. Quantization of the carrier motion in the $z$ direction produces a set of discrete energy levels, dashed lines in Fig. 13.8.

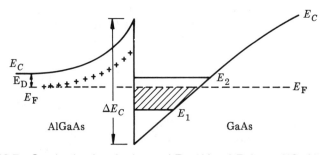

**FIGURE 13.7** Conduction-band edge and Fermi level $E_F$ in an AlGaAs–GaAs heterojunction. $E_1$ denotes the ground state, and electrons occupy states between $E_1$ and $E_F$. The conduction-band discontinuity is $\Delta E_C$.

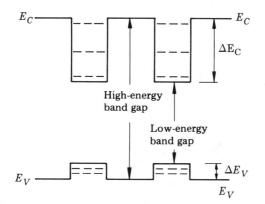

**FIGURE 13.8**  Conduction- and valence-band edges of a multiple quantum well formed by 10-nm-thick layers of AlGaAs and GaAs. The dashed lines show the quantized energy levels in the well.

As discussed in Section 12.4, for a one-dimensional rectangular well infinitely deep, the energies of the series of discrete bound states are given by

$$E_n = \frac{\hbar^2}{2m^*} \left(\frac{n\pi}{L_z}\right)^2 \tag{13.17}$$

with $n = 1, 2, 3 \ldots$, and $m^*$, the effective mass. The energy $E_1$ of the lowest state will decrease as $L_z$ increases. For a three-dimensional system for $L_z \ll L_x$, $L_y$,

$$E = E_n + \frac{\hbar^2}{2m} (k_x^2 + k_y^2) \tag{13.18}$$

where $k_x L_x$ and $k_y L_y$ equal $n_x\pi$ and $n_y\pi$, respectively.

The energy levels in the $k_x$ and $k_y$ directions will be closely spaced (similar to the electron in a cube). For each value of $E_n$, a two-dimensional band in the $k_{xy}$ plane will form leading to an $E$–$k$ diagram as shown in Fig. 13.9. As in Chapter 12, we can write the energy of any of these bands of a given quantum state as

$$E = \frac{h^2}{8m^*L^2} (n_x^2 + n_y^2) = R^2 E_{xy} \tag{13.19}$$

where $E_{xy} = h^2/8m^*L^2$ and $R^2 = n_x^2 + n_y^2$, as before. In a two-dimensional plane, the number $N$ of states within radius $R$ is

$$N = 2 \cdot \frac{1}{4} \pi R^2 = \frac{\pi}{2} \frac{E}{E_{xy}} \tag{13.20}$$

where the factor of 2 is due to the spin states and the $\frac{1}{4}$ is required for the quadrant of positive $n$ values. The density of states $g(E)$ per unit area is

$$g(E) = \frac{1}{L^2} \frac{dN}{dE} = \frac{m^*}{\pi\hbar^2}. \tag{13.21}$$

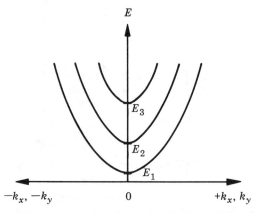

**FIGURE 13.9** $E$–$k$ diagram for a two-dimensional electron gas showing the levels for different values of $E_n$.

The two-dimensional density of states of a given quantum state $E_n$ is independent of energy. The cumulative density of states for a series of bands will be a staircase with steps of $m^*/\pi\hbar^2$ as shown in Fig. 3.10.

Optical absorption measurements at 2 K on multilayer quantum wells show the stairlike nature of the density of states. The values of $E_G$ at 2 K are 1.75 eV for $Al_{0.2}Ga_{0.8}As$ and 1.519 eV for GaAs. Absorption features between these two limits will be due to energy states in the GaAs quantum wells and to absorption by excitons. An exciton is an electron that is excited from the valence band to the conduction band but has a binding energy of 0.004 eV due to the Coulomb attraction between the electron and the hole left in the valence band. The exciton makes a dominant contribution to the band-edge absorption, as shown in Fig. 13.11. For a well thickness of $L_z = 400$ nm, the absorption is similar to that in a bulk sample with a sharp exciton peak. For thinner layers, where quantum size effects appear, there

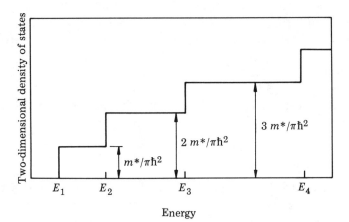

**FIGURE 13.10** Density of states versus energy for a two-dimensional quantum well.

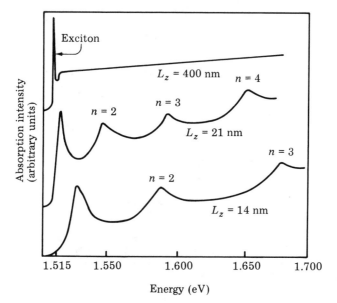

**FIGURE 13.11**   Optical absorption spectra taken at 2 K of 400, 21, and 14-nm-thick layers of GaAs between $Al_{0.2}Ga_{0.8}As$ barriers. [From Dingle (1975).]

are exciton peaks associated with each discrete energy, $E_1, E_2, \ldots$ , and the peaks shift proportional to $L_z^{-2}$, as predicted by eq. (13.17). The absorption spectra for $L_z = 21$ nm and 14 nm have the steplike character that one would deduce from Fig. 13.10 for quantized transitions.

## 13.5 Tunneling Structures

The presence of thin barriers allows the possibility of tunneling between the wells, similar to the situation with the tunnel diode discussed in Chapter 12. Here, too, there is a negative differential resistance because of the discrete bands of available states.

The conduction band for an AlGaAs–GaAs quantum well structure is shown in Fig. 13.12 with $E_1$ shown as the ground level in the GaAs well. The outer layers of GaAs are heavily doped with donors so that $n_n \simeq 10^{18}/cm^3$ and the well has an order-of-magnitude lower concentration of electrons. The three heteroepitaxial layers have a thickness of about 5 nm for efficient tunnel transmission.

When external voltage is applied, the voltage drop is divided equally between the two undoped AlGaAs layers. When $E_1$ lines up with the Fermi level $E_F$ at a voltage of $2E_1/e$, electrons tunnel into the $n = 1$ state at $E_1$ in the GaAs well from the left electrode and out of the well into the right electrode. This is often referred to as resonant tunneling, as transmission of electrons occurs when the Fermi level of the injecting layer is resonant (energy matched) with the quantum state $E_1$ in the confined well. The maximum current occurs at $V = 2E_1/e$ in the symmetric

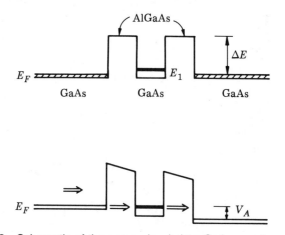

**FIGURE 13.12** Schematic of the energy levels in a GaAs quantum well (with quantum level $E_1$) bounded by undoped AlGaAs layers, which are in turn bounded by doped ($N_D = 10^{18}/cm^3$) layers of GaAs. Resonant tunneling transmission occurs under applied voltage = $2E_1/e$.

structure. At higher applied voltages, the $E_1$ level no longer is aligned with the left electrode Fermi level and the current decreases. At even-higher voltages, the current can rise again due to thermionic emission across the barriers. Peak current-to-valley current ratios of about 3 have been found.

It is also possible to use the tunneling phenomenon to build a transistor, not a conventional *npn* bipolar transistor, but a heterostructure device where the emitter, base, and collector are all made from heavily doped *n*-type GaAs. The emitter and base are separated by a thin (about 10 nm) layer of AlGaAs which is the tunnel barrier (Fig. 13.13). The base and collector are separated by a wider layer of AlGaAs. The AlGaAs layer acts as a barrier between base and collector and prevents the thermal electrons in the base of the transistor from flowing to the collector.

When a voltage is applied, the Fermi level in the base is lowered and electrons can tunnel from the emitter to the base through the thin AlGaAs barrier. With sufficiently high applied voltage, the emitter electrons enter the base with an energy far in excess of the thermal energy of the electrons in the GaAs base. These injected electrons have a high velocity and can traverse the base at a velocity near the hot-electron peak velocity. At high energies, the electrons are scattered predominantly with small angles. The base is thin to minimize scattering events. The high-energy electrons traverse the AlGaAs barrier and reach the collector.

These high-velocity electrons that traverse distances of 10 to 100 nm without large-angle scattering events are referred to as ballistic electrons. The limiting velocity of an electron through GaAs and many other III–V semiconductors is about $10^8$ cm/s (about 1% of the speed of light in the semiconductor) and is determined by the electronic structure of the semiconductor. The electrons may not be truly ballistic, as they may scatter from vibrating lattice atoms, but they only lose a small fraction of their energy in such a collision, which does not significantly

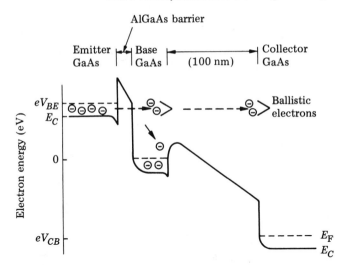

**FIGURE 13.13** Energy diagrams of the conduction band of a tunneling transistor made from heavily $n$-doped GaAs separated by barriers of AlGaAs. Under bias, the electrons in the emitter tunnel into the base at high energy and travel to the collector.

change the original forward direction. The high velocities of the ballistic electrons and the thin layers of the heterostructures lead to transit times less than a picosecond.

# 13.6 Heterojunction Lasers and Light-Emitting Diodes

Light-emitting devices operate by the injection of high concentrations of electrons and holes into the active region where electrons recombine with holes in radiative transitions. This necessitates the use of direct-band-gap materials which have strong radiative recombination cross sections. High-quality material is required to minimize nonradiative recombination caused by defects in the bulk or at interfaces.

In heavily doped GaAs $n$-$p$ junctions operated under forward bias (Fig. 13.14a) electrons are injected at concentrations up to $10^{18}/cm^3$ into the $p$-type materials. The electrons have a lifetime $\tau_n$ of about $10^{-9}$ s in heavily doped material, but because of the high electron mobility, the diffusion length (Chapter 3), $L_n = (D_n \tau_n)^{1/2}$, can be 1 to 3 $\mu$m. To optimize the confinement of the electrons and the optical modes guided along the junction, double-heterojunction structures are used with a 0.1- to 0.2-$\mu$m layer of GaAs—the active region—interposed between layers of AlGaAs (Fig. 13.14b). The spacing $d$ between the heterojunctions is less than a diffusion length, so that the injected carrier density is uniform across $d$. The band discontinuities, $\Delta E_C$ and $\Delta E_V$, confine the injected electrons and holes within the active region.

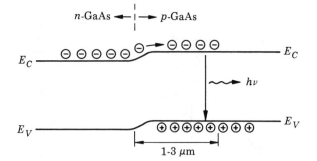

(a) GaAs homojunction laser

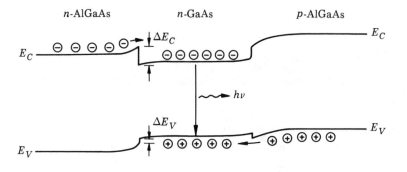

(b) Double heterojunction laser

**FIGURE 13.14** Energy bands of a light-emitting diode under forward bias for (a) a GaAs homojunction and (b) a double heterojunction AlGaAs–GaAs–AlGaAs structure.

The basic concept in the double-heterojunction (DH) structure is to have the recombination region with a direct-band-gap material bracketed on either side by two layers having a higher band-gap energy, to ensure carrier confinement. It is important to use closely lattice matched systems. Misfit dislocations (Chapter 14) have nonradiative centers at the dislocation core and hence degrade the recombination efficiency of the device. For a 1% lattice mismatch at an interface, a dislocation will be generated every 100 atomic planes or every 30 to 50 nm. If $a$ is the lattice spacing and $\Delta a$ the difference in lattice spacings, the value of $\Delta a/a$ should be less than about $10^{-3}$.

Laser- and light-emitting diodes for optical communication use GaAs and AlGaAs for the 0.68- to 0.9-$\mu$m spectral region and InGaAs or GaInAsP for the 1.0- to 1.5-$\mu$m spectral region (Fig. 13.15). The early optoelectronic device development focused on GaAs–AlGaAs emitters, and hence these structures are used for "local" interconnects between computers, processors, and chips. For long-haul communications, GaInAsP emitters on InP substrates are used because of the low-

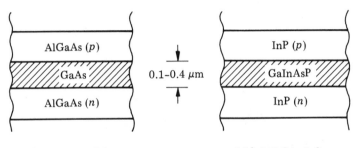

(a) $\lambda \cong 0.68$ to $0.9 \ \mu m$     (b) $\lambda \cong 1.0$ to $1.6 \ \mu m$

**FIGURE 13.15** Schematic cross section of double-heterojunction structures with active light-emitting regions (shaded) between 0.1 and 0.4 nm thick: (a) GaAs; (b) GaInAsP. In (a) the active region can also be AlGaAs as long as its energy gap is less than those of the neighboring layers.

attenuation and dispersion-free windows of optical fibers near 1.3- and 1.55-$\mu$m wavelengths.

As pointed out in Chapter 14, in the $Al_xGa_{1-x}As$ system, the lattice constant shows little variation as $x$ goes from zero to unity. Heterojunctions in the AlGaAs–GaAs system have negligible strain-induced defects. In the GaInAsP system the lattice constant varies by a few percent. InP, $E_G = 1.34$ eV, is used as a substrate and a close lattice match is obtained with $Ga_xIn_{1-x}As_yP_{1-y}$ when $y \cong 2.16x$. In this composition range, the emission wavelength varies from 0.9 $\mu$m ($x = y = 0$) to 1.5 $\mu$m ($x = 0.47$, $y = 1.0$).

Laser diodes and light-emitting diodes can be made from basically the same heterostructure (Fig. 13.16) and can emit nearly the same optical power, typically up to 10 mW. The structures and optical characteristics differ. In light-emitting diodes (a surface-emitting LED is shown in Fig. 13.16a with an optical fiber), the emitted photons bear no fixed phase relationship with each other (incoherent radiation) and their energy can be quite spread out, but centered near the band-gap energy. Lasers emit highly directional coherent radiation by stimulated emission in an optical cavity formed by the two parallel surfaces (parallel crystal planes) at the ends of the diode. The change of refractive index at the semiconductor–air boundaries provides the necessary mirror reflectivity ($\approx 30\%$). The double heterojunction also provides an optical waveguide because the active layer has a higher refractive index than the neighboring layers by about 5%. A typical cavity length is 300 $\mu$m and stripe width $W$ is 10 $\mu$m (Fig. 13.16b). The laser threshold is reached when the gain produced by the injected carriers overcomes the cavity loss and end facet transmission.

The emitted power in a laser diode shows a sharp increase at the threshold current (Fig. 13.17). The current up to threshold is given by the diode equation (Chapter 4), but is modified to include the series resistance $R_s$,

$$I = I_0 \left\{ \exp\left[\frac{e(V_A - IR_s)}{kT}\right] - 1 \right\} \tag{13.22}$$

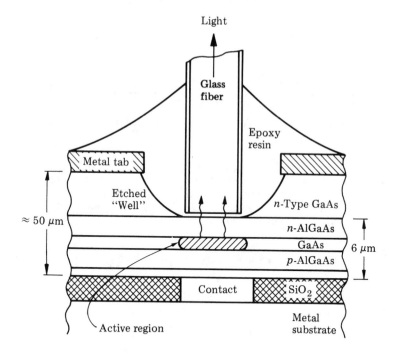

(a)

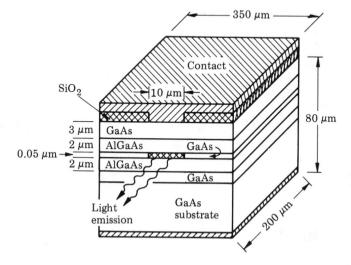

(b)

**FIGURE 13.16** Schematic cross sections of (a) a light-emitting diode and (b) a laser made in double-heterojunction configuration.

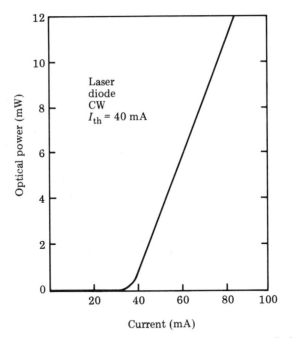

**FIGURE 13.17** Optical power emitted versus current for a laser diode operated under continuous-wave (CW) conditions.

where $V_A$ is the applied voltage, $I_0$ the saturation current at 300 K, $I_0 = J_0$ times diode area with $J_0 \simeq 2 \times 10^3$ A/cm$^2$, and $R_s$ has values between 1 and 10 $\Omega$. Above threshold, the junction voltage drop is at its threshold value, given by about $E_G/e$ for population inversion, so that the current–voltage relation is

$$I = \frac{V_A - E_G/e}{R_s}. \tag{13.23}$$

If we assume an average density $N$ of injected carriers in the active region of width $d$ with a carrier lifetime $\tau$, the current density $J$ is

$$J = \frac{Ned}{\tau}. \tag{13.24}$$

The threshold current density $J_{th}$ will depend on the active width $d$. If we assume that for population inversion the density of injected electrons must equal the effective density of states ($N_C$) (where $N_C$ equals $4.7 \times 10^{17}$/cm$^3$ in GaAs), then

$$J_{th} = 4.7 \times 10^{17} \frac{ed}{\tau} \tag{13.25}$$

which has a value of 3760 A/cm$^2$ times $d$ ($\mu$m) for a value of $\tau = 2 \times 10^{-9}$ s. Threshold current densities of about 500 A/cm$^2$ have been obtained with an active-device thickness of 0.1 $\mu$m.

Our assumption that the injected carrier density is equal to $N_C$ is too low an estimate. The injected electrons and holes have a sufficiently high concentration so that each can be described by an effective or "quasi-" Fermi level in the conduction and valence bands. The energy separation between these quasi-Fermi levels should be greater than $E_G$. In GaAs, this requirement leads to injected carrier densities greater than $10^{18}$ cm$^{-3}$. This is a value that is twice our estimate in eq. (13.25).

Light-emitting diodes (LED) can be operated in pulsed or continuous-wave (CW) mode. The allowable power level depends on heat dissipation and catastrophic optical power limit at the facets where degradation occurs. The attractive feature is the direct modulation of the light output intensity with current. The maximum modulation rate of a light-emitting diode is limited by the carrier recombination lifetime $\tau$ (as is the case for $p$-$n$ junctions, Chapter 4). The modulation bandwidth $\Delta f$, defined as the frequency at which the detected power becomes half (or equivalently 3 dB) of its low-frequency value, can be expressed for an LED as

$$\Delta f = \frac{1}{2\pi\tau}. \tag{13.26}$$

For lifetimes of $10^{-9}$ s, modulation bandwidths of 100 MHz can be achieved. In addition to the carrier lifetime, the external drive circuit, the diode capacitance and the series resistance can limit the response time of an LED.

## 13.7 Photodetectors

Semiconductor devices are used for the detection of light as well as for the emission of light. The absorption of photons at depth $x$ follows the relation

$$N_{\text{ph}} = N_{\text{ph}}(0) \exp(-\alpha x) \tag{13.27}$$

where $N_{\text{ph}}(0)$ is the incident number of photons and $\alpha$ is the optical absorption coefficient in units of cm$^{-1}$. To minimize the thickness of the absorbing material, values of $\alpha$ around $10^4$ cm$^{-1}$ [or 1 ($\mu$m)$^{-1}$] are required. These high absorption coefficients in turn require either direct-band-gap materials or the direct transition component in indirect-gap materials (see Chapter 12). Figure 13.18 gives optical absorption coefficients versus photon wavelength in micrometers for several semiconductors. The conversion between energy $E$ in eV and wavelength $\lambda$ in $\mu$m is $E(\text{eV}) = 1.24/\lambda$ ($\mu$m). The vertical shaded regions are the windows of maximum transmission for silica optical fibers. At the longest wavelength (around 1.55 $\mu$m) GaInAs is required, whereas around 0.8 $\mu$m all the compound semiconductors shown are usable except for GaP. Silicon is marginal, but usable with photoconductors and avalanche photodiodes that have internal gain.

In photodiodes, the objective is to absorb the photons and hence generate hole–electron pairs in the region where there is a high electric field. The photo-generated carriers are swept out, generating current or voltage signals in the external circuitry. The simplest photodiode is a reverse-bias Schottky barrier diode (Fig.

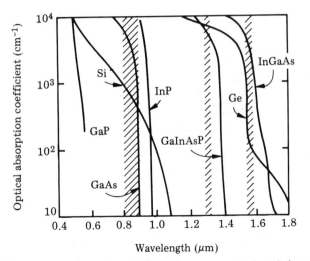

**FIGURE 13.18**   Optical absorption coefficients versus wavelength for several semiconductors. The shaded regions are the maximum transparency of silica fiber optics used in optical communication systems.

13.19) illuminated through the top, semitransparent, thin metal layer. Such structures on GaAs consist of a lightly doped, $n$-type layer of GaAs on a bulk substrate with the metal layer formed on top to form the Schottky barrier. Proton bombardment can be used to form a semi-insulating layer of GaAs around the active area for low leakage currents. Under reverse bias, the electric field is maximum (Chapter 4) at the metal–semiconductor interface. The electric field separates the photo-

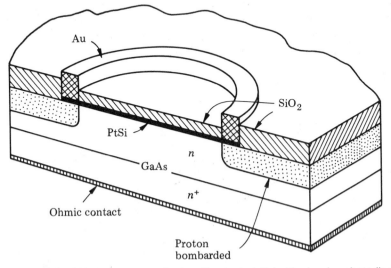

**FIGURE 13.19**   Cross section of a top-illuminated Schottky barrier photodiode.

generated electron–hole pairs; the holes are swept toward the metal contact, where they recombine, and the electrons are swept across the high-field depletion region. As discussed in Section 4.7, the transit time is independent of voltage and is twice the dielectric relaxation time $\tau_{DR} = \rho\epsilon = 1.16 \times 10^{-12}\ \rho$ seconds, where $\rho$ is the resistivity ($\Omega \cdot$ cm). Transit times are in the picosecond regime.

The *p-i-n* (or PIN) diodes have a high resistivity (intrinsic) layer as the active region. Under reverse bias, the high-field region extends across the *i*-region between the *p*- and *n*-type layers (Fig. 13.20). For an intrinsic region (where the difference between donor and acceptor concentrations is zero, $N_D - N_A = 0$) the electric field $\mathscr{E} = V_A/d$ and the transit time, $T_R$, is

$$T_R = \frac{d}{\mu\mathscr{E}} = \frac{d^2}{\mu V_A} \tag{13.28}$$

where $d$ is the width of the *i*-region, $V_A$ the applied voltage, and $\mu$ the mobility of the carrier traversing the region. It is desirable to illuminate the region near the *p-i* interface, top illumination, so that the high-mobility electrons traverse the high-field region. Absorption outside the high-field region, for both the Schottky and the *p-i-n* diode, generates minority carriers, which results in a slow diffusion of carriers to the depletion region. To eliminate absorption in the *p*- and *n*-type layers, these regions are usually made of a wider-band-gap material, AlGaAs, on the intrinsic GaAs.

In a typical long-wavelength planar *p-i-n* diode, GaInAs is the light-absorbing material. Up to 98% of the incident radiation can generate hole–electron pairs in the active region. Top-illuminated *p-i-n* structures with a transparent cap of GaInAsP or InP will have minimal loss of efficiency arising from absorption outside the depletion region.

The response time of the Schottky barrier or *p-i-n* photodiode, which may exceed 10 GHz, is usually limited by the resistance–capacitance (*RC*) time constant of the detector itself and the external receiver circuit. The active area can be made smaller to reduce capacitance and rear illumination used so that the junction area can be matched to the size of the incident light beam.

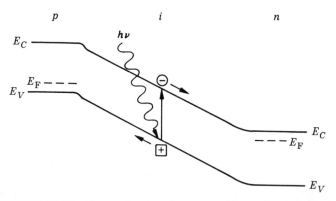

**FIGURE 13.20**  Energy bands of a reverse-bias *p-i-n* photodiode.

Silicon photodetector technologies are well established and use internal gain due to the avalanche process to offset the low optical absorption. Similar to the discussion on reverse-bias breakdown in Section 4.3.2, avalanche photodiodes (APD) are based on the impact ionization effect in semiconductors. Photo-generated carriers are accelerated by the high electric field in the depletion region until they gain sufficient energy to promote an electron from the valence band to the conduction band, thus giving rise to an internal gain. Avalanche gains of 100 are found in Si devices.

The electric field required for impact ionization depends on the mobility (alternatively, the mean free path $l$ between collisions) and the band gap $E_G$. The carrier must acquire enough energy from the field to create a hole–electron pair within a collision distance so that the threshold electric field $\mathscr{E}_{th}$ is approximated by

$$\mathscr{E}_{th} = \frac{E_G}{el}. \tag{13.29}$$

The probability of ionizing collisions is generally greater for electrons than it is for holes. In silicon the ionization coefficient is 40 to 50 times greater for electrons than for holes. The avalanche photodiode structures are designed so that carriers are generated in the region where electrons traverse the high-field region; the avalanche is initiated by the carrier with the highest ionization rate.

Avalanche photodiodes have been made in GaAs. It is more difficult to fabricate avalanche devices for detecting long-wavelength radiation than for short-wavelength radiation. Narrower-band-gap materials, such as GaInAs, are required to absorb the radiation, and these materials have a larger background current than wider-band-gap material under high reverse bias. One solution is to use heterostructure technology, so that the photons are absorbed in a layer of GaInAs and the carriers are swept to an adjacent layer of a wide-band-gap material that contains the avalanche field.

The detector for long-wavelength radiation can be a photoconductor—a simple resistor of GaInAs with ohmic contacts grown on a layer of InP (Fig. 13.21). The layer should be highly resistive to minimize dark current and about 1 to 2 μm thick so that the light can be absorbed. The photo-generated carriers increase the conductivity of the layer, with electrons moving toward the positive contact (anode)

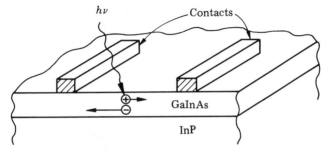

**FIGURE 13.21**  Cross section of a photoconductor made from a 2-μm-thick layer of GaInAs on InP.

and holes toward the negative contact (cathode). The high-mobility electron is collected before the hole; the hole then continues to drift along the conducting layer, resulting in a net positive charge. This excess charge draws an electron into the layer from the ohmic contact at the cathode. Electrons continue to be drawn into the conducting layer until the hole either recombines with an electron or is collected at the cathode. The excess current of electrons is the photocurrent gain and is the ratio of the hole lifetime to the electron transit time. Gains of up to 100 have been obtained in GaInAs devices. Of course, photoconductors have been made from GaAs and AlGaAs as well as Si.

Phototransistors have been made from AlGaAs–GaAs *npn* structures with the emitter being the wide-gap material. The incident photons pass through the emitter and create carriers in the GaAs *p*-type base and *n*-type collector regions. As in the case of the solar cell (Chapter 4), the holes accumulate in the base (the base connection is floating) and decrease the emitter–base junction potential, causing injection of electrons from the emitter into the base. Current gain is the ratio of the lifetime of the injected electrons in the base to their transit time across the base. Current gains of 5000 have been achieved in AlGaAs–GaAs floating base transistors.

Heterostructure technology permits a wide range of photon detection devices. For example, a 10-$\mu$m infrared detector has been made with a superlattice of doped GaAs–AlGaAs quantum wells. The wells were designed to have only two confined states, $E_1$ and $E_2$. Electrons occupy the ground state, $E_1$. An applied voltage establishes a field in the structure. When infrared radiation equal to $E_2 - E_1$ is incident on the sample, electrons are excited to the level $E_2$ and can then tunnel out of the well (Fig. 13.22). The photo-generated and tunneled electron then travels a distance $L$, thereby generating a photocurrent, before being recaptured by one of the wells.

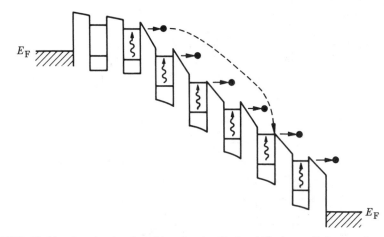

**FIGURE 13.22** Conduction-band levels of a GaAs–AlGaAs multiple-quantum-well structure under applied voltage. The wiggly arrows show the photon-induced transitions from the filled level, $E_1$, to the empty level $E_2$. The photo-excited electrons subsequently tunnel into the high-field domain. [From Levine et al. (1987).]

# 13.8 Optoelectronic Integrated Circuits

Monolithic integration of photon emitter–detector systems and electron transistors provides increased system data capacity and speed. The driving force is the high data rate provided by fiber optic coupling for interchip as well as long-haul (200 km per stage) communication networks. The original development concept was to carry out electronic data processing on large-scale integrated circuits formed on silicon or gallium arsenide substrates. The output of these integrated circuits was used to drive discrete photon emitters coupled to fiber optics and coupled to discrete photodetectors.

The approach was natural because laser diodes are 100 to 300 $\mu$m long, require parallel facets for optical cavities, and generally have a vertical height of 1 or 2 $\mu$m. Optical emitters require more space than electronic transistors. Further, laser diodes have threshold current densities of kA/cm$^2$ and require sinks for heat dissipation. Local heating is a major source of performance degradation. Even at an injection level of 50 mA, a laser diode with a 3-$\Omega$ series resistance will operate more than 10°C higher in temperature than an adjacent heat sink.

The materials requirements for optical emitters differ from those for field-effect transistors. Both emitters and transistors require defect-free material and controlled process technology for high production yields. However, laser diodes require direct-band-gap material and high injection levels (carrier densities of $10^{18}$/cm$^3$), which in turn require heavily doped layers. Heterojunctions are required for carrier confinement in optical emitters, whereas homojunctions are common in transistors, which are, however, based on the use of epitaxial layers grown on an inactive substrate.

The last statement, epitaxial growth on inactive substrates, is the key concept behind integrated optoelectronics. From the onset of high-circuit-density integrated-circuit development, the substrate—100 to 200 $\mu$m thick—provided an easily handled, large-area, perfect crystal lattice on which 1-$\mu$m-thick epitaxial layers could be grown to contain the active circuit elements. Processing—lithography, ion implantation, diffusion, and metallization—focused on the top 1 $\mu$m with the substrate treated as a passive component. At one point sapphire was used as the substrate with silicon layers grown epitaxially (despite the poor lattice match) on top; this silicon-on-sapphire (SOS) technology still has applications in radiation-hard circuits.

Heteroepitaxy is the key to integrated optoelectronics and is described in Chapter 14. There are basically two approaches: GaAs grown on silicon substrates or AlGaAs grown on GaAs substrates. We use GaAs as the generic term to include III–V compound semiconductors which have high carrier mobilities and direct band gaps that can be tailored for light emission and detection at the appropriate wavelengths for fiber optics.

The easiest to implement is AlGaAs on GaAs. Here the substrate can be a semi-insulating layer of GaAs on GaAs which provides a lattice-matched system for heteroepitaxy. The major problems lie in process technology to ensure that the manipulation of the light-emitting materials with their heavy doping and optical cavities do not degrade the electronic properties of the driving, logic, and memory components in the adjacent portions of the chip.

The most elegant approach is GaAs on silicon. Silicon wafers can be grown to large size—6- to 8-in wafer diameters—and the processing technology is low cost. As a substrate material silicon is ideal because of its high mechanical strength and perfect lattice. The thermal conductivity of silicon is three times greater than that of GaAs: 1.5 and 0.46 W/cm · °C, respectively. This is an important factor when considering the heat-dissipation requirements for optical emitters. The major problem is the differences between Si and GaAs: the 4% mismatch in lattice constant, 0.543 and 0.565, respectively, and the factor of 2.5 between the thermal expansion coefficients, $2.6 \times 10^{-6}$ and $6.86 \times 10^{-6}$ per °C, respectively. Despite these differences, it is possible to grow high-quality GaAs layers on silicon and to fabricate field-effect transistors and light-emitting diodes.

Semiconductor chips using both technologies have been made with integrated optoelectronic transmitters, with both lasers and support circuitry, for driving and shaping the light pulses of the laser. Receiver chips contain the photodetectors as well as the amplifiers and pulse-shaping circuitry. Cost and specific application dictate the degree of optoelectronic integration used. The standard silicon integrated circuit with megabit memory and the discrete light emitters and detectors are useful in many applications. Integrated optoelectronics with optical interconnects between transistor circuits are now available.

# REFERENCES

AGRAWAL, G. P., and N. K. DUTTA, *Long-Wavelength Semiconductor Lasers,* Van Nostrand Reinhold, New York, 1986.

BEAN, J. C., "Silicon-Based Semiconductor Heterostructures," Ch. 11 in *Silicon-Molecular Beam Epitaxy* ed. E. Kasper and J. C. Bean, (CRC Press, Boca Raton, Fla., 1988).

CASEY, JR., H. C., and M. B. PANISH, *Heterostructure Lasers,* Part A, *Fundamental Principles,* Part B. *Materials and Operating Characteristics,* Academic Press, Orlando, Fla., 1978.

DINGLE, R., "Confined Carrier Quantum States in Ultrathin Semiconductor Heterostructures," in *Festkörperprobleme 15,* ed. H. J. Queisser, Pergamon Press, Braunschweig, West Germany, 1975, p. 21.

DINGLE, R., ed., *Applications of Multiquantum Wells, Selective Doping and Superlattices,* Vol. 24 in *Semiconductors and Semimetals,* ed. R. K. Willardson and A. C. Beer, Academic Press, Orlando, Fla., 1987.

DINGLE, R., W. WIEGMANN, and C. H. HENRY, *Phys. Rev. Lett. 33,* 827 (1974).

EINSPRUCH, N. G., and W. R. WISSEMAN, *GaAs Microelectronics,* Vol. II in *VLSI Electronics for Microstructure Science,* ed. N. G. Einspruch, Academic Press, Orlando, Fla., 1985.

HUNSPERGER, R. G., *Integrated Optics: Theory and Technology,* 2nd ed., Springer-Verlag, Berlin, 1984.

KROEMER, H., "Heterojunction Bipolar Transistors and Integrated Circuits" *Proc. IEEE*, 70, 13 (1982).

KRESSEL, H., ed., *Semiconductor Devices for Optical Communication,* Springer-Verlag, Berlin, 1980.

LEVINE, B. F., K. K. CHOI, C. G. BETHEA, J. WALKER, and R. J. MALIK, *Appl. Phys. Lett. 50,* 1092 (1987).

MILNES, A. G., and D. L. FEUCHT, *Heterojunctions and Metal–Semiconductor Junctions,* Academic Press, New York, 1972.

MORKOÇ H., in *Molecular Beam Epitaxy and Heterostructures,* ed. L. L. Chang and K. Ploog, NATO ASI Series, Series E, No. 387, Martinus Nijhoff, Dordrecht, The Netherlands, 1985, p. 625.

MORKOÇ H., and H. UNLU, "Factors Affecting the Performance of (Al, Ga)As/GaAs and (Al, Ga)As/InGaAs Modulation-Doped Field-Effect Transistors: Microwave and Digital Applications," in *Semiconductors and Semimetals,* Vol. 24, *Applications of Multiquantum Wells, Selective Doping, and Superlattices,* ed. Raymond Dingle, 1987, Chapter 2.

PARKER, E. H. C., ed., *The Technology and Physics of Molecular Beam Epitaxy,* Plenum Press, New York, 1985.

YU, P. K. L., and P. C. CHEN, "GaAs Opto-Electronic Device Technology," in *Introduction to GaAs Technology,* ed. C. Wang, Wiley, New York, 1988, Chapter 7.

## PROBLEMS

**13.1** For a AlGaAs–GaAs *npn* transistor at 300 K, where the emitter is AlGaAs with $5 \times 10^{17}$ electrons/cm$^3$ and the base is GaAs with $5 \times 10^{18}$ holes/cm$^3$ and a width of 0.2 μm, the value of $\Delta E_G = 0.2$ eV.

(a) Calculate the ratio of injected electrons to injected holes for a forward-bias voltage.

(b) Calculate the diffusion lengths for carrier lifetimes of $10^{-9}$ s if the electron mobility in GaAs is 5000 cm$^2$/V · s and the hole mobility in AlGaAs is 500 cm$^2$/V · s.

(c) Calculate the ratio of electron diffusion current densities in a long base ($d = 100$ μm) to that in a short base ($d = 0.5$ μm) for a forward applied voltage.

**13.2.** A modulation-doped, heterostructure FET is formed by AlGaAs on GaAs at 300 K with a sheet carrier concentration of $5 \times 10^{11}$ electrons/cm$^2$ in the channel in the undoped GaAs. The channel length is 0.8 μm and the width is 15 μm, the electron mobility in the channel is 8000 cm$^2$/V · s.

(a) In the low-voltage region, $V_D = 0.1$ V, calculate $I_D$.

(b) Calculate the gate capacitance $C_G$ for AlGaAs thickness $d = 50$ nm and $\epsilon_r = 12$.

(c) Calculate the mutual transconductance in the velocity-saturated regime, $g_m(\text{sat})$.

(d) Calculate the transit time.

**13.3.** For a one-dimensional quantum well 20 nm wide bounded by infinite potential barriers formed in GaAs, with $m^* = 0.067$ m:

(a) Calculate the energy $E_1$ of the first level.

(b) Calculate the density of states in that level.

(c) Calculate the energy difference between levels $E_1$ and $E_2$ and the wavelength of light equal to this difference.

**13.4.** A double heterojunction laser at 300 K with a GaAs active region 0.2 μm wide has an area of $10^{-4}$ cm$^2$ and a series resistance of 3 Ω. The carrier lifetimes $= 10^{-9}$ s. The threshold current requires an injected carrier density of $10^{18}$/cm$^3$.

(a) Calculate the threshold current $I_{th}$.

(b) What applied voltage $V_A$ is required?

(c) What wavelength of light would be emitted?

**13.5.** For a Schottky barrier on $n$-type GaAs (doped with $10^{16}$ electron/cm$^3$) and a barrier height of 0.8 V at 300 K:

(a) Calculate the reverse voltage required to obtain a depletion width of 2 μm.

(b) Calculate the transit time for electrons using (1) the dielectric relaxation time and then (2) the saturated velocity $v_s$.

(c) What is the maximum and average electric field in the depletion region, and would you expect the electrons to be in the saturated velocity regime?

**13.6.** Consider photodetectors at 300 K with a gain of 100, carrier lifetimes of $10^{-9}$ s, an electron mobilities of 5000 cm$^2$/V · s.

(a) In a photoconductor with 5 V across the contacts, what contact spacing would be required?

(b) In a phototransistor, what base width would be required?

# CHAPTER 14

# Heteroepitaxy

## 14.1 Introduction

Heteroepitaxy is the epitaxial growth of a layer of one semiconductor on another semiconductor where the lattice structures are registered with each other. In Si technology, the epitaxial layer is Si; this kind of epitaxy is termed homoepitaxy. In GaAs and other III–V compound technology, heteroepitaxy is the growth of a layer such as $Al_xGa_{1-x}As$ on GaAs or $Ga_xIn_{1-x}As$ on InP. Several epitaxial growth techniques are now available to grow thin layers of different materials on top of each other in sequence, resulting in double-heterojunction and multi-quantum-well structures. Many modern electronic devices (described in Chapter 13) are based on the epitaxial growth of heterojunctions. While epitaxy offers additional control in defects and doping profiles in the layers, the defects at or near the interface between different layers become increasing more significant as the epitaxial layers become thinner. Interfacial structure and defects ultimately control the performance of the device.

Epitaxial methods can be categorized generally into (1) vapor-phase epitaxy, where the epitaxial layer is condensed from a vapor phase produced by evaporation [molecular beam epitaxy (MBE)] or as a result of chemical reactions between various gaseous reactants near or at the vapor–substrate surface [chemical vapor deposition (CVD)], and (2) liquid-phase epitaxy (LPE), where the epitaxial layer is precipitated out of a liquid solution in a temperature gradient. The processes of molecular beam epitaxy, metal–organic vapor-phase epitaxy (sometimes referred to as MOVPE), and liquid-phase epitaxy are usually used for the growth of hetero-junctions and quantum well structures. Acronyms are listed in Table 13.1.

In Chapter 13 we showed that certain electronic properties are unique only to heterostructures and are not possible to achieve in conventional homostructures. These unique electronic properties depend in large part on how the band edges line up when a heterojunction is formed (i.e., the conduction-band offset $\Delta E_C$ and the valence-band offset $\Delta E_V$). In this chapter we show how two semiconductors join electronically and structurally. If we know how they join together, we can then design heterostructures to give us the desired electronic properties. Heterostructures of interest here are often made with a combination of III–V compounds. The tendency is to match semiconductor pairs with identical or similar lattice constants that would lead to desired electronic or optoelectronic properties. One of the most extensively studied combinations is the $Al_xGa_{1-x}As$–GaAs system. We will use this combination as an example to address some of the issues of heteroepitaxy. We then discuss strained layer superlattices and growth methods.

## 14.2 Ternary Solid Solutions

Figure 14.1 shows the ternary phase diagram of As–Ga–Al at 900°C. The way to read this phase diagram has been explained in Section 10.5. We note that GaAs is connected to AlAs by a solid tie line, indicating that GaAs is in equilibrium with AlAs. In addition, GaAs and AlAs are completely miscible, forming solid solutions with any proportions. This is indicated by the label "$Al_xGa_{1-x}As$" above the solid tie line. Any composition inside the triangle As–GaAs–AlAs has two phases in equilibrium: solid As and $Al_xGa_{1-x}As$. Any composition below the $Al_xGa_{1-x}As$ line and above the 900°C liquidus line also has two phases in equilibrium. A dashed tie line is drawn from $x = 0.01$ (a very Al poor and very Ga rich liquid) to the solid with a composition of $Al_{0.63}Ga_{0.37}As$ in equilibrium. Any composition lying on the dashed tie line has a liquid phase (Al poor) in equilibrium with a solid phase ($Al_{0.63}Ga_{0.37}As$). Any composition lying along and below the liquidus line is a single-phase ternary liquid solution (very As poor). One should recognize that inside the liquid–solid two-phase region, there is a solid solution in equilibrium with any given liquid solution.

The energy gap of ternary alloys depends on the composition. For $Al_xGa_{1-x}As$, the energy gap increases with Al content and changes from a direct

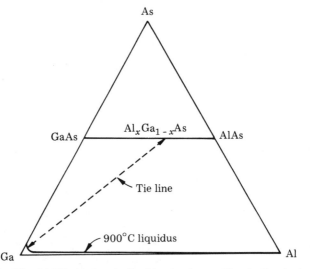

**FIGURE 14.1** The 900°C Al–Ga–As liquidus isotherm. The tie line is drawn from $x = 0.01$ to the equilibrium solid $Al_{0.63}Ga_{0.37}As$. [After Casey and Panish (1978).]

**TABLE 14.1**  Compositional Dependence of the Energy Gaps of the Binary III–V Ternary Solid Solutions at 300 K

| Compound | Direct Energy Gap $E_G$ (eV) |
|---|---|
| $Al_xIn_{1-x}P$ | $1.351 + 2.23x$ |
| $Al_xGa_{1-x}As$ | $1.424 + 1.247x^a$ |
| | $1.424 + 1.455x^b$ |
| $Al_xIn_{1-x}As$ | $0.360 + 2.012x + 0.698x^2$ |
| $Al_xGa_{1-x}Sb$ | $0.726 + 1.129x + 0.368x^2$ |
| $Al_xIn_{1-x}Sb$ | $0.172 + 1.621x + 0.43x^2$ |
| $Ga_xIn_{1-x}P$ | $1.351 + 0.643x + 0.786x^2$ |
| $Ga_xIn_{1-x}As$ | $0.36 + 1.064x$ |
| $Ga_xIn_{1-x}Sb$ | $0.172 + 0.139x + 0.415x^2$ |
| $GaP_xAs_{1-x}$ | $1.424 + 1.150x + 0.176x^2$ |
| $GaAs_xSb_{1-x}$ | $0.726 - 0.502x + 1.2x^2$ |
| $InP_xAs_{1-x}$ | $0.360 + 0.891x + 0.101x^2$ |
| $InAs_xSb_{1-x}$ | $0.18 - 0.41x + 0.58x^2$ |

*Source:* After Casey and Panish (1978).
[a] $(0 < x < 0.45)$.
[b] $(0 < x < 0.37)$; Kuech et al. (1987).

gap to an indirect gap for $x$ greater than about 0.4. Casey and Panish (1978) report

$$E_G = 1.424 + 1.247x \qquad \text{with } 0 < x < 0.45.$$

Kuech et al. (1987) report

$$E_G = 1.424 + 1.455x \qquad \text{with } 0 < x < 0.37. \qquad (14.1)$$

The difference in the two determinations presumably lies in the measurement of the Al content. The compositional dependence of the energy gap in some III–V compounds at 300 K is listed in Table 14.1 and the temperature dependence of the band gap is listed in Table 14.2.

## 14.3 Lattice Matching

To form high-quality heterostructures, it is often necessary to have a semiconductor pair with very similar or identical lattice parameters. When this happens the heterostructure is said to be lattice matched. Heterostructures are often formed by growing ternary alloys (or quaternary alloys) on binary alloys with properly chosen composition to provide lattice matching, and more important, desired energy gap and

**TABLE 14.2** Energy Gaps of the Binary III–V Compounds

| Compound | Type of Energy Gap | Experimental Energy Gap $E_G$ (eV) | | Temperature Dependence of Energy Gap $E_G(T)$ (eV) |
|---|---|---|---|---|
| | | 0 K | 300 K | |
| AlP | Indirect | 2.52 | 2.45 | $2.52-3.18 \times 10^{-4}T^2/(T + 588)$ |
| AlAs | Indirect | 2.239 | 2.163 | $2.239-6.0 \times 10^{-4}T^2/(T + 408)$ |
| AlSb | Indirect | 1.687 | 1.58 | $1.687-4.97 \times 10^{-4}T^2/(T + 213)$ |
| GaP | Indirect | 2.338 | 2.261 | $2.338-5.771 \times 10^{-4}T^2/(T + 372)$ |
| GaAs | Direct | 1.519 | 1.424 | $1.519-5.405 \times 10^{-4}T^2/(T + 204)$ |
| GaSb | Direct | 0.810 | 0.726 | $0.810-3.78 \times 10^{-4}T^2/(T + 94)$ |
| InP | Direct | 1.421 | 1.351 | $1.421-3.63 \times 10^{-4}T^2/(T + 162)$ |
| InAs | Direct | 0.420 | 0.360 | $0.420-2.50 \times 10^{-4}T^2/(T + 75)$ |
| InSb | Direct | 0.236 | 0.172 | $0.236-2.99 \times 10^{-4}T^2/(T + 140)$ |

*Source:* After Casey and Panish (1978).

band offset. Figure 14.2 shows the lattice constant as a function of composition for ternary III–V solid solutions. The lattice constants for binary alloys are also labeled (see also Table 14.3). When the lattice-constant line for a ternary solution crosses the horizontal line for a binary compound, the composition at the intersec-

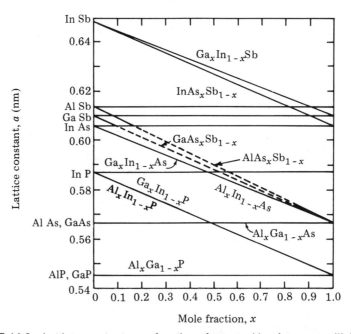

**FIGURE 14.2** Lattice constant as a function of composition for ternary III–V crystalline solid solutions in which the end components differ by less than 0.05 nm. Vegard's law is assumed to be obeyed in all cases. The dashed lines show regions where miscibility gaps are expected. [After Casey and Panish (1978).]

tion provides lattice matching at room temperature. For example, $Al_xIn_{1-x}As$ is lattice matched with InP at $x \simeq 0.5$. At about this same $x$ value, $Al_xIn_{1-x}P$ is lattice matched with GaAs. One can see that the line for AlAs is similar to that of GaAs, indicating that $Al_xGa_{1-x}As$ is closely lattice matched to GaAs for any values of $x$. This fact makes the AlGaAs–GaAs system one of the most interesting systems for the formation of heterostructures.

If one combines the information shown in Fig. 14.2 and the information listed in Table 14.1, an energy gap versus lattice constant plot may be constructed as shown in Fig. 14.3. The boundaries joining the binary compounds give the ternary energy gap and lattice constant. The solid boundaries denote a direct band gap and the dashed boundaries denote an indirect band gap. The area inside the boundaries denotes quaternary alloys. From this figure one can see that GaAs is lattice matched to AlAs and to any ternary solid solution along the boundary between GaAs and AlAs. The ternary alloy can have energy gap from about 1.4 to 2.2 eV. For the epitaxial growth of $In_xGa_{1-x}As$ on InP, lattice matching is possible only at $x = 0.53$ ($E_G \sim 0.8$ eV). Quaternary alloys (GaInPAs) along the GaAs–AlAs boundary but inside the area bounded by GaP–GaAs–InAs–InP are also lattice matched to GaAs. On the other hand, a series of quaternary solid solutions, GaInAsP, inside the area bounded by GaP–GaAs–InAs–InP can be lattice matched to InP. Figure 14.4 shows the $x$-$y$ compositional plane for $Ga_xIn_{1-x}P_yAs_{1-y}$ at 300 K. This figure is another representation of the area bounded by GaP–GaAs–InP–InAs shown in Fig. 14.3. It is more convenient to use Fig. 14.4 to obtain the composition of the

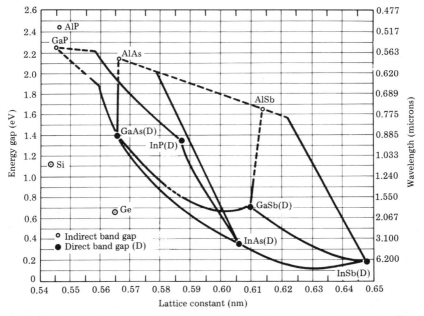

**FIGURE 14.3** Energy gap and lattice constant for several III–V compounds. The boundaries joining the binary compounds give the ternary energy gap and lattice constant. [After Cho (1985).]

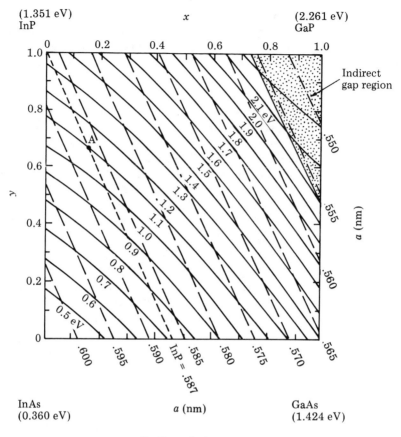

FIGURE 14.4 The *x-y* compositional plane for $Ga_xIn_{1-x}P_yAs_{1-y}$ at 300K. The *x-y* coordinate of any point in the plane gives the composition. The solid lines are constant direct-energy-gap values. The dashed lines are constant lattice constants. The lattice constant of InP is shown to illustrate the range of *x* and *y* that lattice match InP. [After Casey and Panish (1978).]

quaternary solid solution with a desired energy gap lattice matched to a given substrate. For example, if a quaternary alloy with a $E_G$ of about 1.1 eV and lattice matched to InP is desired, the composition of this alloy should be $Ga_{0.15}In_{0.85}P_{0.65}As_{0.35}$, as shown by point A in Fig. 14.4. Table 14.3 lists some of the lattice constants of binary III–V compounds. Table 14.4 lists some of the binary to ternary III–V lattice-matched systems, and Table 14.5 lists some of the binary to quaternary III–V lattice-matched systems applicable to heterostructure lasers.

## 14.4 Band-Edge Lineup of Abrupt Heterojunctions

When two different semiconductors are placed together forming an abrupt hetero-junction [i.e., semiconductor A is grown epitaxially on semiconductor B with a

**TABLE 14.3** Lattice Constants of the
Binary III–V Compounds

| Compound | Lattice Constant (nm) | $T$ (°C) |
|----------|----------------------|----------|
| AlP | 0.5451 | — |
| AlAs | 0.56605 | 0 |
| AlSb | 0.61355 | 18 |
| GaP | 0.545117 | 25 |
| GaAs | 0.565325 | 27 |
| GaSb | 0.609593 | — |
| InP | 0.586875 | 18 |
| InAs | 0.60584 | 18 |
| InSb | 0.647937 | 25 |

*Source:* After Casey and Panish (1978).

very sharp compositional transition from A to B at the interface (within a few lattice constants)], one of the central issues is how the bands are lined up at the interface. There are generally three types of band lineups (Kroemer, 1985): (1) straddling, (2) staggered, and (3) broken-gap (see Fig. 14.5). If the semiconductor pair are lattice matched, it is often possible to calculate or estimate the band offsets.

**TABLE 14.4** Binary-to-Ternary III–V Lattice-Matched Systems for Heterostructure Lasers

| System | Laser Emission | | |
|--------|----------------|---|---|
| | Energy $E$ (eV) | Wavelength $\lambda$ ($\mu$m) | Temperature (K) |
| AlSb–Ga$_{0.9}$In$_{0.1}$Sb | 0.63 | 2.0 | 300 |
| | 0.71 | 1.75 | 0 |
| AlSb–InAs$_{0.82}$Sb$_{0.18}$ | 0.38 | 3.26 | 0 |
| GaAs–Al$_x$Ga$_{1-x}$As | 1.42 | 0.87 | 300 |
| GaAs–Al$_x$Ga$_{1-x}$As– Al$_y$Ga$_{1-y}$As,  $y < x$ | 1.42–1.61 | 0.87–0.773 | 300 |
| GaAs–Ga$_{0.51}$In$_{0.49}$P | 1.42 | 0.87 | 300 |
| GaP–Al$_x$Ga$_{1-x}$P | Indirect $E_G$ | — | — |
| GaSb–AlAs$_{0.08}$Sb$_{0.92}$ | 0.72 | 1.72 | 300 |
| GaSb–InAs$_{0.91}$Sb$_{0.09}$ | 0.347 | 3.57 | 0 |
| InP–Ga$_{0.47}$In$_{0.53}$As | 0.86 | 1.44 | 300 |
| InP–Al$_{0.47}$In$_{0.53}$As | 1.35 | 0.92 | 300 |
| InAs–AlAs$_{0.16}$Sb$_{0.84}$ | 0.42 | 3.0 | 0 |
| InAs–GaAs$_{0.08}$Sb$_{0.92}$ | 0.42 | 3.0 | 0 |

*Source:* After Casey and Panish (1978).

**TABLE 14.5** Binary to Quaternary III–V Lattice-Matched Systems for Heterostructure Lasers

| Quaternary | Lattice-Matching Binary |
|---|---|
| $Al_xGa_{1-x}P_yAs_{1-y}$ | GaAs for low $y$ |
| $Al_xGa_{1-x}P_ySb_{1-y}$ | GaAs, InP, InAs |
| $Al_xGa_{1-x}As_ySb_{1-y}$ | InP, InAs, GaSb |
| $Al_xIn_{1-x}P_yAs_{1-y}$ | InP |
| $Al_xIn_{1-x}P_ySb_{1-y}$ | GaAs, InAs, AlSb, GaSb |
| $Al_xIn_{1-x}As_ySb_{1-y}$ | InP, GaSb, AlSb |
| $Ga_xIn_{1-x}P_yAs_{1-y}$ | InP, GaAs |
| $Ga_xIn_{1-x}P_ySb_{1-y}$ | GaAs, InP, InAs, AlSb |
| $Ga_xIn_{1-x}As_ySb_{1-y}$ | InP, GaSb, AlSb |
| $(Al_xGa_{1-x})_yIn_{1-y}P$ | GaAs, $Al_xGa_{1-x}As$ |
| $(Al_xGa_{1-x})_yIn_{1-y}As$ | InP |
| $(Al_xGa_{1-x})_yIn_{1-y}Sb$ | AlSb |
| $Al(P_xAs_{1-x})_ySb_{1-y}$ | InP |
| $Ga(P_xAs_{1-x})_ySb_{1-y}$ | InP |
| $In(P_xAs_{1-x})_ySb_{1-y}$ | AlSb, GaSb, InAs |

*Source:* After Casey and Panish (1978).

If the semiconductor system is lattice mismatched, the lineups may depend on how the mismatch is accommodated at the interface; theoretical prediction of the band lineups becomes difficult and may not be very meaningful. Before we discuss some of the theories for the prediction of band lineups, let us consider certain correlations and rules that are generally observed in energy-band lineups.

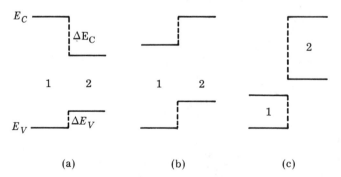

**FIGURE 14.5** Types of energy-band lineups: (a) straddling (i.e., AlGaAs/GaAs); (b) staggered (i.e., InP/InSb); (c) broken-gap (i.e., InAs/GaSb). [After Kroemer (1985).]

## 14.4.1 The Anion Correlation Rule

There is a general observation that in systems such as AlGaAs/GaAs and InAs/GaAs where the anion atoms are the same (i.e., As in these two systems) the valence-band offset $\Delta E_V$ is smaller than the conduction-band offset, $\Delta E_C$. For example, the valence-band offset, $\Delta E_V$, of the $Al_xGa_{1-x}As/GaAs$ was measured experimentally and can be expressed as

$$\Delta E_V (Al_xGa_{1-x}As/GaAs) = 0.45x \text{ eV}. \qquad (14.2)$$

The sign convention used here is that $\Delta E_V$ (A/B) is positive if semiconductor A has the lower valence band. For $x = 0.3$, $\Delta E_V = 0.135$ eV. Then

$$\Delta E_C = \Delta E_G - \Delta E_V = 0.436 - 0.135 = 0.301 \text{ eV}.$$

One can see that $\Delta E_V$ is about 30% of $\Delta E_G$, whereas $\Delta E_C$ is about 70% of $\Delta E_G$ for this pair of semiconductors with a common anion of As. ($\Delta E_G$ is calculated using Table 14.1.) The reason for this correlation is related to the fact that the valence-band wave functions derive largely from the anion atom wave function. The band lineup for the $Al_xGa_{1-x}As/GaAs$ heterostructure for $x < 0.4$ is of the straddling type (Fig. 14.5).

For semiconductor pairs with common cations X, the valence-band energies at the interface correlate with anion electronegativities. The electronegativities for P, As, and Sb are 2.19, 2.18, and 2.05, respectively. For III–V compounds this suggests that

$$E_V(XP) < E_V(XAs) < E_V(XSb). \qquad (14.3)$$

The relationship shown in eq. (14.3) implies that the valence-band edge moves up as one crosses the interface from GaP to GaAs for such a pair of semiconductors with a common cation Ga.

## 14.4.2 Linearity and Transitivity

Many band-edge lineup theories assume that there is a specific absolute energy associated with the band edges of every semiconductor, and that the band offsets are simply the difference between the relevant absolute band energies of the semiconductor pair. This means that these theories assume that

$$\Delta E_V [A/B] = E_V(B) - E_V(A). \qquad (14.4)$$

Such theories are called linear theories. If linear theories are valid, we have the following relationship:

$$\Delta E_V (A/B) + \Delta E_V (B/C) + \Delta E_V (C/A) = 0. \qquad (14.5)$$

This property is referred to as "transitivity" by Frensley and Kroemer (1977). The transitivity property may be used to test the validity of theories. If the calculated band offsets for three pairs of semiconductors do not pass the transitivity test, the validity of the theory is therefore in doubt. The transitivity property may also be used to test for experimental accuracy and consistency. For example, for three pairs of lattice-matched semiconductors, Ge on ZnSe, ZnSe on GaAs, and Ge on GaAs, the measured $\Delta E_V$ values follow the transitivity property:

$$\Delta E_V(\text{Ge/ZnSe}) + \Delta E_V(\text{ZnSe/GaAs}) - \Delta E_V(\text{Ge/GaAs})$$
$$= -1.52 \text{ eV} + 0.96 \text{ eV} + 0.53 \text{ eV} = -0.03 \text{ eV}.$$

The algebraic sum of $-0.03$ eV is very small and is below the accuracy of measurements. It appears that transitivity is a sound assumption for well-prepared heterostructures.

### 14.4.3  The Electron Affinity Rule

Electron affinities are the energies required to remove an electron from the bottom of the conduction band to the outside of the semiconductor (commonly referred to as the vacuum level). If band bending is present near the surface, the electron affinity is the energy to remove an electron near the surface, as shown in Fig. 14.6. The electron affinity rule asserts that the conduction-band offset at an abrupt heterojunction is equal to the difference in electron affinities between the two semiconductors, with the signs chosen such that the semiconductor with the smaller affinity has the higher conduction band at the interface:

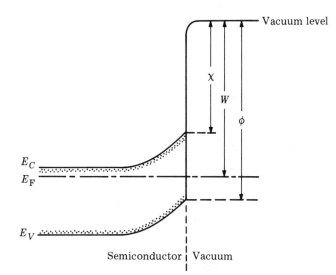

**FIGURE 14.6**   The electron affinity $\chi$, the work function $W$, and the ionization energy $\phi$ all represent the work required to remove an electron to the outside of the semiconductor, just beyond the range of dipole and image forces but from different starting levels. For the electron affinity, the initial energy is the bottom of the conduction band; for the ionization energy (or photothreshold) it is the top of the valence band. In both cases the band energies are taken at the surface, not deep in the bulk. For the work function the starting energy is the Fermi level. [After Kroemer (1985).]

$$\Delta E_C = E_{C_2} - E_{C_1} = \chi_1 - \chi_2 \qquad (14.6)$$

where $\chi$ is the electron affinity. Since the ionization energy $\phi$ is the sum of $\chi$ and $E_G$, it follows that

$$\Delta E_V = E_{V_2} - E_{V_1} = \phi_1 - \phi_2 \qquad (14.7)$$

or

$$\Delta E_C + \Delta E_V = \Delta E_G. \qquad (14.8)$$

An example of band lineup for an $n$-GaAs/$p$-Ge lattice-matched heterojunction using the electron affinity rule is shown in Fig. 14.7; the parameters used in this example are listed in Table 14.6. The band diagrams for the two semiconductors before joining together to form a heterojunction are shown in Fig. 14.7a. There is no band bending involved and the vacuum levels line up. When the two semiconductors are placed together to form a heterojunction, the Fermi levels line up under thermal equilibrium, resulting in the band-lineup diagram shown in Fig. 14.7b. The vacuum levels are now displaced by an amount called the built-in potential.

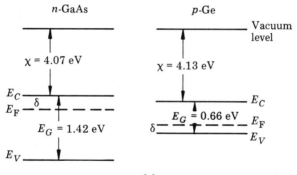

(a)

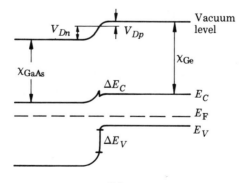

(b)

**FIGURE 14.7** Construction (not drawn to scale) of the energy-band diagram for $n$-$p$ GaAs/Ge. [After Milnes and Feucht (1972).]

**TABLE 14.6**   Values Assumed in Construction of the $n$-$p$ GaAs/Ge Heterojunction

|  | GaAs | Ge |
|---|---|---|
| Energy gap, $E_G$ (eV) | 1.42 | 0.66 |
| Electron affinity, $\chi$ (eV) | 4.07 | 4.13 |
| Net donor doping, $N_D - N_A$ (cm$^{-3}$) | $10^{16}$ | — |
| Net acceptor doping, $N_A - N_D$ (cm$^{-3}$) | — | $3 \times 10^{16}$ |
| $E_C - E_F = \delta_{GaAs}$ (eV) | 0.1 | — |
| $E_F - E_V = \delta_{Ge}$ | — | 0.14 |
| Lattice constant, $a$ (nm) | 0.5654 | 0.5658 |
| Dielectric constant, relative | 13.1 | 16 |

*Source:* After Milnes and Feucht (1972)

The built-in potential, $V_0$, is the difference in vacuum levels after the Fermi levels on both sides of the heterojunction line up. In the notation of Fig. 14.7b, the built-in potential equals $V_{Dn} + V_{Dp}$. The conduction band offset is

$$\Delta E_C = \chi_{Ge} - \chi_{GaAs} = (4.13 - 4.07)\text{ eV} = 0.06\text{ eV}.$$

The corresponding valence-band offset $\Delta E_V$ is

$$\Delta E_V = \Delta E_G - \Delta E_C = (1.42 - 0.66) - 0.06 = 0.7\text{ eV}.$$

The energy "spike" in $\Delta E_C$ shown in Fig. 14.7b is due to the band bendings near the interface conforming to the built-in potential. The energy spike may also appear in $\Delta E_V$, depending on the heterojunction. For example, an energy spike in $\Delta E_V$ will appear in a $p$-GaAs/$n$-Ge heterojunction.

The electron affinity rule is simple and widely used. The question is: How good it is compared to experimental values? This question was addressed by R. S. Bauer and his co-workers (Bauer et al., 1983). They compared the experimentally measured $\Delta E_C$ values with those predicted by the electron affinity rule. Figure 14.8 shows the comparison. As one can see, the agreement between predicted values and experimental value is not good. One should recognize that electron affinity is a property of the free surface for a small test charge. When two semiconductors are joined together, there may be other forces associated with the interface, which cause the energy required to remove electrons from the bottom of the conduction band to deviate from that of a free surface. Table 14.7 lists the electron affinities and other properties of commonly encountered semiconductors.

### 14.4.4 Harrison Atomic-like Orbital Theory

The Harrison atomic-like orbital (HAO) theory is a more successful theory for the prediction of band lineups. The idea here is to calculate the potential difference at the interface of a semiconductor pair using potentials for free atoms (Harrison, 1977). We shall follow the presentation of Kroemer (1985) for a discussion of this approach. Kroemer expressed the results of the HAO model in terms of band-lineup

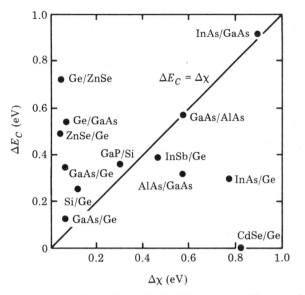

**FIGURE 14.8**  Conduction-band discontinuity $\Delta E_C$ measured by core-level photo-electron spectroscopy using the band gaps given in Milnes and Feucht (1972). The materials are listed with the substrate semiconductor written first, and the deposited overlayer written second. While the growth conditions vary widely among these results, no systematic trend is obtained when the electron affinity rule $\Delta E_C = \Delta \chi$ is applied by plotting the offsets as a function of electron affinity difference $\Delta \chi$ [from tables in Milnes and Feucht (1972)]. [After Bauer et al. (1983).]

tables, as shown in Table 14.8. The bottom entry in each box represents the valence-band edge, the top the conduction-band edge. The valence-band edge of GaAs is taken to be 0.0 eV and serves as the reference point for all other entries. The designation X, $\Delta$, or L preceding the conduction-band value indicates an indirect band (see Chapter 12); the corresponding direct gap at $k = 0$ (or $\Gamma$ in Fig. 12.21a) is given in parentheses. Let us demonstrate the use of these tables with an example.

**EXAMPLE 1:** _____

Estimate the band offset of the InAs/GaSb system using the HAO lineup tables.
From eq. (14.4), we have

$$\Delta E_V(\text{InAs/GaSb}) = E_V(\text{GaSb}) - E_V(\text{InAs})$$
$$= (+0.84) - (+0.32) \text{ eV}$$
$$= +0.52 \text{ eV}.$$

The values of $E_V$ for GaSb and InAs are given in Table 14.8a. Following the sign convention, the valence band moves up from InAs to GaSb across the abrupt interface of this heterostructure. The HAO tables also give the minimum band-gap energies: for example, the minimum band-gap energy of GaAs: $\Delta E_G = +1.42 + 0.0 = 1.42$ eV. This is the algebraic sum of the conduction-

**TABLE 14.7** Properties of Some Semiconductors Used in Heterojunctions

| Material | Energy Gap 300°K (eV)[a] | Gap Transition | Mobility 300 K (cm²/V · s) | | Temperature Coefficient of Expansion at 300 K ($\times 10^{-6}$ °C$^{-1}$) | Dielectric Constant (Relative)[a] | Electron Affinity (eV) |
|---|---|---|---|---|---|---|---|
| | | | Electron | Hole | | | |
| Si | 1.12 | Indirect | 1,350 | 480 | 2.33 | 11.9 | 4.01 |
| Ge | 0.66 | Indirect | 3,600 | 1,800 | 5.75 | 16.0 | 4.13 |
| AlAs | 2.15 | Indirect | 280 | — | 5.2 | 10.1 | — |
| AlSb | 1.58 | Indirect | 900 | 400 | 3.7 | 14.4 | 3.65 |
| GaP | 2.26 | Indirect | 300 | 150 | 5.3 | 11.1 | 4.3 |
| GaAs | 1.42 | Direct | 5,000–8,000 | 300 | 5.8 | 13.1 | 4.07 |
| GaSb | 0.72 | Direct | 5,000 | 1,000 | 6.9 | 15.7 | 4.06 |
| InP | 1.35 | Direct | 4,500 | 100 | 4.5 | 12.4 | 4.38 |
| InAs | 0.36 | Direct | 30,000 | 450 | 4.5 (5.3) | 14.6 | 4.9 |
| InSb | 0.17 | Direct | 80,000 | 450 | 4.9 | 17.7 | 4.59 |
| ZnS (hex.) | 3.68 | Direct | 120 | — | 6.2–6.5 | 5.2 | 3.9 |
| ZnSe | 2.67 | Direct | 530 | — | 7.0 | — | 4.09 |
| ZnTe | 2.26 | Direct | 530 | 130 | 8.2 | — | 3.5 |
| CdS (hex.) | 2.42 | Direct | 340 | — | 4.0 | 5.4 | 4.5 |
| CdSe (hex.) | 1.7 | Direct | 600 | — | 4.8 | 10.0 | 4.95 |
| CdTe | 1.56 | Direct | 700 | 65 | — | 10.2 | 4.28 |
| SiC (hex.) | 3.0 | Indirect | 60–120 | 10–20 | 5.7 | 10.0 | — |
| PbTe | 0.31 | Indirect | 2,500 | 1,000 | — | 30.0 | — |

*Source:* After Milnes and Feucht (1972).

[a]From Sze (1981).

**TABLE 14.8** Band-Edge Values from the HAO Theory[a]

| | (a) III–V Compounds | | | | (b) II–VI Compounds | | |
|---|---|---|---|---|---|---|---|
| | P | As | Sb | | S | Se | Te |
| Al | X: 1.95 (3.1) −0.50 | X: 2.17 (2.52) −0.04 | X: 2.44 (3.08) +0.86 | Zn | +1.93 −1.87 | +1.82 −1.05 | +2.42 +0.03 |
| Ga | X: 1.79 (2.31) −0.47 | +1.42 0.00 | +1.57 +0.84 | Cd | +0.97 −1.59 | +1.02 −0.82 | +1.81 +0.21 |
| In | +1.24 −0.11 | +0.68 +0.32 | +1.29 +1.12 | | | | |

(c) The cuprous halides (only the valence-band energies are given)

| CuCl | CuBr | CuI |
|---|---|---|
| −3.58 | −2.37 | −1.09 |

(d) Si and Ge

| | Si | Ge |
|---|---|---|
| | $\Delta$: +1.15 (4.21) +0.03 | L: +1.08 (1.22) +0.41 |

*Source:* After Kroemer (1985).

[a]See text for details.

band edge energy and the valence-band edge energy. For Ge, the minimum band-gap energy is $E_G = (+1.08 - 0.41) = 0.67$ eV. Since this minimum energy occurs along $k = <111>$ direction, it is an indirect gap. Germanium has a direct gap with an energy of 1.22 eV, quoted in parentheses in Table 14.8d.

Figure 14.9 shows a plot of $\Delta E_V$(experimental) versus $\Delta E_V$(HAO). The agreement is substantially better than that shown in Fig. 14.8. There are other theories for the prediction of band offsets. The HAO theory appears to be the most successful theory at present. Interested readers are referred to a review by Kroemer (1985) for more details.

# 14.5 Graded Junctions

In previous discussions we considered only abrupt junctions where the compositional transition across the interface from one semiconductor to the other changes within a few lattice constants. Energy spikes in $\Delta E_C$ and $\Delta E_V$ may be present at such an interface (see Fig. 14.7). These energy spikes may not be desirable in a heterostructure, as they present an additional energy barrier to current transport. Energy spikes can be smoothed out by compositional grading across the junction. In the preparation of the heterostructure, graded junctions are more often encountered in liquid-phase epitaxy (LPE), although abrupt junctions can be grown by LPE. Molecular beam epitaxy generally produces abrupt junctions unless graded junctions are deliberately prepared. The question we now ask is: What happens to the band offset and energy spikes across a graded junction? A simplified treatment of this issue by Cheung et al. (1975) is given here. Let us use the GaAs/$Al_{0.3}Ga_{0.7}As$

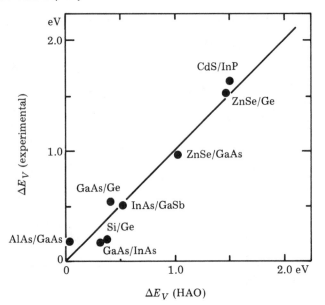

**FIGURE 14.9**   Experimental valence-band offsets in eV for the seven reference systems, plotted as a function of the theoretical valence-band offsets as predicted by the Harrison theory. The value for CdS/InP is also shown. The AlAs/GaAs value is extrapolated from $Al_xGa_{1-x}As/GaAs$ for $x = \sim 0.3$. [After Kroemer, (1985).]

$p$-$n$ heterojunction as an example. The energy-band diagram for an abrupt junction is shown in Fig. 14.10. There is an energy spike in $\Delta E_C$; $\Delta E_V$ is assumed to be small (only 15% of $\Delta E_G$). The total built-in potential, $V_0$, is shown to be 1.643 eV in the figure with the given parameters. This band diagram is quite similar to that for a $p$-$n$ homojunction (shown in Fig. 4.1) except for the presence of $\Delta E_C$, $\Delta E_V$, and $\Delta E_G$ across the junction. In fact, the band diagram of a heterojunction may be approximately constructed from a homojunction band diagram using the principle of superposition (i.e., adding $\Delta E_C$, $\Delta E_V$, and $\Delta E_G$) onto a band diagram of a homojunction. To apply the principle of superposition, we start by expressing the spatial variation of the valence band, $E_V(x)$, across a junction at $x = 0$ by a sum of two parts:

$$E_V(x) = E_s(x) + \Delta E_V(x) \tag{14.9}$$

where $E_s(x)$ is related to the built-in potential, $V_0$ [eq. (14.12)]. For the conduction band, we may write

$$E_C(x) = E_s(x) + E_{G_1} + \Delta E_C(x) \tag{14.10}$$

where $E_{G_1}$ is the band gap of the smaller-band-gap material. Note that if we set $\Delta E_C(x) = \Delta E_V(x) = 0$ and $E_{G_1} = E_{G_2}$, eqs. (14.9) and (14.10) are then applicable to $p$-$n$ homojunctions (i.e., material 1 = material 2).

Figure 14.11 shows such a homojunction with the same parameters as those in Fig. 14.10. One should be aware of the possible difference in effective mass (see

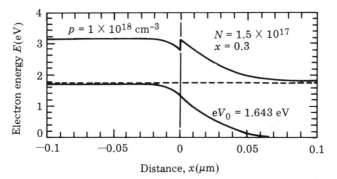

**FIGURE 14.10** Energy-band diagram for GaAs/Al$_{0.3}$Ga$_{0.7}$As $p$-$n$ heterojunction at 297 K. The built-in potential, $V_0$, is 1.643 eV. The valence-band offset, $\Delta E_V$, is small compared to $\Delta E_C$. [After Casey and Panish (1978).]

Chapter 12) between the two semiconductors to correct for the Fermi levels [eq. (3.10)] in constructing a diagram such as that shown in Fig. 14.11. On the $p$-side (the GaAs side) of the "homojunction,"

$$E_s(x) = V_0 \quad \text{for } -\infty < x < -d_p, \tag{14.11}$$

This equation indicates that $E_s(x)$ is equal to the full built-in potential outside the depletion region on the $p$-side. Within the depletion region $d_p$, $E_s(x)$ varies according to eq. (14.12):

$$E_s(x) = V_0 - \frac{eN_A}{2\epsilon_1}(d_p + x)^2 \quad \text{for } -d_p < x < 0. \tag{14.12}$$

On the $n$-side (the AlGaAs side) within the depletion region $d_n$,

$$E_s(x) = \left(V_0 - \frac{eN_A}{2\epsilon_1}d_p^2\right) - \frac{qN_D}{2\epsilon_2}[d_n^2 - (d_n - x)^2] \quad \text{for } 0 < x < d_n. \tag{14.13}$$

Outside the depletion region,

$$E_s(x) = 0 \quad \text{for } d_n < x < \infty. \tag{14.14}$$

While eqs. (14.12) and (14.13) may look complicated, they essentially reflect the parabolic dependence of the built-in potential on distance within the depletion regions and may be derived in a similar manner as that for eq. (4.12) taking two semiconductors into consideration. The intent here is to demonstrate the principle of superposition without going into the details of band diagram calculations of heterojunctions. Interested readers are referred to Chapter 4 in Casey and Panish (1978) for a more detailed discussion. The spatial variation of $E_s(x)$ is plotted in Fig. 14.11a in accordance with eq. (14.11) to (14.14). Addition of $\Delta E_V(x)$ and $\Delta E_C$ to $E_V(x)$ and $E_C(x)$ are done in Fig. 14.11b in accordance with the principle of superposition.

Now suppose that we gradually grade the composition across the interface; one would intuitively speculate that the energy spike in $\Delta E_C$ in Fig. 14.11b could be "smeared out." Cheung et al. (1975) modeled the graded junction using the prin-

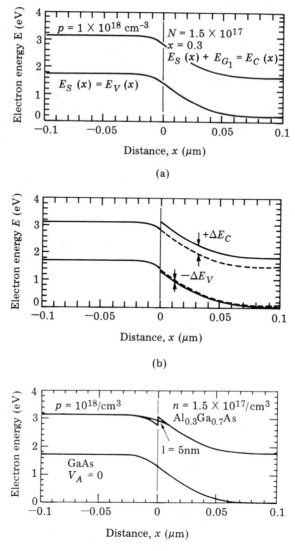

**FIGURE 14.11**   Energy-band diagram by superposition for GaAs/Al$_{0.3}$Ga$_{0.7}$As $p$-$n$ heterojunction of Fig. 14.10: (a) "homojunction" portion of energy-band diagram; (b) addition of $\Delta E_V(x)$ to $E_V(x)$ and $\Delta E_C(x)$ to $E_C(x)$ to obtain heterojunction energy-band diagram; (c) energy-band diagram for graded GaAs/Al$_{0.3}$Ga$_{0.7}$As $p$-$n$ heterojunction with $\ell = 5$ nm. [After Casey and Panish (1978).]

ciple of superposition by adding simple expressions for $\Delta E_C(x)$ to $E_s(x)$. A hyperbolic function was chosen in this case so that we have

$$\Delta E_C(x) = \frac{\Delta E_C}{2} \left[ 1 + \frac{\tanh (x - x_0)}{l} \right] \qquad (14.15)$$

where $x_0$ is the center of the grading region and $l$ is the characteristic grading length, which may be taken to be the compositional transition distance. We recall that the tanh function has a value of $-1$ for large negative arguments, a value of 0 when the argument is zero, and value of $+1$ for large positive arguments. Figure 14.11c shows the energy-band diagram for the graded GaAs/Al$_{0.3}$Ga$_{0.7}$As *p-n* heterojunction. One can see that the energy spike is "smoothed out" over the distance of the graded junction.

Graded-band-gap layers of Ga$_{1-x}$In$_x$As epitaxially grown on GaAs can be used to make ohmic contacts to *n*-type GaAs, as shown in Fig. 14.12. The barrier height of metal–Schottky barriers on *n*-type GaAs (Fig. 14.12a) is about 0.8 eV, and it is difficult to achieve electron concentrations greater than about $5 \times 10^{18}/\text{cm}^3$ in

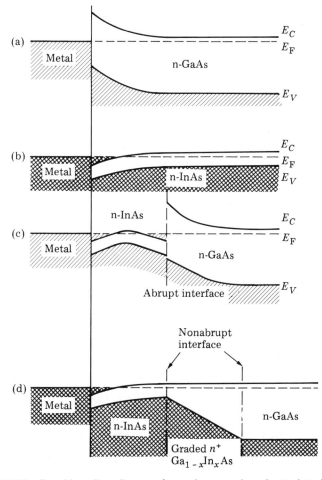

**FIGURE 14.12** Band-bending diagram for various semiconductor interfaces: (a) metal on *n*-GaAs; (b) metal on *n*-InAs; (c) metal on *n*-InAs on *n*-GaAs; (d) metal on $n^+$-InAs on graded $n^+$–Ga$_{1-x}$In$_x$As on *n*-GaAs. [After Woodall et al. (1981).]

the substrate. Consequently, low-resistance ohmic contacts utilizing electron tunneling through the barrier are not feasible, in contrast to the case in Si (Section 12.5).

On metal-to-InAs contacts, the Fermi level is located at or in the conduction band on InAs surfaces (Fig. 14.12b). However, the lattice mismatch of InAs ($a = 0.606$ nm) on GaAs ($a = 0.565$ nm) is large (as shown in Table 14.3) and would lead to a barrier at the interface (Fig. 14.12c). Woodall et al. (1981) demonstrated that an epitaxial layer could be grown on GaAs of $n$-type $Ga_{1-x}In_xAs$ which is graded in composition from $x = 0$ at the GaAs interface to $0.8 \leq x \leq 1$ at the surface (Fig. 14.12d). The barrier height for the metal/$Ga_{1-x}In_xAs$ interface was nearly zero and the graded-band-gap structure on $n$-type GaAs gave a low contact resistance.

## 14.6 Lattice-Mismatched Interfaces

In the discussion of heterostructures above, emphasis was placed on obtaining lattice-matched interfaces for the production of high-quality epitaxial layers. As

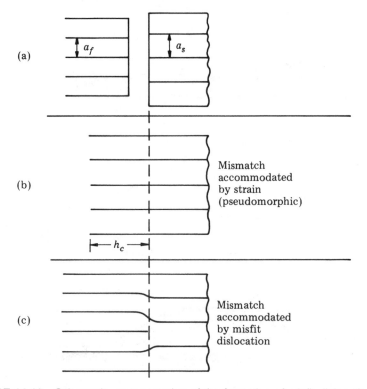

**FIGURE 14.13** Schematic representation of the formation of misfit dislocations: (a) unstrained lattice; (b) thickness of the film is less than $h_c$; (c) thickness of the film is greater than $h_c$ misfit dislocations are generated. [After Woodall et al. (1983).]

the lattice mismatch increases, high-quality epitaxial growth cannot continue indefinitely because the strain energy of the layer is eventually completely or partially relieved by the generation of dislocations at the interface (misfit dislocations).

When a film with an unstrained lattice constant $a_f$ is deposited on a substrate with a different lattice constant, $a_s$ (Fig. 14.13a), it initially grows with a lattice constant equal to that of the substrate. The mismatch (misfit strain), $(|a_f - a_s|)/a_{avg}$, is accommodated by strain in the layer. This is known as a pseudomorphic film. This continues until the film reaches some critical thickness $h_c$ (Fig. 14.13b). When the film thickness exceeds $h_c$, the misfit is accommodated by the formation of misfit dislocations, and the lattice constant of the film relaxes toward the unstrained value (Fig. 14.13c). The strain energy necessary to form these dislocations is approximately proportional to the product of the misfit strain and the film thickness (Woodall et al., 1983). The generation of misfit dislocations may be observed directly using transmission electron microscopy. Figure 14.14 shows an electron micrograph for a layer of $Ga_{0.93}In_{0.07}As$ about 1 μm thick grown on a GaAs substrate with molecular beam epitaxy. The misfit between the two materials is about 0.5%. The perpendicular lines in the image correspond to misfit dislocations at the interface. The thickness of the GaInAs layer exceeded the critical thickness for dislocation generation.

If the thickness of the lattice-mismatched epitaxial layer is below a certain critical thickness, $h_c$, dislocations are not generated because it is energetically more favorable to accommodate the strain energy by stretching the epitaxial layer than to create dislocations. To understand the energetics of the situation, let us consider an epitaxial layer grown on a substrate material. We define $a_f$ and $a_s$ to be the lattice constants in the film and the substrate, respectively. The lattice mismatch $f$

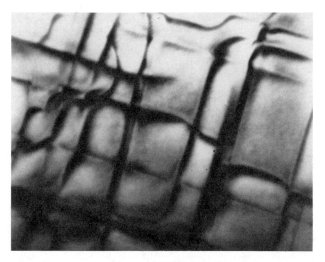

**FIGURE 14.14**  Transmission electron micrograph of the interface between $Ga_{0.93}In_{0.07}As$ about 1 μm thick on GaAs containing misfit dislocations. Misfit is about 0.5%; average dislocation spacing = 140 nm. [After Woodall et al. (1983).]

may be defined as $(a_f - a_s)/a_{avg}$. If the entire mismatch is taken up by accommodation in the film, the elastic strain energy per unit volume of the film is $(Y/2)f^2$, where $Y$ is Young's modulus. The total strain energy $\sigma_s$ in the film per unit area is

$$\sigma_s = \frac{Y}{2} h f^2 \tag{14.16}$$

where $h$ is the thickness of the epitaxial film. Compare this to the energy of a grid of edge dislocations. The energy of an edge dislocation $\Gamma_e$ is given by (assuming $Y \simeq G$, the shear modulus, in this approximation).

$$\Gamma_e \simeq \alpha G b^2 \simeq \alpha Y b^2 \tag{6.36}$$

where $Y$ is Young's modulus, $b$ is a Burgers vector (about the same length as $a_f$ and $a_s$), and $\alpha$ is a geometrical factor (about equal to 1). If we assume that the dislocation grid is a square with length $D$, the dislocation energy $\sigma_D$ per unit area is

$$\sigma_D = \frac{2D\Gamma_e}{D^2} = \frac{2\Gamma_e}{D} = \frac{2\alpha Y b^2}{D} \tag{14.17}$$

remembering that the area of the grid is $D^2$ and the length of the dislocation associated with a square grid is $2D$ (each dislocation is shared by two neighboring grids). For the case where the mismatch is entirely taken up by dislocations, the spacing of the grid, $D$, is given by

$$D = \frac{a_{avg}}{f} \tag{14.18}$$

so that the dislocation energy is

$$\sigma_D = \frac{2\alpha Y b^2 f}{a_{avg}} \simeq 2\alpha Y b f \tag{14.19}$$

where $f$ is the lattice mismatch and $b \simeq a_{avg}$.

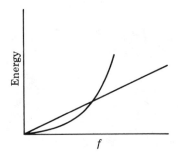

**FIGURE 14.15** Comparison of energy versus mismatch. The straight line represents eq. (14.19) (mismatch is taken up by dislocations). The curve represents eq. (14.16) (the mismatch is taken by strain).

Figure 14.15 shows a schematical plot of eq. (14.16) (mismatch taken up by the strain in the film) and eq. (14.19) (mismatch relaxed by dislocations). For small mismatch, it is energetically more favorable to accommodate the mismatch by straining the film. For large mismatch, the generation of misfit dislocations is favored. The crossing point of these two lines depends on the thickness of the epitaxial film. For a given mismatch $f$ there is a critical thickness $h_c$ at which the crossover occurs. Equating $\sigma_D = \sigma_s$ at $h_c$, we have

$$h_c = \frac{4\alpha b}{f} \tag{14.20}$$

A more sophisticated derivation (Matthews and Blakeslee, 1974) shows that for single heterojunctions

$$h_c = \frac{b(1 - \mu \cos^2\theta)}{2\pi f(1 + \mu) \cos \phi} \ln \left( \frac{h_c}{b} + 1 \right) \tag{14.21}$$

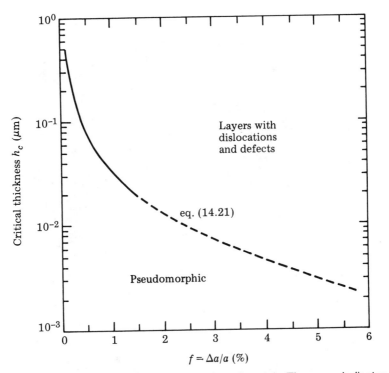

**FIGURE 14.16**  Critical thickness versus lattice mismatch. The curve indicates the critical thickness for misfit dislocation generation. For lattice mismatch smaller than ∼ 1.5%, pseudomorphic two-dimensional growth prevails in layers thinner than $h_c$. For lattice mismatch larger than ∼ 1.5% (indicated by the dashed line), pseudomorphic two-dimensional growth can continue only to a very limited thickness (one or two monolayers); three-dimensional growth follows in subcritical films. For layers thicker than $h_c$, dislocations and staking faults are generated in the epitaxial layer to relax the strain. (Courtesy of J. Woodall, IBM, 1987.)

where $\mu$ the Poisson ratio, $\theta$ the angle between the dislocation line and $b$, and $\phi$ is the angle between the slip direction and that direction in the film plane which is perpendicular to the line of intersection of the slip plane and the interface (for $<001>$ growth direction, $\theta = \phi = 45°$).

Figure 14.16 shows the relationship between $h_c$ and the lattice mismatch, $f$. For small $f$ and thicknesses below $h_c$, two-dimensional pseudomorphic growth without dislocations is possible. Above $h_c$, misfit dislocations are generated to relieve the mismatch. For large $f$ ($\geq 2\%$) and below thicknesses $h_c$, very thin two-dimensional pseudomorphic growth is possible; however, it is more likely to have three-dimensional island growth on the substrate. These islands coalesce, leading to possible stacking fault formation.

For thicknesses greater than $h_c$, misfit dislocations and other defects are present to accommodate the lattice mismatch. Figure 14.17 shows a cross-sectional transmission electron micrograph of an interface between GaAs and $Ga_{0.1}In_{0.9}As$. The

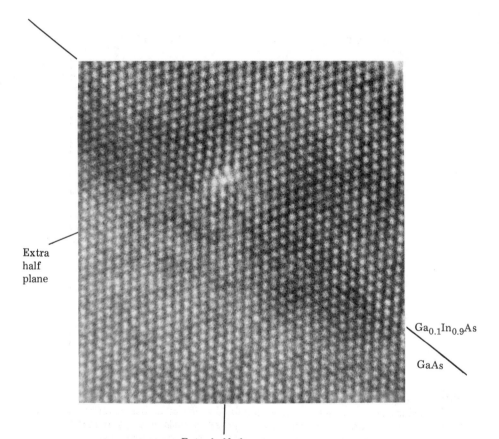

Extra
half
plane

$Ga_{0.1}In_{0.9}As$

GaAs

Extra half plane

**FIGURE 14.17** Cross-sectional transmission electron micrograph of an interface between GaAs and $Ga_{0.1}In_{0.9}As$. The dark areas may be considered as atomic position. (Courtesy of P. Kirchner, M. Chisholm, and J. Woodall, IBM, 1988.)

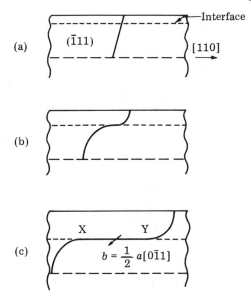

**FIGURE 14.18**    Generation of a length X Y of misfit dislocation by the glide of a threading dislocation. [Adapted from Matthews (1975).]

critical thickness $h_c$ has been exceeded and a misfit dislocation is generated at the interface as shown by the extra half-planes in the figure. Misfit dislocations can also be generated by the glide of a threading dislocation shown in Fig. 14.18. A threading dislocation is a dislocation line that extends from the substrate, across the interface, and through the overgrowth layer (Fig. 14.18a). As a result of the mismatch between the two materials there are shear forces acting on this dislocation. These forces increase as the overgrowth layer increases and cause the dislocation to bow as shown in Fig. 14.18b. Eventually, the shear forces are strong enough to cause the dislocation to glide along the interface and leave a length of misfit dislocation line at the interface. In Fig. 14.18c the interfacial dislocation has a Burgers vector of magnitude $a/2$ directed along $[0\bar{1}1]$.

From the discussion above we see that it is possible to grow dislocation-free epitaxial layers even if there is mismatch involved in the heterostructure system. The critical thickness for dislocation generation ranges between 1 to 10 nm for mismatch in the range 1 to 5%.

## 14.7  Strained-Layer Superlattice

A strained-layer superlattice (SLS) is a series of alternating epitaxial layers of two mismatched materials, each layer having a thickness below the critical thickness $h_c$. This produces a thick layer of dislocation-free material with unusual electronic and optical properties. The growth of these structures was first reported by Matthews and Blakeslee (1974). These strained-layer superlattices (SLSs) provide the

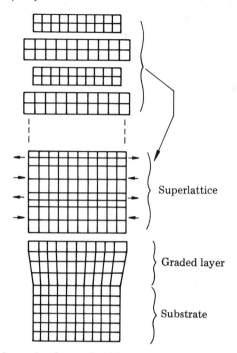

**FIGURE 14.19** Schematic of a strained-layer superlattice structure (SLS) grown on a lattice-matched graded layer (same average composition) and a lattice-mismatched substrate. [After Osbourn et al. (1987).]

freedom to combine materials of different band structures without regard to lattice matching. This unique flexibility has great potential for the fabrication of novel devices. Most ternary semiconductors, while having a continuous range of lattice constants, can lattice match to only a few binary substrates. On the other hand, strained-layer superlattice structures have the capability of varying the band gap ($E_G$) and the lattice constant ($a$) independently.

The preparation of SLS structures is illustrated in Fig. 14.19. We start by growing a graded layer with a desired lattice constant and follow by the growth of alternate thin (below $h_c$) mismatched layers. This results in a superlattice containing tetragonally strained layers in the SLS structure. For example, a SLS structure can be grown starting with a GaAs substrate with a graded layer of $In_{0.1}Ga_{0.9}As$, and finishing with alternate layers of $In_{0.2}Ga_{0.8}As$ and GaAs on top of the graded layer. The typical thickness of these layers is about or less than 10 nm and the strain is about a few percent.

The lattice constant of the superlattice that determines its lattice matching to other materials is the one in the plane parallel to the superlattice interfaces ($a^{||}$). The $a^{||}$ of SLS is given by (Osbourn et al., 1987)

$$a^{||} = a_1 \left[ 1 + \frac{f}{1 + \beta \, (h_1/h_2)} \right] \qquad (14.22)$$

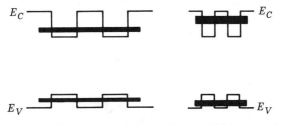

**FIGURE 14.20** Schematic quantum-well structures of thick-layered and thin-layered superlattices formed by the bulk conduction band minima ($E_C$) and the bulk valence band maxima ($E_V$) of the layers. The horizontal bands represent the new superlattice conduction bands and valence bands formed by the quantum wells. [After Osbourn (1983).]

where $a_1$ is the unstrained lattice constant of layer 1 of the superlattice, $h_1$ and $h_2$ are the thicknesses of the individual superlattice layers, $f$ is the lattice mismatch of the unstrained layer materials, and $\beta$ depends on the shear moduli and the elastic constants of materials 1 and 2. Inspection of eq. (14.22) reveals that there are four degrees of freedom in designing a SLS without changing $a^{11}$ (i.e., the compositions and the thicknesses of the individual layers). Consequently, there are a large number of different SLS structures with the same lattice constant. This suggests the possibility of varying material properties independently of the lattice constant.

The SLS band gap is an example of such a property. As with conventional superlattices grown from lattice-matched materials, the SLS band gap can be controlled by the layer composition and layer thicknesses. These quantities determine the multiple-quantum-well (MQW) properties as shown schematically by the solid lines labeled $E_C$ and $E_V$ in Fig. 14.20. It should be recognized that when the spacing between individual quantum wells is large ($>$ a few tens of nanometers), an MQW structure is just a series of heterojunctions joined together. As the spacing between the quantum wells decreases, the electron in a well can tunnel across the energy barrier (see Section 12.9) and communicate with the electron in the next quantum well. At this point the multiple-quantum-well structure becomes a superlattice, and collectively there is a new band gap associated with the superlattice.

The conduction and valence bands of a superlattice are shown schematically in Fig. 14.20 as a broad band running horizontally through the wells. The separation of these bands, $E_G$, can be varied by changing the width as shown in Fig. 14.20 or by changing the layer compositions. The band gap generally increases with decreasing individual layer width. This can be associated with the increased energy of a particle in a quantum well as the size of the wells decreases. The bandwidth of both the conduction band and the valence band also increases with decreasing layer thickness, as shown schematically in Fig. 14.20. The effective mass [eq. (12.7)] along a particular direction is determined by the curvature of the band. The curvature in turn is related to the bandwidth along a particular direction (Chapter 12). Therefore, the effective mass of electrons and holes perpendicular to the superlattice interfaces decrease with decreasing layer width. The mobility of these carriers can be modulated with layer width and layer composition independently of $a^{11}$ and $E_G$.

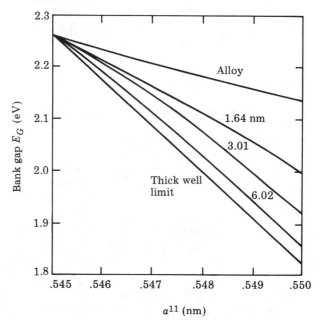

**FIGURE 14.21** GaAs$_x$P$_{1-x}$/GaP(100) SLS band gaps versus SLS lattice constant for various layer thicknesses of GaP (equal to the GaAs$_x$P$_{1-x}$ thickness in each structure). The SLS lattice constant is varied by linearly varying $x$. For comparison, the energy gap of bulk GaAsP versus lattice constant and the limit on obtainable SLS band gaps (in the limit of thick SLS layers) are also shown. [After Osbourn (1983).]

Figure 14.21 shows the calculated band gap of a GaAs$_x$P$_{1-x}$/GaP strained-layer superlattice as a function of $a^{11}$. It is shown that as the layer width (equal layer thicknesses) decreases, the band gap increases. The upper limit is for a homogeneous alloy where the layer width is one atomic layer. On the other hand, when the layer width increases to greater than 6 nm (60 Å), the band gap is no longer sensitive to layer width. With increasing individual layer width, the superlattice eventually turns into a multiple-quantum-well structure. The collective effect of the electrons and holes in a superlattice no longer applies and the MQW structure has two band gaps, characteristic of the two materials.

Strained-layer superlattices may be dislocation-free but not strain-free. In Chapter 3 we see that the band gap of a semiconductor can be changed by an applied stress (or strain). In a SLS structure, strain may be varied by changing the ratio of the layer thicknesses and is an additional factor in tailoring the band gap. Summarizing the discussion on the tailoring of SLS band gaps, there are three ways that $E_G$ may be altered: (1) layer width, (2) layer composition, and (3) strain which may be varied by layer widths. Appropriately designed SLS structures could have one or more optical and transport properties which are advantageous for optimized device performance. To date, SLS structures have been successfully used for high-speed devices such as MESFETs and HEMTs and for optoelectronic devices such as photodetectors and double-heterostructure lasers. The research on SLS structures is a rapidly expanding field. This class of unusual materials has added a new dimension in the science and technology of future semiconductor devices.

## 14.8 Growth Techniques

The heterostructures and superlattice structures discussed above are often prepared by two epitaxial growth techniques, molecular beam epitaxy (MBE) and metal–organic chemical vapor deposition (MOCVD), and less often by liquid-phase epitaxy (LPE).

### 14.8.1 Molecular-Beam Epitaxy

Molecular-beam epitaxy can best be described as a highly controlled ultrahigh-vacuum deposition process. Epitaxial layers are grown by impinging thermal beams of molecules or atoms on a heated substrate. The apparatus used for the MBE growth of GaAs and $Al_xGa_{1-x}As$ is shown schematically in Fig. 14.22. The principal components in this system are the resistance-heated source furnace (called effusion or Knudsen cells), the shutters, and the heated substrate holder. There are usually some surface analysis equipment attached to the growth chamber to monitor surface cleanliness and surface crystallography before and during growth. The effusion cells have small apertures (radius about 1 cm) for the vapor to escape, thus forming a molecular beam. The flux $F_c$ of total number of atoms escaping through the aperture per second is

$$F_c = \frac{PAN_A}{\sqrt{2\pi MRT}} = 3.51 \times 10^{22} \frac{PA}{\sqrt{MT}} \qquad \text{molecules/s} \qquad (14.23)$$

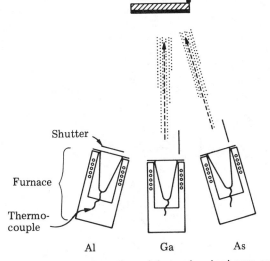

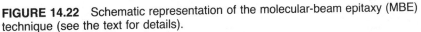

**FIGURE 14.22** Schematic representation of the molecular-beam epitaxy (MBE) technique (see the text for details).

where $A$ = area of the aperture

$P$ = pressure (torr) inside the effusion cell

$N_A$ = Avogadro's number

$M$ = molecular weight

$R$ = gas constant

$T$ (K) = temperature of the cell

If the substrate is positioned at a distance $l$ from the aperture and is directly in line with the aperture, the flux $F_s$ or total number of molecules per second striking the substrate of unit area is

$$F_s = 1.118 \times 10^{22} \frac{PA}{l^2\sqrt{MT}} \qquad \text{molecules/cm}^2 \cdot \text{s.} \qquad (14.24)$$

Take Ga as an example. At $T = 970°C$ (1243 K) the vapor pressure is 2.2 $\times$ $10^{-3}$ torr and $M = 70$. For $A = 5 \text{ cm}^2$ and $l = 12$ cm, the value of $F_s$ is 2.94 $\times$ $10^{15}/\text{cm}^2 \cdot \text{s}$. This arrival rate is about 3 monolayers, or 1.5 nm (15 Å) per second. For the growth of GaAs, the arrival rate of arsenic as $As_2$ or $As_4$ is about 10 times greater than that of Ga. Arsenic atoms only stick onto a Ga surface; therefore, the rate-limiting step is controlled by the arrival rate of the Ga atoms. At the Ga arrival rate of 3 monolayers per second ($2.94 \times 10^{15}/\text{cm}^2 \cdot \text{s}$), a 1-$\mu$m ($10^4$ Å)-thick layer of GaAs takes less than 10 minutes. With a more typical arrival rate of $10^{12} - 10^{14}/\text{cm}^2 \cdot \text{s}$, it may take hours to grow a 1-$\mu$m-thick GaAs layer. Figure 14.23 shows a list of semiconducting materials grown by MBE.

For the growth of AlGaAs, the sticking coefficient of Al and Ga are both close to unity at the usual substrate temperature. Therefore, the Ga-to-Al ratio in the grown layer is the same as the ratio of Ga to Al flux at the substrate surface.

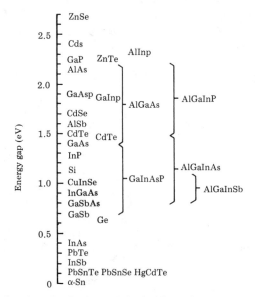

**FIGURE 14.23** Semiconducting materials that have been grown by MBE. [After Cho (1985).]

Metal–organic molecular beam epitaxy (MOMBE) can be carried out in an MBE chamber with a gas source of trialkyl group III metals [$Ga(CH_3)_3$] replacing the group III elemental sources. A source of $As_2$ or $P_2$ beam can be provided by thermal decomposition of the hydrides ($AsH_3$ or $PH_3$) in the furnace cells. The organometallic compounds, the alkyls, decompose after impinging on the substrate surface. The chemistry that takes place at the growing surface in MOMBE is more complex than with conventional MBE. Although MOMBE is carried out in an MBE chamber [the dopants are supplied from conventional MBE effusion (or Knudsen) cells], the pressure is higher, between $10^{-3}$ and $10^{-4}$ torr. This is still in the regime of molecular flow and the minimum mean free path is greater than the source-to-substrate distance. MOMBE techniques are still in the early stage of development, although high-quality epitaxial layers have been produced.

### 14.8.2 Metal–Organic Chemical Vapor Deposition

The growth of epitaxial materials in metal–organic chemical vapor deposition (MOCVD) or metal–organic vapor-phase epitaxy (MOVPE) techniques is typically accomplished by the coreaction of reactive metal alkyls with a hydride of the nonmetal component. A diversity of chemical growth precursors and growth system designs has allowed for the successful growth of a large number of materials and structures, despite the complex nature of the growth process. Figure 14.24 shows

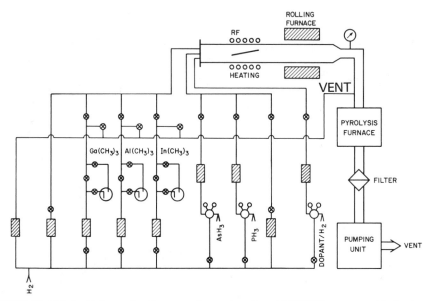

**FIGURE 14.24** Schematic diagram of a conventional MOVPE gas panel indicates the use of both liquid and gas sources. The liquid sources are contained in temperature-controlled baths. The use of a "to vent" or "to reactor" gas valving arrangement facilitates the formation of abrupt heterojunction interfaces. The rolling furnace could be used to vapor etch the reactor between growth runs. The pyrolysis furnace is one of several means of removing the toxic and pyrophoric compounds from the effluent of the reactor. [After Kuech (1987).]

a schematic diagram of a conventional MOVPE system. In the growth of III–V compounds, a general overall reaction is given by

$$MR_3 \text{ (a metal alkyl)} + XH_3 \text{ (a hydride)} \longrightarrow MX + 3RH \qquad (14.25)$$
$$RH = CH_3, C_2H_5, \ldots$$

The most common example of this reaction is found in the growth of GaAs and $Al_xGa_{1-x}As$:

$$Ga(CH_3)_3 + AsH_3 \longrightarrow GaAs + 3CH_4 \qquad (14.26)$$

or

$$xAl(CH_3)_3 + (1 - x)Ga(CH_3)_3 + AsH_3 \longrightarrow Al_xGa_{1-x}As + 3Ch_4. \qquad (14.27)$$

   A simplified description of the growth process occurring near and at the substrate surface is shown in Fig. 14.25 for the case of GaAs growth from $Ga(CH_3)_3$ and $AsH_3$ (Kuech, 1987). The growth ambient generally has a large excess of the group V constituent over the metal alkyl. Several steps must occur for epitaxial growth to proceed. Mass transport of the reactants to the growth surface, their reaction at or near the surface, incorporation of the new material into the growth front, and removal of the reaction by-products must all take place. The slowest step in this sequence will determine the growth rate step. The substrate temperature is typically held at a value substantially higher than the pyrolysis temperature of the metal alkyl, ensuring its rapid decomposition at the growth surface. The growth rate is then rate limited by the mass transport of the group III reactant to the growth surface. While the growth rate is determined by the mass transport of reactants to the surface, the surface and gas-phase chemical reactions can influence the materials properties. The MOVPE growth technique has been proven capable of fabricating materials of high purity and excellent morphology. Artificial microstructures such

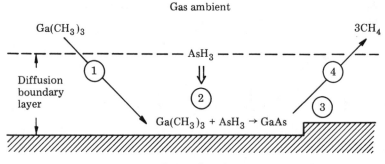

**FIGURE 14.25**   Several processes must take place in the vapor deposition of material: step 1, mass transport of nutrients to surface; step 2, surface reaction; step 3, incorporation of molecule into the crystalline substrate; and step 4, removal of reaction products. The slowest of these serial process steps will be the growth-rate-limiting step. MOVPE growth usually is undertaken in a mass-transport-limited (step 1) regime. [After Kuech (1987).]

as superlattices, quantum wells, and atomically abrupt interfaces can be produced routinely by the MOVPE technique.

### 14.8.3 Liquid-Phase Epitaxy

In the most general terms, liquid-phase epitaxy (LPE) is the growth of an oriented crystalline layer of material from a saturated or supersaturated liquid solution onto a crystalline substrate that has a similar enough crystal structure and lattice constant to the growing layer to permit continuation of the coherent crystal structure. Most frequently, the major constituent of the liquid solution is one of the major components of the solid, and the phase equilibria are such that the liquid solution from which growth occurs is relatively dilute in all components but one.

Let us use a hypothetical binary system AC of Fig. 14.26 to illustrate the thermodynamic basis of LPE. In this phase diagram we have a congruently melting solid AC with a melting temperature $T_F$. Congruent melting is the melting of a compound at a fixed temperature at which the solid and liquid have the same composition. The temperature $T_1$ and $T_2$ are the temperatures for the equilibrium between the solid phase and the composition $X_C(T_1)$ and $X_C(T_2)$ on the liquidus curve. At a given temperature $X_A + X_C = 1$. Let us assume that a group III element is A and a group V element is C; a III–V compound AC can be in

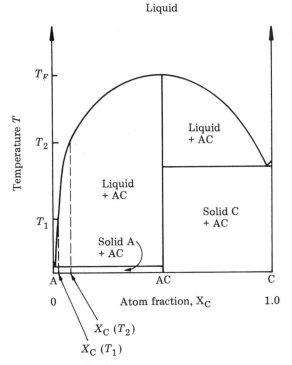

**FIGURE 14.26** Schematic representation of the liquidus–solidus equilibrium in a system with a large AC primary phase field. [After Casey and Panish (1978).]

equilibrium with liquid solutions that are very dilute in C. The growth process in LPE is a near-equilibrium situation. The epitaxial growth of the binary compound AC onto a substrate is equivalent to the loss of $[X_C(T_1) - X_C(T_2)]$ atom fraction of C and the same amount of A from the liquid solution (see Fig. 14.26) as a result of slow cooling from $T_2$ to $T_1$. This technique of growth is called the uniform cooling technique (Casey and Panish, 1978) and is commonly used for the growth of GaAs/Al$_x$Ga$_{1-x}$As heterostructures. The GaAs substrate crystal is brought into contact with the equilibrated solution at $T_2$ and then slowly cooled, together with the solution, to $T_1$. LPE techniques are commonly used to grow heterostructures for lasers and photodetectors, whereas MBE and MOVPE are used to grow multiple-quantum-well structures and superlattice structures.

## 14.9 Gallium Arsenide on Silicon

The primary semiconductor in use today is Si because of the availability of large wafers of good quality and the existence of a sophisticated processing technology. Since GaAs on Si offers possibilities for new types of devices and ICs with the advantages of both semiconductors, it is not surprising that the heteroepitaxial growth of GaAs on Si is an interesting area for investigation. Since the lattice constant of GaAs (0.565 nm) is about 4% larger than that of Si (0.543 nm), pseudomorphic growth is difficult. Three-dimensional island growth leading to the formation of stacking faults, threading, and misfit dislocations frequently results. The dominant stacking faults in this case are antiphase domains where either a plane of Ga atoms grows on a Ga surface or a plane of As atoms grows on a As surface. To circumvent these difficulties, a two-step growth technique has been found to be successful in growing a single-domain GaAs layer on Si (Wang, 1984). In this technique a thin GaAs layer of less than 20 nm is deposited on silicon at low temperatures (200 to 450°C) followed by heating the substrate up to 600°C (for MBE) or 700 to 750°C (for MOCVD) and then a second layer is grown (Kaminishi, 1987). The first layer of GaAs is amorphous in structure. As the substrate temperature is raised for the growth of the second layer, the amorphous layer crystallizes via the process of solid-phase epitaxy [(SPE); see Chapter 8]. Further growth at normal temperatures on the crystallized layer can lead to epitaxial layers with a single domain. It has also been found that (100) Si wafers with a small offset angle toward the [110] direction facilitates single-domain growth of GaAs. It is believed that such an offset produces steps two atomic layers in height on the Si surface; these bi-atomic layer steps are conducive to the growth of a single-domain GaAs layer. Figure 14.27 shows a cross-sectional transmission electron micrograph with atomic resolution of an island of GaAs grown on (100) Si. The growth temperature is about 500°C, slightly higher than the conventional first-step growth temperature of 200 to 450°C. The GaAs island is crystalline in structure and contains planar defects (twins). The surface of the island is faceted. In the second step of growth, the temperature of growth is increased to about 600 to 700°C. Most of the planar defects anneal out at that temperature (Nieh, 1988).

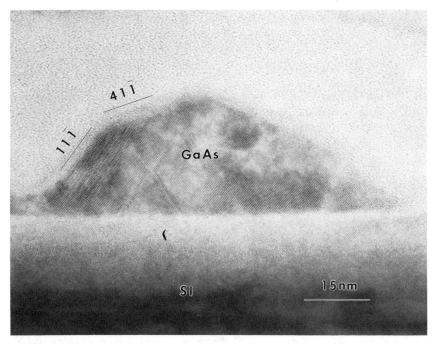

**FIGURE 14.27** Cross-sectional transmission electron micrograph with atomic resolution of a GaAs island grown on <100> Si at about 500°C. (Courtesy of S. Nieh, Caltech, 1988.)

Electronic and optoelectronic devices have been fabricated with reasonable success using GaAs–Si heterostructures. These experimental results has demonstrated the feasibility of monolithic integration of Si ICs and optical devices.

## REFERENCES

### General References

CASEY, JR., H. C., and M. B. PANISH, *Heterostructure Lasers,* Part A, *Fundamental Principles,* Part B, *Materials and Operating Characteristics,* Academic Press, New York, 1978.

CHO, A. Y., "Growth and Properties of III–V Semiconductors by Molecular Beam Epitaxy," in *Molecular Beam Epitaxy and Heterostructures,* ed. Leroy L. Chang and Klaus Ploog, NATO ASI Series, Series E, No. 87, Martinus Nijhoff, Dordrecht, The Netherlands, 1985.

KASPER, E., and J. C. BEAN ed. *Silicon-Molecular Beam Epitaxy,* Vol. I and II, CRC Press, Boca Raton, Florida, 1988.

KROEMER, H., "Theory of Heterojunctions: A Critical View," in *Molecular Beam Epitaxy and Heterostructures,* ed. Leroy L. Chang and Klaus Ploog, NATO ASI Series E, No. 87, Martinus Nijhoff, Dordrecht, The Netherlands, 1985.

KUECH, T. F., "Metal–organic vapor phase epitaxy of compound semiconductors" *Mater. Sci. Rep. 2*,(1) (1987).

MATTHEWS, J. W., "Coherent Interfaces and Misfit Dislocations," *Epitaxial Growth* Parts A and B, ed. J. W. Matthews, Academic Press, New York, 1975.

MILNES, A. G., and D. L. FEUCHT, *Heterojunctions and Metal–Semiconductor Junctions,* Academic Press, New York, 1972.

STOWELL, M. J., "Defects in Epitaxial Deposits," in *Epitaxial Growth,* Part A and B, ed. J. W. Matthews, Academic Press, New York, 1975.

## Specific References

BAUER, R. S., PETER ZURCHER, and HENRY W. SANG, JR., *Appl. Phys. Lett. 43*(7), 663 (1983).

CHEUNG, D. T., S. Y. CHIANG, and G. L. PEARSON, *Solid State Electron. 18,* 263 (1975).

FRENSLEY, W. R., and H. KROEMER, *Phys. Rev. B16,* 2642 (1977).

HARRISON, W. A., *J. Vac. Sci. Technol. 14,* 1016 (1977).

KAMINISHI, K. *Solid State Technol.,* Nov. 1987, p. 91.

KUECH, T. F., D. J. WOLFORD, R. POTOEMSKI, J. A. BRADLEY, K. H. KELLEHER, D. YAN, J. PAUL FARRELL, P. M. S. LESSER, and F. H. POLLAK, *Appl. Phys. Lett. 51*(7), 505 (1987).

MATTHEWS, J. W., and A. E. BLAKESLEE, *J. Cryst. Growth 27,* 118 (1974).

NIEH, S., private communication, Caltech, 1988.

OSBOURN, G. C., *J. Vac. Sci Technol. B1*(2), 379, 1983.

OSBOURN, G. C., P. L. GOURLEY II, J. FRITZ, R. M. BIEFELD, L. R. DAWSON, and T. E. ZIPPERIAN, "Principles and Applications of Semiconductor Strained-Layers Superlattices," in *Semiconductors and Semimetals—Applications of Multiquantum Wells, Selective Dopings, and Superlattices,* Vol. 24, ed. Raymond Dingle, Academic Press, Orlando, Fla., 1987.

WANG, W. I., *Appl. Phys. Lett. 44*(12), 1149 (1984).

WOODALL, J. M., J. L. FREEOUF, G. D. PETIT, T. N. JACKSON and P. KIRCHNER, *J. Vac. Sci. Technol, 19,* 626 (1981).

WOODALL, J. M., G. D. PETTIT, T. N. JACKSON, C. LANZA, K. L. KAVANAGH, and J. W. MAYER, *Phys. Rev. Lett. 51*(19), 1783 (1983).

# PROBLEMS

**14.1.** Use the electron affinity rule to calculate the band lineups for the Ge/GaAs systems. Compare this lineup with that predicted by the HAO theory.

**14.2.** Use the band-edge values from the HAO theory (Table 14.7) to calculate the band lineups for the following systems: (a) InP/InAs, (b) InP/InSb, (c) InAs/GaSb, and (d) Si/Ge. Which of these will have the (1) straddling, (2) staggered and (3) broken-up lineups?

**14.3.** Calculate the critical thickness $h_c$ of an epitaxial Ge layer on Si using eq. (14.20) and compare with Figure 14.16.

**14.4.** Discuss the structural and the electronic aspects of a graded junction.

**14.5.** (a) An epitaxial layer of $Ga_xIn_{1-x}As$ is grown on an InP substrate. Find $x$ for the lattice-match condition.
(b) An epitaxial layer of $Al_xIn_{1-x}As$ is also grown on InP. What is $x$ in this case for the lattice-match condition?
(c) What are the energy gaps for these two ternary compound under the lattice-match condition?

**14.6.** The lattice constant of Si is 0.543 nm and that of GaAs 0.565 nm at room temperature.
(a) What would be the critical thickness, $h_c$, for the pseudomorphic growth of GaAs on Si under the normal one-step growth technique?
(b) Why does the two-step growth discussed in Section 14.9 allow for the growth of a good quality GaAs layer thicker than $h_c$?

# Assembly and Packaging

## 15.1 Introduction

The final stages in the fabrication of integrated circuits are the processes that transform the wafer to electronic components mounted on a printed circuit board. These processes are called packaging and include separating the wafer into chips, wire bonding, mounting the chips in ceramic or plastic packages, and sealing the package to protect it from moisture or contaminants.

In this book we have discussed Si and GaAs transistors and the process steps from diffusion to metallization. Figure 15.1 shows a cross section of an $n$-channel MOSFET that is part of a CMOS circuit. The FET was formed by growing an epitaxial layer of $p$-type Si on the substrate, by oxidizing and patterning the surface, and by implanting and diffusing the source and drain regions. Silicide contacts are made to the source and drain as well as to the polycrystalline Si (poly Si) gate that is contained in a sidewall oxide. Further metallization consists of Al lines with 3

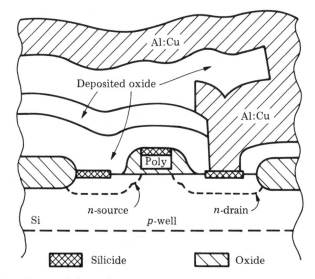

**FIGURE 15.1**   Cross section of a portion of an integrated circuit on Si.

to 4 atom % of Cu to reduce electromigration and of tungsten plugs in vias between interconnect lines. The interconnects are separated by deposited oxide layers. For a megabit dynamic random access memory (DRAM), the minimum feature size of the lines are 1 μm and the chip size is 10 mm on an edge (area of 100 mm$^2$).

At this stage in the fabrication we would like to make contact with the outside world, and to do that we enter into the field of packaging. As described below, there are many different ways to package a chip, ranging from plastic to ceramic. Cost and performance are determined by the final package configuration, and the final packaging strategy often involves a trade-off between cost and performance.

There are many packaging options available, so we will concentrate on common technologies involving the packaging of a single-chip made from a silicon wafer. Even then there are multiple options, from front-mount to flip-chip configurations and from dual-in-line packages to pin-grid arrays. The options dictate the circuit layout and configuration of the chip, so there is an interdependence between wafer processing and packaging. Commonly used acronyms are listed in Table 15.1 (other lists of acronyms are given in Tables 5.1 and 13.1).

Although we focus on single-chip packaging, there are important areas involving hybrid packaging where one mounts individual components such as capacitors together with integrated-circuit chips on a single substrate. Hybrid packaging would also involve optoelectronic integrated circuits as well as individual photodetectors and light-emitting diodes.

We start as an illustration with the simplest package, the TO-5 header, with a single bipolar transistor as shown in Fig. 15.2. The TO-5 package is a metal cup with 8 to 10 or 12 insulated pins (Fig. 15.2 shows the original three-pin design). The cup is covered with a metal can in an inert, dry atmosphere and then top and bottom are joined to form a hermetic seal for protection against the environment.

The silicon bipolar transistor has been mounted on a Kovar (an Fe–Ni–Co alloy with a thermal expansion coefficient close to that of Si) tab or paddle which is

**TABLE 15.1**  Acronyms Commonly Used in Packaging Technology

| | |
|------|--------------------------------------|
| CTE  | Coefficient of thermal expansion     |
| DIP  | Dual-in-line package                 |
| IC   | Integrated circuit                   |
| I/O  | Input/output                         |
| LCC  | Leadless chip carrier                |
| LSI  | Large-scale integration              |
| MLC  | Multilayer ceramic                   |
| PCB  | Printed circuit board                |
| PGA  | Pin grid array                       |
| PTH  | Plated-through holes (on circuit board) |
| PWB  | Printed wiring board                 |
| SMT  | Surface-mount technology             |
| TAB  | Tape-automated bonding               |
| TCM  | Thermal conduction module            |
| VLSI | Very large scale integration         |

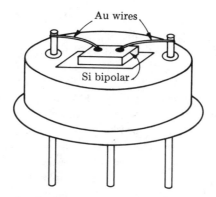

**FIGURE 15.2**   Schematic of a single bipolar transistor mounted on a TO-5 header.

attached to the TO-5 bottom cup. The connections to the transistor are made by 25-μm-diameter (0.001 in. or 1 mil) Au wires that are bonded to the transistor and pin. Although primitive by today's integrated-circuit packages, the TO-5 header package does require many of the packaging steps required in modern packages: slicing of chips from the wafer, chip attachment, wire bonding, and encapsulation.

## 15.2 Chip Separation and Back-Side Bonding

At the end of the integrated-circuit process line, the wafers, typically 150 mm in diameter, are mounted in a test fixture so that the electrical performance of the individual chips can be evaluated. Figure 15.3 shows a wafer after inspection; the flat indicates the crystal orientation (Fig. 6.1). The wafer is then cut or separated into individual chips (sometimes referred to as "die") by cutting with a diamond-impregnated saw blade. Chip separation (also called dicing) can also be done with a pulsed laser or a diamond-tipped scribing tool. Diamond sawing is often used for full wafer separation because it leads to straighter edges with less mechanical damage. The chips are then sorted according to the results of the electrical inspection.

For face-up configuration (illustrated in Fig. 15.2) the active components of the integrated circuit are on the top side of the chip and the back side is bonded to a metal lead frame or metallized ceramic (usually alumina) substrate which has a thermal expansion coefficient close to that of silicon. The linear coefficient of thermal expansion of alumina is $6.5 \times 10^{-6}/°C$, while that of Si is $2.6 \times 10^{-6}/°C$.

One of the common techniques for chip attachment makes use of a Au–Si eutectic alloy (Chapter 9). Eutectic die bonding metallurgically attaches the chip to the substrate material. The source of the gold in the alloy can be plating on a metal tab or paddle, a layer on a ceramic substrate, or a thin (usually less than 0.05 mm) alloy preform placed between the chip and the substrate (Fig. 15.4).

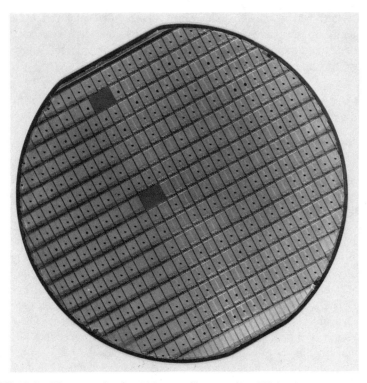

**FIGURE 15.3**  Photograph of a 100-mm silicon wafer. (Wafer furnished by R. D. Pashley, Intel Corporation, 1987.)

The Au–Si preform generally is 98 wt % Au and 2 wt % Si and the bonding temperature is usually around 400°C, a temperature well above the Au–Si eutectic temperature. At the bonding temperature, the preform reacts with the Si chip and forms a liquid layer containing some extra Si dissolved from the chip. During bonding, the unit is usually ultrasonically agitated or scrubbed to ensure uniform contact between preform and chip (reducing void formation) and is surrounded with an inert atmosphere to prevent the formation of an oxide.

Polymers are also used for chip attachment. Silver-filled epoxies are electrically conducting and are cured between 125 and 175°C to cross-link the bonds so that

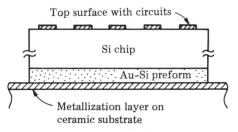

**FIGURE 15.4**  Schematic showing a Au–Si preform used to attach a Si integrated-circuit chip to a metallized ceramic substrate.

the chip is attached to the substrate. Such adhesive die bonding requires lower process temperatures than those for preform alloys and can be used except in applications that require high temperatures or high currents through the chip–substrate bond.

## 15.3 Wire Bonding

The integrated-circuit chip has now been bonded to a substrate and the next step is to make electrical connections to the top surface. The top-surface metallization is aluminum lines containing Cu to reduce electromigration. The metallization pattern leads the Al lines to square pads, typically 150 to 200 $\mu$m on an edge, located at the periphery of the chip (Figs. 1.1 and 1.15). These pads, then, provide access between the outside world and the integrated circuit.

Wire bonding—Au or Al wire—is used to make the connections to front-surface integrated circuits. The Au-wire bonds are shown in Fig. 1.16 and the procedure is illustrated in Fig. 15.5. The Au wires, typically 25 $\mu$m in diameter, are fed through a fine tube (capillary) and a ball is formed on the end of the wire by a flame tip. The chip is heated to about 150°C and the ball is pressed onto the Al-bonding pad on the chip. The combination of pressure and heat is called thermo-compression wire bonding. Ultrasonic agitation is used to ensure good contact. During the bonding process intermetallic Au–Al compounds are formed as described in Section 11.5. After the wire is bonded to the integrated-circuit pad (Fig. 15.5), the capillary head is moved and the wire is positioned over a Au-coated lead on the package frame. The second bond is formed, a wedge bond in this case, by the pressure of the capillary tip along with ultrasonic agitation. The wire is pulled and broken, leaving a tail-less, wedge bond. The combination is called ball-wedge bonding. The Au wire is ball bonded on the chip and wedge bonded on the package substrate.

Gold is the common material for wire bonds because it is ductile and has a long tradition of use. The disadvantage is cost and the formation of intermetallic com-

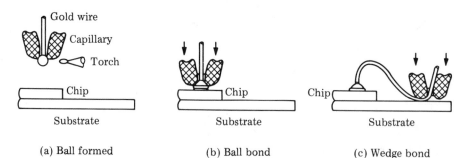

(a) Ball formed       (b) Ball bond       (c) Wedge bond

**FIGURE 15.5**  Thermocompression ball-wedge bonding of a Au wire: (a) ball formation at the tip of the bonder capillary; (b) ball bonding on the chip; (c) wedge bonding on the substrate.

**FIGURE 15.6**  Wedge-bonded Al wires on an integrated-circuit chip. (Chip supplied by R. D. Pashley, Intel Corporation, 1982; photo by Scott Pesarcik, 1988.)

pounds. Aluminum wire bonding is used to obtain Al/Al interconnections. Figure 15.6 shows two Al wires wedge bonded to the Al pads of an integrated-circuit chip.

## 15.4  Flip-Chip Technology

With wire bonding, the bonding pads are located around the edges of the chip. An alternative bonding method is to use an area array of solder bumps on the face of the chip—the interconnect pads are located on any point on the active chip surface, allowing over 100 interconnect pads with solder bumps on the surface. The chip is then turned over (flipped over) and mated with a metallized ceramic substrate having a matching footprint of input/output (I/O) terminals. The flip-chip concept is shown in Figs. 1.17 and 15.7 and described by Fried et al. (1982).

In wire bonding the temperatures are well below any liquidus and the intermetallics are formed in solid-phase reactions. In solder ball technology, the alloy melts and dissolves some of the coating on the interconnect pads. For example, in Pb–Sn alloys, a Pb-rich alloy is used with a liquidus temperature somewhat above 300°C (the Pb–Sn eutectic temperature is 183°C). The solder connection is between metals, typically Cu, which have a higher melting point. A glass dam surrounding the solder ball limits the flow of the liquid solder to the tip of the substrate metallization.

Heat dissipation in a flip-chip configuration is through the solder joints to the substrate, in contrast to the case where the back side of the chip is eutectic bonded

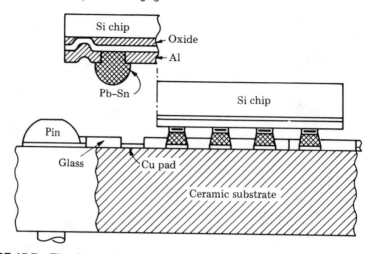

**FIGURE 15.7** Flip-chip solder bonded to a ceramic substrate. Enlarged view at left shows Pb-Sn solder before bonding.

to the substrate. The amount of heat dissipation is dependent on the thermal cross section, which is determined by the number and size of the solder joints.

If we consider one-dimensional heat flow through $N$ solder connections of area $A$, the flux $F_{th}$ of thermal energy (heat) transmitted per unit time per unit area (see Section 2.10) in a temperature gradient $dT/dx$ through a total area $A_T$ is

$$F_{th} A_T = -K_{th} N A \frac{dT}{dx} \qquad (15.1)$$

where $K_{th}$ is the thermal conductivity in W/cm · K. The negative sign indicates a positive flow of heat toward lower temperature. Table 5.2 lists values for the thermal conductivity and thermal expansion coefficient of several materials.

**TABLE 15.2** Thermal Conductivities of Materials and Coefficients of Thermal Expansion $dL/L \, dT$

| | $K_{th}$ (W/cm · K) | $\alpha$ (K$^{-1}$) |
|---|---|---|
| Alumina (~90% Al$_2$O$_3$) | 0.17 | 6.5 |
| Solder: Pb–5% Sn | 0.63 | 29 |
| Silicon (Si) | 1.5 | 2.6 |
| Gold (Au) | 3.45 | 14.3 |
| Au–3% Si | 0.27 | 12.3 |
| Epoxy (Ag loaded) | 0.008 | 53 |

*Source:* Sze, ed. *VLSI Technology,* McGraw-Hill, New York, 1983.

The solder balls have a diameter of about 100 μm (area of about $0.75 \times 10^{-4}$ cm²) and height $\Delta X$ of about 75 μm. If heat flow is only through 100 solder balls,

$$F_{th}A_T = -K_{th}NA \frac{\Delta T}{\Delta X}$$

$$= -K_{th} \times 100 \times 0.75 \times 10^{-4} \times \frac{\Delta T}{75 \times 10^{-4} \text{ cm}}$$

$$= -K_{th} \Delta T$$

in watts, where 1 watt = 1 joule per second. The thermal resistance $R_{th}$ in $K$ per watt can be written in analogy with electrical resistance as

$$R_{th} = -\frac{\Delta T}{F_{th}A_T} \tag{15.2}$$

where the negative sign accounts for the sign of the temperature difference $\Delta T$. For 100 solder balls with $K_{th} = 0.63$,

$$R_{th} = -\frac{\Delta T}{F_{th}A_T} = \frac{\Delta X}{K_{th}NA} \tag{15.3}$$

so that

$$R_{th} = \frac{1}{0.63} \simeq 1.6 K/\text{W}.$$

The constraint imposed by heat dissipation can be relieved by use of a thermal conduction module (TCM) where heat-conducting pistons are in contact with the back surface of the chip. Helium gas is used as the encapsulated cooling atmosphere to improve thermal conduction. Three to four watts per chip are dissipated, with maximum junction temperatures of less than 80°C, by the use of the piston back-side contact and the high-thermal-conductivity helium atmosphere in the enclosure.

## 15.5 Tape-Automated Bonding

Tape-automated bonding (TAB) is a concept for an interconnection that is similar to wire bonding except that the wires are now flexible copper-film fingers that are placed on a polymer (usually polyimide) tape. As in wire bonding, the Al metallization pads are on the periphery of the chip; however, the Al pads are "bumped" by depositing Au pads on top of the Al interconnect pads. The Cu fingers are then thermocompression bonded to the Au bumps on the chip (Fig. 15.8).

Tape bonding is used because it offers another approach to the automation of chip assembly. Even though wire bonding is automated and requires less than 0.2 s/wire, tape bonding allows simultaneous bonding of all the interconnects. Tapes with over 100 input/output terminals are made for chips with 100- to 200-μm center spacing of the Al interconnect pads around the perimeter of the chip.

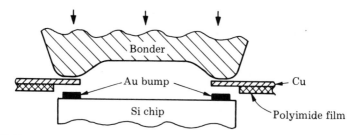

**FIGURE 15.8** Tape-automated bonding showing the bonder head, the Cu fingers on the polyimide tape, and the Au bumps on the chip.

The flexible leads, typically Cu, are made by photopatterning and etching a layer of the Cu on a polymer (polyimide) carrier tape. The film tape is strong enough to support the thin Cu fingers, but flexible enough to minimize stresses to the interconnection joints. The fingers are photoprinted and held in place by the polymer carrier tape to allow automatic joining to the high density of input/output (I/O) connections.

As in any packaging technology, many options are available. The bonding bumps can be placed on the conducting fingers rather than on the chip interconnection pads. Tin-coated ($\simeq 0.5$ μm of Sn) Cu fingers can be eutectically soldered to the Au bumps on the chip. The Au–Si eutectic temperature at 30 at % Sn is at 280°C, a temperature that is higher than that used in thermocompression bonding. Various polymer tapes can be used with 50-μm-wide bond fingers on 100- to 200-μm centers.

After bonding of the Cu fingers to the chip, the tape Cu leads can then be bonded to a variety of plastic or ceramic substrates, such as dual-in-line packages or pin-grid arrays.

# 15.6 Package Types

Single-chip integrated circuits are mounted in either ceramic packages or on metal lead frames in plastic packages. Ceramic packages are hermetically sealed from the environment. Plastic-encapsulated packages (see Fig. 1.18, for example) are lower in cost and are used where hermeticity is not required. Again, there are many package configurations and combinations of ceramic, metal, and plastic.

## 15.6.1 Dual-In-Line Package

A schematic of a dual-in-line package (DIP) is shown in Fig. 15.9, where the electrical connections between the chip and the package leads are made by wire bonding. The chips themselves are usually eutectic bonded to the lead frame. Tape-automated bonding can be used, as well as wire bonding in this package.

The pins of the package are on 2.54-mm (100-mil) centers and are inserted and soldered to plated-through holes on printed circuit boards. Alternatively, they can

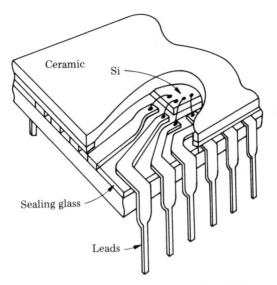

**FIGURE 15.9**  Dual-in-line package with the connections between chip and package made by wire bonds.

be bent outward (or inward) and mounted directly on the surface of a printed circuit board. This surface mount technology (SMT) requires solder joints to connect the package to the board but does not require plated-through holes on the circuit board.

In plastic dual-in-line packages, the package and lead frame are molded first and then the chip and interconnects are added. Finally, molding compounds, under pressure and heat ($\simeq$ 175°C), flow around the chip and connections to provide encapsulation. Mismatch between the expansion coefficient of the plastic molding compound and the silicon chip with surface passivation and bonds leads to stresses after molding, which can result in loss of electrical connection.

### 15.6.2 Quadpacks

The quadpack has interconnections on all four sides of the package, with an internal configuration similar to that of the dual-in-line package. Again, leaded quadpacks can be mounted with pins through the circuit board or on the surface of the board. Leadless quadpacks with contacts on the side can be soldered to the circuit board in surface-mount technology.

### 15.6.3 Pin-Grid Arrays

The pin-grid array (PGA) is a ceramic, usually sintered-alumina package with terminals on four sides but nests a second row of pins inside the outer row, giving a total of over 100 pins. The integrated circuit chip can be mounted on the array by wire bonding, tape-automated bonding, or by flip-chip technology. The flip-chip mount is shown in Fig. 15.10.

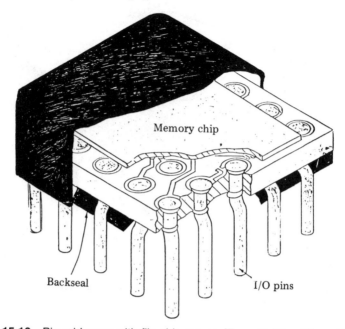

**FIGURE 15.10**   Pin-grid array with flip-chip mount. [From C. Woychik and R. Senger in Seraphim et al. (1988).]

The pins can be locked into the holes on the ceramic substrate and are soldered to the connecting metallization lines on the ceramic. The pins are then inserted into holes on the circuit board.

A single-layer alumina substrate can be used with metallized lines. However, multilayer ceramics (MLC) are more common, in which there are many layers of ceramic with a metal layer pattern for custom wiring on each ceramic layer. These layers are aligned with holes for interconnects between the ceramic levels and then fired at about 1600°C to form a monolithic sintered body. The combination of the multilayer metallization on the integrated-circuit chip itself and the multilayer ceramic with buried redistribution lines within the ceramic layers provides the high circuit density required in very large scale integration.

## 15.7 Pressure Sensors

In micromechanical devices such as pressure sensors, silicon is used as the mechanical material. As discussed in Section 3.10, the relative change in resistivity $\Delta\rho/\rho$ is proportional to the stress $\sigma_s$:

$$\frac{\Delta\rho}{\rho} = P_s\sigma_s$$

where $P_s$ is the piezoresistance constant, which depends on crystal orientation. If one forms a $p$-type resistor $R$ (ohms) in $n$-type Si by implanting boron ions, for example, the relative change in resistance $\Delta R/R$ is

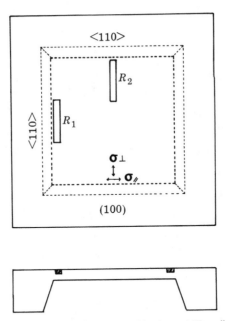

**FIGURE 15.11** Resistors $R_1$ and $R_2$ oriented in the $<110>$ direction on an $n$-type Si square diaphragm. [From Tanigawa et al. (1985) (© 1985 IEEE).]

$$\frac{\Delta R}{R} = P_s(l)\sigma_s(l) + P_s\sigma_s(t)$$

where $l$ and $t$ denote longitudinal and transverse directions, respectively.

For $p$-type resistors $R_1$ and $R_2$ arranged in a $<110>$ direction on $n$-type Si with (100) surface orientation as shown in Fig. 15.11 (Tanigawa et al., 1985), the resistance variations of the two resistors are equal and opposite:

$$\frac{\Delta R_1}{R_1} = -\frac{\Delta R_2}{R_2}$$

since $P_s(l) = -P_s(t)$; that is, under applied pressure one is increased and the other decreased. The resistors can then be connected in a bridge circuit. The output is about 35 mV per $10^5 \text{N/m}^2$ (1 atm is about $10^5 \text{ N/m}^2$; Chapter 1) for $\pm 5$ V applied voltage.

The ratio $\Delta R/R$ for a given pressure $P$ in $\text{N/m}^2$ is proportional to the inverse of the square of the substrate thickness $t_s$ ($\Delta R/R \propto P/t_s^2$). The pressure sensitivity of a silicon pressure sensor is increased by thinning the Si to form a diaphragm. The resistors are formed directly on the diaphragm, as shown in Fig. 15.11. Uniform diaphragm layers 20 to 30 μm thick have been made.

Silicon resistors are also temperature sensitive. A resistor formed on the Si substrate outside the diaphragm region can be used as a pressure-independent temperature sensor to compensate for the temperature dependence of resistors $R_1$ and $R_2$. A feedback circuit compensates the bridge output of the pressure-sensitive circuit for temperature changes. An $n$-channel metal–oxide–Si FET (NMOS) is

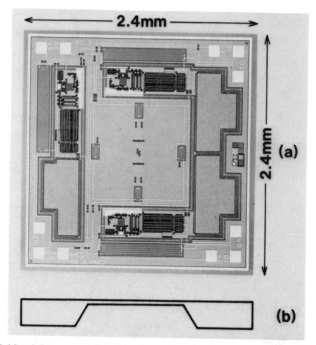

**FIGURE 15.12** (a) top, and (b) side view of an integrated-circuit pressure sensor with temperature compensation and amplifier circuits. [From Tanigawa et al. (1985) (© 1985 IEEE).]

fabricated for signal amplification so that the output signal can then be connected directly to analog-to-digital (A/D) convertors. Figure 15.12 shows the layout of the resistors and the integrated circuits.

In order to utilize the circuit and diaphragm, the Si chip has to be mounted in a package. As a first step, the chip is mounted on a glass die of the same size which has a pressure inlet hole. The glass is chosen to have a thermal expansion

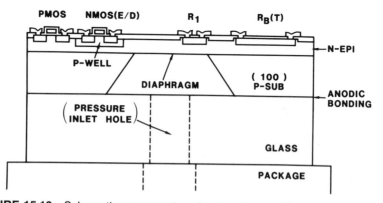

**FIGURE 15.13** Schematic cross section showing piezoresistors in the diaphragm and the pressure-independent resistor for temperature compensation in the Si web. [From Tanigawa et al. (1985) (© 1985 IEEE).]

**FIGURE 15.14** Pressure sensor mounted in a TO-5 header with a pressure inlet tube. [From Tanigawa et al. (1985) (© 1985 IEEE).]

coefficient nearly equal to that of silicon (Chapter 1). The glass mount minimizes any unwanted stress due to the thermal coefficient mismatch between the silicon and the package header. A schematic of the assembly cross section is shown in Fig. 15.13. The Si and glass dies are mounted on TO-5 headers with a pressure tube at the center. A photo of the final assembly is shown in Fig. 15.14.

## 15.8 Alpha-Particle-Induced Soft Errors

Soft errors are random, nonrecurring, single-bit errors in memory devices. A source for these errors in dynamic random access memories (DRAM) and charge-coupled devices (CCDs) is the passage of α-particles or cosmic rays through the memory array (May and Woods, 1979). While cosmic rays are a ubiquitous component in the environment, α-particles are emitted from the radioactive decay of polonium, thorium, and uranium which are present in parts per million level in chemical etches and packaging materials.

In radioactive decay, the emitted α-particles have energies between 5 and 9 million electron volts (MeV). Alpha-particles are doubly ionized helium ions

($^4$He$^{2+}$). When an $\alpha$-particle penetrates silicon it loses energy in electronic collisions (see Section 8.4). Hole–electron pairs are produced along the path of the $\alpha$-particle. The average energy $\epsilon_\alpha$ required to produce a hole–electron pair in silicon is 3.6 eV (Feldman and Mayer, 1986), so that the number of pairs $N_{pair}$ produced by an $\alpha$-particle with energy $E_\alpha$ is

$$N_{pair} = \frac{E_\alpha}{\epsilon_\alpha}. \tag{15.4}$$

For a 7.2-MeV $\alpha$-particle incident on Si, $2 \times 10^6$ hole–electron pairs will be produced.

The penetration depth of a 5 to 10-MeV $\alpha$-particle incident normal to the Si surface is about 25 to 80 $\mu$m. Although this depth is deeper than the thickness of active devices on the surface of Si (the devices are fabricated in the upper 1 $\mu$m of the surface), the carrier diffusion length $L$ is larger than this length. From Section 3.9, the diffusion length $L$ is

$$L = (D\tau)^{1/2} \tag{15.5}$$

where $D$ is the carrier diffusion constant and $\tau$ is the lifetime. For $D = 25$ cm$^2$/V · s and $\tau = 10^{-5}$ s, $L$ has a value of 158 $\mu$m. Consequently, the hole–electron pairs created by an $\alpha$-particle can diffuse to the surface region and can accumulate in the potential well under a gate electrode in a field-effect transistor.

The integrated circuits are made on epitaxial layers with appropriate doping to form potential barriers, so that only charge created within a few micrometers of the surface will be collected. Alpha-particles lose energy at a rate of 0.18 MeV/$\mu$m and create about $0.5 \times 10^5$ pairs/$\mu$m, so that for a 4-$\mu$m-thick epitaxial layer we assume that about $2 \times 10^5$ pairs are collected. The remainder of the pairs created in the silicon cannot be collected, due to the potential barrier at the interface.

From Chapter 5 the amount of charge $\mathcal{Q}_s$ that can be stored under a gate on an SiO$_2$ layer of thickness $t$ at a voltage $V$ is

$$\mathcal{Q}_s = \frac{V\epsilon_{ox}A}{t} \tag{15.6}$$

where $A$ is the area of the device and for SiO$_2$ $\epsilon_{ox} = 0.33 \times 10^{-12}$ F/cm. If the electric field $\mathcal{E} = V/t$, the charge density $\mathcal{Q}_s/A = Q_s$ is

$$Q_s = \epsilon_{ox}\mathcal{E}. \tag{15.7}$$

For an electric field strength of $10^3$ V/m,

$$Q_s = 3.3 \times 10^{-3} \text{ C/m}^2$$

or

$$Q_s \approx 2 \times 10^4 \text{ electrons/}\mu\text{m}^2.$$

For a gate with an area of 10 $\mu$m$^2$, this gives a stored charge of $2 \times 10^5$ electrons.

Thus the carriers created by an $\alpha$-particle could match the charge stored under a gate and would lead to a soft error. For this reason, particle-induced soft errors, it is important to monitor packaging materials for radioactive material. For high-density integrated circuits with a stored charge of $2 \times 10^5$ electrons per gate, every $\alpha$-particle that strikes the memory array can produce a soft error.

## 15.9 Electromigration and Failure Rate

Integrated circuits fail during operation. Large-scale integrated circuits are operated near the limit of their capabilities in terms of current densities in interconnects, voltages across gate oxides and moisture penetration in non-hermetic packages. We are concerned with when (not if) a circuit fails and how to establish device reliability criteria.

In failure analysis, we consider an integrated circuit (IC) to be a single device even though it may consist of many individual components (FET's, resistors, diodes, etc.). The failure of any one of these components can lead to a failure of the IC. The unit of failure rate is the FIT (*Failure unit* = FIT) with

$$1 \text{ FIT} = \frac{1 \text{ failure}}{10^9 \text{ device-hour}}.$$

If a system (e.g., a computer) is made up of a collection of $10^4$ IC's, each with an intrinsic failure rate of 10 FIT, then one would expect approximately one system failure per year (without redundancy),

$$\frac{\text{failures}}{\text{system year}} = \left(\frac{10 \text{ failure}}{10^9 \text{ device} \cdot \text{hr}}\right)\left(\frac{10^4 \text{ devices}}{\text{system}}\right)\left(\frac{8760 \text{ hr}}{\text{year}}\right) = 0.876.$$

Electromigration (Section 10.3.2) in the thin-film interconnects, the current conductors, provides an example of failure under operating conditions. Electromigration is transport of atoms in the conductor due to the current flowing through the conductor. Failure occurs due to the formation of open circuits in the conductors in areas where atomic transport results in material depletion, voids (Fig. 15.15).

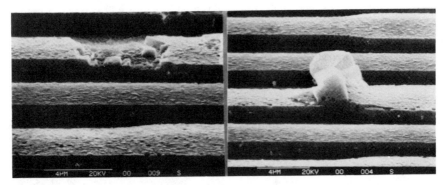

**FIGURE 15.15** Electromigration-induced void formation and Al extrusions in metallization lines formed by Al(Si)/Ti/Al(Si) sandwich structures. In these structures the electrons flow from the left to right side. (Photo from Joe McPherson, Texas Instruments.)

The same figure also shows a hillock of Al formed by mass accumulation in a localized area. Hillocks can cause short circuits between conductors or provide corrosion paths by breakage of protective layers.

Atomic transport in electromigration is directly proportional to the current density $J$ (neglecting heating effects) because the drift velocity

$$v_a = \text{(atomic mobility)} \times \text{(driving force)}$$

$$= \left(\frac{D}{kT}\right) \times (eZ^* \, \mathscr{E}) \tag{15.18}$$

$$= \frac{D}{kT} \cdot eZ^* \cdot \rho J$$

where $D$ is the diffusivity, $eZ^*$ the effective charge on the atom and $\rho$ the resistivity of the conductor. Temperature effects can not be excluded because joule heating (Section 2.10) given by $J^2\rho$ increases the diffusivity, $D = D_0 \exp(-E_A/kT)$. Electromigration becomes all the more severe in high density integrated circuits where the design effort is to reduce the dimensions of devices and conductors. The current density increases as the square of the reduction factor.

The current density and the temperature are two critical factors in failure due to electromigration. In one model (d'Heurle and Ho, 1978), the median time, $t_{50}$, to failure (the time when 50% of the lines have failed) is given by

$$t_{50} = A \, J^{-n} \exp(E_A/kT) \tag{15.19}$$

where $A$ depends on the conductor and its structure, and the current density exponent $n$ values are often between 2 and 3. Clearly failure times decrease strongly with an increase in temperature and current density. This is a consideration in high density integrated which operate hot and with current densities approaching $10^6$ A/cm$^2$.

Reliability can be expressed in terms of distribution functions (Bertram, 1983). The simplest is the exponential distribution function with a constant failure rate. The reliability function $R(t)$ is the probability that a device will survive to time $t$ and the cumulative distribution function $F(t)$ is the probability that the device will fail on or before time $t$. They are related by $1 - F(t) = R(t)$. For an exponential function with failure rate $\lambda$,

$$R(t) = e^{-\lambda t}. \tag{15.20}$$

The probability density function $f(t)$ is the derivative of $F(t)$ so that the mean time to failure (MTTF) is given by

$$\text{MTTF} = \int_0^\infty t \, f(t)dt = \int_0^\infty \lambda t \, e^{-\lambda t} \, dt = \frac{1}{\lambda}. \tag{15.21}$$

Electromigration is a wearout phenomenon (increasing failure rate) and, as such, cannot be described by the constant failure rate distribution (Equation 15.20). To describe wearout, the lognormal distribution is commonly used,

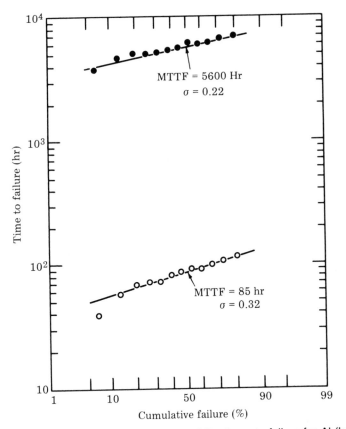

**FIGURE 15.16** Log normal probability plots of the times to failure for Al (lower curve) and Al-4 wt % Cu-1.7 wt % Si (upper curve). (From A. J. Learn, *J. Electron. Mater., 3,* 531 (1974).

$$f(t) = \frac{1}{\sigma t \sqrt{2\pi}} \exp \left\{ -\frac{1}{2} \left( \frac{\text{Int} - \text{Int}_{50}}{\sigma} \right)^2 \right\} \qquad (15.22)$$

where $t_{50}$ is the median time to failure and the dispersion parameter $\sigma \simeq \ln(t_{50}/t_{16})$ with $t_{16}$ the time when 15.866% of the devices have failed. As shown in Fig. 15.16, one can obtain a straight line fit when one plots the logarithm of the time to failure versus the cumulative percent of failures. This figure also shows the two order of magnitude increase in time to failure that is obtained by the addition of Cu to the Al. This addition reduces grain boundary diffusion and consequently Al interconnects commonly contain a few wt. % of Cu as well as Si if the interconnect is in contact with Si (Section 10.3).

In this section we have used failure due to electromigration as an example of a time dependent failure mechanism. There are a number of mechanisms that degrade the reliability of integrated circuits (Bertram, 1983). In the packaging area alone

one is concerned about the integrity of a hermetically-sealed package due to seal leaks and fatigue in the bonds and other mechanical interfaces.

## 15.10 Overview

This chapter provides a description of the bonding techniques and packages used in integrated-circuit technology. The processing steps used to produce integrated circuits in Si and GaAs are well in hand, with a drive to produce increasingly dense arrays of circuits on the chips. The number of devices per integrated-circuit chip and their performance must be met with compatible package designs. Chips with higher input/output counts require more package pins, larger packages, and more heat dissipation. Multilayer metallization in the chip leads to multilayer connections in the package.

Packaging lies in the area between fabrication of the integrated circuit and operation of the electronic system. Both of these must be taken into consideration when designing the optimum package. This book focuses on the principles underlying the fabrication of the integrated circuit. A proper appreciation of the constraints imposed on package design also requires analysis of cost and performance criteria imposed by the electronic system.

## REFERENCES

BERTRAM, W. J., in "VLSI Technology," edited by S. M. Sze (McGraw-Hill, New York, 1983), chapter 14.

COLCLASSER, R. A., *Microelectronics: Processing and Device Design,* Wiley, New York, 1980.

D'HEURLE, F. M., and P. S. Ho in "Thin Films—Interdiffusion and Reactions," edited by J. M. Poate, K. N. Tu and J. W. Mayer (Wiley-Interscience, New York, 1978), Chapter 8.

DUNN, C. F., F. R. BROTZEN, and J. W. MCPHERSON, "Electromigration and Microstructural Properties of Al-Si/Ti/Al-Si VLSI Metallization," *J. Electronic Materials 15(5),* 273 (1986).

FELDMAN, L. C. and J. W. MAYER, *Fundamentals of Surface and Thin Film Analysis,* North-Holland, Amsterdam, 1986.

FRIED, L. J., J. HAVAS, J. S. LECHATON, J. S. LOGAN, G. PAAL, and P. A. TOTTA, "A VLSI Bipolar Metallization Design with Three-Level Wiring and Area Array Solder Connections," *IBM J. Res. Devel. 26,* 362 (1982).

MAY, T. C., and M. H. WOODS, "Alpha-Particle Induced Soft Errors in Dynamic Memories," *IEEE Trans. Electron Devices ED-26,* 2 (1979).

SERAPHIM, D. P., R. C. LASKY and C.-Y. LI, ed. *Principles of Electronic Packaging,* McGraw-Hill, New York, 1989.

TANIGAWA, H., T. ISHIHARA, M. HIRATA, and K. SUZUKI, *IEEE Trans. Electron Devices ED-32,* 1191 (1985).

# Answers to Selected Problems

**Chapter 1**

**1.1.**   (a) $4.0 \times 10^{17}$ Si/cm²; $0.8 \times 10^{-5}$ cm
      (b) $2.7$ g/cm³; $1.54 \times 10^{15}$ Al atoms/cm²

**1.2.**   (a) $3.93 \times 10^{21}$ atoms/sec
      (b) stress $= 1.79 \times 10^{6}$ N/m²; strain $= 1.79 \times 10^{-5}$

**1.3.**    —

**1.4.**   (a) $\Delta l = 9 \times 10^{-4}$ cm
      (b) relative $\Delta l = 9 \times 10^{-4}$ cm; 1/10 solder ball dia.

**1.5.**    —

**1.6.**    $T_R = 2.5 \times 10^{-12}$ sec, $T_{\text{light}} = 7.5 \times 10^{-2}$ cm

**1.7.**    $2.0 \times 10^{4}$ transistors on 2 mm pin head

**Chapter 2**

**2.1.**   (a) $R = 10 \ \Omega$
      (b) $V = 0.1$ volt; $10^{-3}$ Watt
      (c) resistivity unchanged, $R = 7.5 \ \Omega$
      (d) resistance unchanged
      (e) $R_s = 2.5 \times 10^{-2} \ \Omega/\square$

**2.2.**   (a) $v_d = 1.25 \times 10^{2}$ cm/sec
      (b) flux $= 1.25 \times 10^{18}$/cm² sec; $J = 2 \times 10^{-1}$ A/cm²
      (c) less than
      (d) $D = 1.6$ cm²/sec
      (e) $dn/dx = 7.7 \times 10^{17}$ electrons/cm⁴

**2.3.**   (a) $8.46 \times 10^{18}$ Cu/cm²
      (b) $\mu = 36.8$ cm²/volt-sec
      (c) $R = 5.1 \ \Omega$
      (d) $J = 3.92 \times 10^{5}$ Amp/cm²; $\mathscr{E} = 0.666$ volt/cm
      (e) $R_s = 1.7 \times 10^{-2} \ \Omega/\square$ and $1.7 \times 10^{-1} \ \Omega/\square$

**2.4.**   (a) $\tau = 4.8 \times 10^{-12}$ sec; $l = 5.66 \times 10^{-7}$ m
      (b) Energy $= 4.1 \times 10^{-2}$ eV

**2.5.**    —

**2.6.**   **(a)** resistivity = 50 $\Omega$-cm
    **(b)** mobility = 3.33 × $10^3$ cm$^2$/V-sec
    **(c)** $n$ = 3.75 × $10^{13}$/cm$^3$
    **(d)** $V_x$ = 0.5 volt; $V_H$ = 0.05 volt

**2.7.**   **(a)** C = 6.9 × $10^{-15}$ farad, Q = 3.45 × $10^{-15}$ Coulomb
    **(b)** 1.5 × $10^6$ pairs and 3.5 volts
    **(c)** V = 4.8 volt, A = 1.45 × $10^{-5}$ cm$^2$

**2.8.**   **(a)** $J$ = 61.92 A/cm$^2$

**2.9.**   **(a)** $\Delta T$ = 5.88 × $10^2$ °C
    **(b)** $F_{th}$ = 0.4 J/cm$^2$ sec; $\Delta T$ = 0.67 × $10^{-2}$ °C

**2.10.**   **(a)** $R$ increase, $V_H$ constant
    **(b)** $R$ unchanged, $V_H$ constant

**2.11.**   **(a)** hot end positive
    **(b)** heavily doped end has same voltage


## Chapter 3

**3.1.**   **(a)** Si, **(b)** Si, **(c)** Si, **(d)** Al, **(e)** same, **(f)** neither

**3.2.**   —

**3.3.**   **(a)** Si: $n$: 4.17 $\Omega$-cm; $p$: 13.9 $\Omega$-cm
        GaAs: $n$: 0.735 $\Omega$-cm; $p$: 15.6 $\Omega$-cm
    **(b)** $n$-type, 4.17 × $10^4$ $\Omega$/□; $p$-type, 13.9 $\Omega$/□
    **(c)** sheet resistance = 1.48 × $10^3$ $\Omega$/□

**3.4.**   **(a)** 2.71 × $10^{-8}$ cm, 1.357 × $10^{15}$/cm$^2$, 2.71 × $10^{-7}$ cm
    **(b)** $r$ = 6.3 × $10^{-8}$ cm, $N$ = 5 × $10^{20}$/cm$^3$
    **(c)** $E_D$ = 0.09 eV

**3.5.**   **(a)** $n$ = 2 × $10^{16}$/cm$^3$; $n_i$ = 1.88 × $10^{13}$/cm$^3$; $p$ = 1.78 × $10^{10}$/cm$^3$,
        $E_F$ = 0.26 eV
    **(b)** $n$ = 1 × $10^{16}$/cm$^3$; $n_i$ unchanged, $p$ and $E_F$ change

**3.6.**   **(a)** away
    **(b)** toward
    **(c)** no change
    **(d)** away

**3.7.**   Si: $E_F$ = 0.204 eV; GaAs: $E_F$ = 0.099 eV

**3.8.**   **(a)** $L_n$ = 1.97 × $10^{-3}$ cm (Si); $L_n$ = 4.7 × $10^{-3}$ cm (GaAs)
    **(b)** J = 31.4 × $10^{-2}$ A/cm$^2$ (Si); J = 75 × $10^{-2}$ A/cm$^2$
    **(c)** E = 4.4 V/cm (Si); E = 11.7 V/cm (GaAs)

**3.9.**   **(a)** greater in intrinsic
    **(b)** decrease
    **(c)** smaller in GaAs

**3.10.**   **(a)** 0.178 eV
    **(b)** $n_i$ = 2.78 × $10^{13}$/cm$^3$, $p$ = 7.76 × $10^9$/cm$^3$

**3.11.** (a) $E_G = 1.24$ eV
(b) $\mu = 2000$ cm$^2$/V-sec
(c) $m^*/m = 0.037$
(d) $N_c = 2.5 \times 10^{16}$/cm$^3$
(e) $n_i = 3.83 \times 10^8$/cm$^3$
(f) $p = 1.6 \times 10^4$ holes/cm$^3$

## Chapter 4

**4.1.** —

**4.2.** (a) $\rho(n\text{-type}) = 0.42$ $\Omega$-cm; $\rho(p\text{-type}) = 1.4 \times 10^{-2}$ $\Omega$-cm
(b) $E_C - E_F = 0.205$ eV, $E_F - E_V = 6.04 \times 10^{-2}$ eV
(c) $V_0 = 0.855$ eV
(d) $p_n = 4.05 \times 10^3$/cm$^3$
(e) $d_n = 0.33$ micron

**4.3.** (a) $n_p = 2.6 \times 10^{10}$/cm$^3$; $p_n = 2.61 \times 10^{12}$/cm$^3$
(b) $L_p = 5.08 \times 10^{-3}$ cm
(c) $I = 2.12 \times 10^{-7}$ Amp

**4.4.** (a) $d_n = 6.6 \times 10^{-5}$ cm
(b) $I = 8.12 \times 10^{-16}$ A
(c) $I = 5.28 \times 10^{-12}$ A

**4.5.** (a) $\mathcal{E}(n\text{-type}) = 1.32 \times 10^{-3}$ V/cm; $\mathcal{E}(p\text{-type}) = 1.32 \times 10^{-5}$ V/cm
(b) $V(n\text{-type}) = 2.64 \times 10^{-4}$ V; $V(p\text{-type}) = 2.64 \times 10^{-6}$ V
(c) $I (V = 0.5) = 4 \times 10^{-4}$ A

**4.6.** $T_R = d/v_d = 1.77 \times 10^{-12}$ sec

**4.7.** (a) $V_0 = 0.5$ V
(b) $d_n = 8.5 \times 10^{-5}$ cm, $C = 1.24 \times 10^{-12}$ F
(c) $J = 1.78 \times 10^{-5}$ A/cm$^2$

**4.8.** —

**4.9.** —

**4.10.** —

**4.11.** (a) $n_p = 7.9 \times 10^7$ e/cm$^3$
(b) $J_n = 8.1 \times 10^{-2}$ A/cm$^2$
(c) $C = 8.7 \times 10^{-11}$ F

## Chapter 5

**5.1.** —

**5.2.** (a) $p_n = 2.38 \times 10^7$/cm$^3$
(b) $p_n = 5.67 \times 10^{11}$/cm$^3$
(c) $J = 0.41$ A/cm$^2$
(d) $J = 8.1 \times 10^{-4}$ A/cm$^2$

**5.3.** —

**5.4.** —

**5.5.** **(a)** —

**(b)** $E_F(e) = 0.044$ eV, $E_F(b) = 0.138$ eV, $E_F(c) = 0.264$ eV

**(c)** $V_0(e-b) = 0.938$ V; $V_0(b-c) = 0.718$ V

**(d)** $d(base) = 1.6 \times 10^{-5}$ cm, d(c) $= 9.7 \times 10^{-5}$ cm

**(e)** $n_i = 1.45 \times 10^{10}/cm^3$; $p_n(e) = 42/cm^3$; $n_p(b) = 4.2 \times 10^3/cm^3$; $p_n(c) = 2.1 \times 10^5/cm^3$

**5.6.** **(a)** $J = 0.619$ A/cm$^2$

**(b)** $J = 3.14 \times 10^{-3}$ A/cm$^2$

**(c)** $V(e-b) = 0.557$ V (from 5.5(e))

**(d)** $p_n = 10^{11}/cm^3$ (from 5.5(e))

**(e)** $J = 1.72 \times 10^{-4}$ A/cm$^2$

**5.7.** **(a)** $J = 3.6 \times 10^{-11}$ A/cm$^2$ (use values from 5.5(e))

**(b)** $d_n$ (c) $= 2.73 \times 10^{-4}$ cm

**(c)** $J = 3.17 \times 10^{-8}$ A/cm$^2$

**(d)** $J = 6.16 \times 10^{-1}$ A/cm$^2$

**5.8.** —

**5.9.** —

**5.10.** —

**5.11.** —

**5.12.** —

# Chapter 6

**6.1.** **(a)** 8.92 gm/cm$^3$

**(b)** 3.5 Å

**(c)** 2.83 Å

**6.2.** **(a)** 1.09 Å

**(b)** $5.85 \times 10^{22}/cm^3$

**(c)** $5.85 \times 10^{18}/cm^2$

**6.3.** **(a)** $d_{100}$

**(b)** (100)

**(c)** 2.9 Å

**6.4.** **(a)** B (FCC)

**(b)** B (FCC)

**(c)** A (simple cubic)

**6.5.** **(a)** $1.2 \times 10^{23}$

**(b)** 3.27 Å

**(c$_1$)** $4/\sqrt{14}$ Å $= 1.07$ Å

**(c$_2$)** $6.25 \times 10^{14}$ atoms/cm$^2$ on each side

**6.6** —

**6.7** —

**6.8**     —

**6.9**     —

**6.10**    —

**6.11**    —

## Chapter 7

**7.1.**    (a)   $5 \times 10^{14}$ cm$^{-2}$
        (b)   $1.8 \times 10^{4}$ sec.
        (c)   $1.12 \times 10^{11}$ atoms/cm$^{2}$ sec

**7.2.**    (a)   $2.1 \times 10^{-12}$ cm$^{2}$/sec
        (b)   $4 \times 10^{19}$ cm$^{-3}$
        (c)   $1.2 \times 10^{12}$ atoms/cm$^{2}$ sec
        (d)   $8 \times 10^{15}$ cm$^{-2}$
        (e)   1.15 eV

**7.3.**    (a)   Boron
        (b)   Arsenic
        (c)   temperature
        (d)   4.2 eV

**7.4.**    (a)   0.22 μm, $1.095 \times 10^{14}$ Os/cm$^{2}$ (linear approx.)
        (b)   $10^{4}$ V/cm
        (c)   1208°C

**7.5**        —

**7.6**        —

**7.7**        —

**7.8**        —

**7.9.**       0.59 μm

**7.10**    (a)   $1.09 \times 10^{-5}$ cm
        (b)   $3.6 \times 10^{-5}$ cm
        (c)   yes

**7.11**      —

**7.12**      —

## Chapter 8

**8.1.**    (a)   $1.9 \times 10^{16}$/cm$^{2}$
        (b)   $R_p = 0.05$ micron; $\Delta R_p = 0.022$ micron
        (c)   $3.4 \times 10^{21}$/cm$^{3}$
        (d)   0.117 micron

**8.2.**    (a)   $3.6 \times 10^{7}$ cm/sec
        (b)   2.8 Weber/m$^{2}$
        (c)   $R_p = 0.025$ micron

**8.3.** (a) As: $1.68 \times 10^3$ eV/nm; Sb: $2.9 \times 10^3$ eV/nm
(b) $R_p(\text{As}) = 5 \times 10^{-2}$ micron; $R_p(\text{Sb}) = 4 \times 10^{-2}$ micron
(c) $\Delta R_p(\text{As}) = 2.13 \times 10^{-2}$ micron; $\Delta R_p(\text{Sb}) = 1.47 \times 10^{-2}$ micron

**8.4.** (a) $k = 1$
(b) —
(c) —

**8.5.** (a) $1.47 \times 10^{-5}$ cm and $10^{20}/\text{cm}^3$
(b) $0.54 \times 10^{20}/\text{cm}^3$
(c) $2.2 \times 10^{-5}$ cm, $3.04 \times 10^{-5}$ cm

**8.6.** (a) $6.25 \times 10^2$ $\Omega/\square$
(b) $7.2 \times 10^{-2}$ A
(c) $4.5 \times 10^{-1}$ V

**8.7.** (a) $5 \times 10^3$ sec
(b) $1.25 \times 10^3$ sec
(c) $14.3 \times 10^3$ sec

# Chapter 9

**9.1.** (a) $4.6 \times 10^{14}$ atoms/cm$^2$
($b_1$) increase
($b_2$) decrease
(c) $2.53$ eV
(d) top half

**9.2.** (a) $88.6$ sec, $450$ Å
(b) accumulate
(c) $37.7$ min.
(d) —

**9.3** As consumed and diffuses away

**9.4.** 1 hr 35 min

**9.5.** curvature is $1.15$ m$^{-1}$.

**9.6.** $1.4 \times 10^{11}$ cm$^{-2}$

**9.7** —

# Chapter 10

**10.1** —

**10.2** —

**10.3** —

**10.4.** (a) At $\approx 1300°C$, the mixture starts to solidify.
(b) Si will solidify first.
(c) $\approx 12.2$ at $\cdot \% \cdot$ Si.

**10.5** —

**10.6** —

**10.7.** (a) Au is not stable in contact with InP in a closed or open system.
  (b) $AuIn_2$, $AuIn$, $Au_2P_3$ and $\gamma'$.
  (c) $AuIn_2$.
  (d) —

**10.8.** —

# Chapter 11

**11.1.** (a) $T < 750°C$
  (b) 680 Å
  (c) substrate at ground, $Pd_2Si$ at $+0.3$ volt.
  (d) increase

**11.2.** (a) $2.2 \times 10^{-9}$ cm$^2$/sec
  (b$_1$) unchanged
  (b$_2$) unchanged
  (b$_3$) decrease
  (c) 2.5 eV.

**11.3.** $Ni_2Si$, 900 Å; $NiSi$, 1800 Å; $NiSi_2$, 3600 Å

**11.4.** —

**11.5.** (a) $Si/Pd_2Si/VSi_2$
  (b) $Si/Pd_2Si/(Pd_2Si + VSi_2)$

**11.6** —

**11.7.** —

**11.8.** —

# Chapter 12

**12.1.** (a) 9.4 eV, 37.5 eV
  (b) $9.4 \times 10^6$

**12.2.** (a) 4.25 eV
  (b) $1.05 \times 10^{10}$/m
  (c) $1.2 \times 10^6$ m/sec

**12.3.** (a) $T = 379$ K
  (b) Si: $6.9 \times 10^{16}$/cm$^3$ and $9.5 \times 10^{18}$/cm$^3$; GaAs: $1.16 \times 10^{15}$/cm$^3$, $1.6 \times 10^{17}$/cm$^3$

**12.4.** (a) $1.5 \times 10^5$/m
  (b) $1.02 \times 10^{10}$/m
  (c) $1.76 \times 10^{-2}$ m$^2$/V sec
  (d) $0.36 \times 10^{23}$/cm$^3$
  (e) $9.9 \times 10^{-7}$ Ω-cm

**12.5.** —

**12.6.** (a) $5.41 \times 10^{19}/\text{cm}^3$
(b) $0.48 \times 10^{19}/\text{cm}^3$
(c) away

**12.7.** —

**12.8.** —

**12.9.** (a) 0.119 and 0.0179
(b) 0.135 and 0.0183

**12.10.** ($a_1$) $3 \times 10^{11}/\text{cm}^3$
($a_2$) $10^5$ $\Omega$-cm
($a_3$) middle
($a_4$) $n_i$ increase, $E_F$ unchanged
($b_1$) $5 \times 10^{16}/\text{cm}^3$
($b_2$) $n_i$ unchanged
($b_3$) $1.95 \times 10^6/\text{cm}^3$
($b_4$) $1.25$ $\Omega$-cm

**12.11.** (a) Si: $2.84 \times 10^{19}$ and $1.03 \times 10^{19}/\text{cm}^3$
GaAs: $4.67 \times 10^{17}$ and $0.82 \times 10^{19}/\text{cm}^3$
(b) 0.17 (Si) and 0.064 (GaAs)
(c) below (Si) and above (GaAs)

**12.12.** —

**12.13.** (a) 0.119 eV
(b) just at
(c) $1.4 \times 10^{11}/\text{cm}^3$
($d_1$) decrease
($d_2$) increase
($d_3$) increase

**12.14.** (a) 150 K
(b) 555 K

**12.15.** (a) $1.48 \times 10^{11}/\text{cm}^3$
(b) $10^6/\text{cm}^3$
(c) 0.119 eV

**12.16.** (a) $2.6 \times 10^6/\text{cm}^3$
(b) closer to $E_V$
(c) 1.79 m
(d) holes
(e) $E > E_G$
(f) Si on Al site, Te on P site

**12.17.** (a) $2.2 \times 10^5$ $\Omega$-cm
($b_1$) 0.264 eV
($b_2$) $2.1 \times 10^5/\text{cm}^3$
($b_3$) 4.17 $\Omega$-cm
(c) —
(d) $3.41 \times 10^{-2}$ cm

## Chapter 13

**13.1.** **(a)** $2.3 \times 10^2$
**(b)** GaAs: $3.6 \times 10^{-4}$ cm, AlGaAs: $1.1 \times 10^{-4}$ cm
**(c)** 0.14

**13.2.** **(a)** $I_D = 1.2 \times 10^{-3}$ A
**(b)** $2.55 \times 10^{-14}$ F
**(c)** $g_m = 6.375 \times 10^{-3}$ S
**(d)** $4 \times 10^{-12}$ sec

**13.3.** **(a)** $1.4 \times 10^{-2}$ eV
**(b)** $2.8 \times 10^{13}/\text{eV cm}^2$
**(c)** $4.2 \times 10^{-2}$ eV

**13.4.** **(a)** 0.32 A
**(b)** 0.97 V
**(c)** 0.87 micron

**13.5.** **(a)** 26.8 V
**(b)** $1.7 \times 10^{-13}$ sec and $10^{-11}$ sec
**(c)** $1.3 \times 10^5$ V/cm (ave.)

**13.6.** **(a)** 5 micron
**(b)** 0.36 micron

## Chapter 14

**14.1.** $\Delta E_C = .06$ eV; $\Delta E_V = -0.7$ eV, according to the electron affinity rule.
$\Delta E_C = 0.34$; $\Delta E_V = -0.41$ eV, according to the HAO theory.

**14.2.** **(a)** $\Delta E_C = -0.56$ eV; $\Delta E_V = 0.43$ eV, straddling type.
**(b)** $\Delta E_C = 0.05$ eV; $\Delta E_V = 1.23$ eV, staggered type
**(c)** $\Delta E_C = 0.89$ eV; $\Delta E_V = 0.52$ eV, broken gap type
**(d)** $\Delta E_C = -0.07$ eV; $\Delta E_V = 0.38$ eV, straddling type

**14.3.** $h_c \sim 540$ Å from eq. 14.20. $h_c \sim 45$ Å from Figure 14.16.

**14.4** —

**14.5.** **(a)** $x = 0.47$

**14.6.** $h_c \simeq 40$ Å from Fig. 14.6

# Index

# Properties of Si, GaAs and SiO$_2$ at 300 K

| Properties | Si | GaAs | SiO$_2$ |
|---|---|---|---|
| Atoms/cm$^3$, molecules/cm$^3$ $\times$ 10$^{22}$ | 5.0 | 4.42 | 2.27[a] |
| Structure | diamond | zincblende | amorphous |
| Lattice constant (nm) | 0.543 | 0.565 | — |
| Density (g/cm$^3$) | 2.33 | 5.32 | 2.27[a] |
| Relative dielectric constant, $\epsilon_r$ | 11.9 | 13.1 | 3.9 |
| Permittivity, $\epsilon = \epsilon_r \epsilon_0$ (farad/cm) $\times$ 10$^{-12}$ | 1.05 | 1.16 | 0.34 |
| Expansion coefficient (dL/LdT) $\times$ (10$^{-6}$ K) | 2.6 | 6.86 | 0.5 |
| Specific Heat (joule/g K) | 0.7 | 0.35 | 1.0 |
| Thermal conductivity (watt/cm K) | 1.48 | 0.46 | 0.014 |
| Thermal diffusivity (cm$^2$/sec) | 0.9 | 0.44 | 0.006 |
| Energy Gap (eV) | 1.12 | 1.424 | ~9 |
| Drift mobility (cm$^2$/volt-sec) | | | |
|     electrons | 1500 | 8500 | — |
|     holes | 450 | 400 | — |
| Effective density of states (cm$^{-3}$) $\times$ 10$^{19}$ | | | |
|     conduction band | 2.8 | 0.047 | — |
|     valence band | 1.04 | 0.7 | — |
| Intrinsic carrier concentration (cm$^{-3}$) | 1.45 $\times$ 10$^{10}$ | 1.79 $\times$ 10$^6$ | |

From Sze, "Physics of Semiconductor Devices" (1981) and from Beadle et al., "Quick Reference Manual for Silicon Integrated Circuit Technology" (1985).
[a]Formed under dry oxidation conditions.

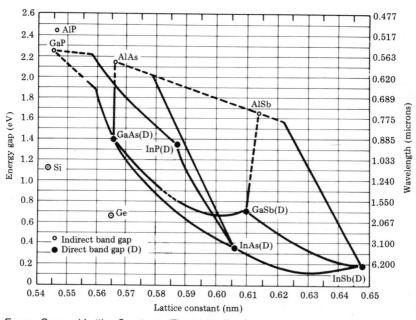

Energy Gap and Lattice Constants (Fig. 14.3, after Cho, 1985).